中国科学院科学出版基金资助出版

项目策划

向安全　林　鹏　王春香　吕　虹　范庆奎　王　璐

执行编辑

范庆奎

国外数学名著系列（影印版） 2

Basic Notions of Algebra

代数学基础

Igor R. Shafarevich

科学出版社
北京

图字：01-2005-6732

图书在版编目 (CIP) 数据

代数学基础＝Basic Notions of Algebra/(俄罗斯)沙法列维奇(Shafarevich,I. R.)著. —影印版. —北京:科学出版社,2006

(国外数学名著系列)

ISBN 978-7-03-016691-3

Ⅰ.代… Ⅱ.沙… Ⅲ.代数-英文 Ⅳ.015

中国版本图书馆 CIP 数据核字(2005)第 154387 号

科学出版社 出版
北京东黄城根北街 16 号
邮政编码：100717
http://www.sciencep.com
涿州市般润文化传播有限公司印刷
科学出版社发行 各地新华书店经销
*
2006 年 1 月第 一 版 开本:B5(720×1000)
2024 年 9 月第八次印刷 印张:16 3/4
字数：316 000
定价:118.00元
(如有印装质量问题,我社负责调换)

《国外数学名著系列》(影印版) 序

要使我国的数学事业更好地发展起来，需要数学家淡泊名利并付出更艰苦地努力。另一方面，我们也要从客观上为数学家创造更有利的发展数学事业的外部环境，这主要是加强对数学事业的支持与投资力度，使数学家有较好的工作与生活条件，其中也包括改善与加强数学的出版工作。

从出版方面来讲，除了较好较快地出版我们自己的成果外，引进国外的先进出版物无疑也是十分重要与必不可少的。从数学来说，施普林格(Springer)出版社至今仍然是世界上最具权威的出版社。科学出版社影印一批他们出版的好的新书，使我国广大数学家能以较低的价格购买，特别是在边远地区工作的数学家能普遍见到这些书，无疑是对推动我国数学的科研与教学十分有益的事。

这次科学出版社购买了版权，一次影印了 23 本施普林格出版社出版的数学书，就是一件好事，也是值得继续做下去的事情。大体上分一下，这 23 本书中，包括基础数学书 5 本，应用数学书 6 本与计算数学书 12 本，其中有些书也具有交叉性质。 这些书都是很新的，2000 年以后出版的占绝大部分，共计 16 本，其余的也是 1990 年以后出版的。这些书可以使读者较快地了解数学某方面的前沿，例如基础数学中的数论、代数与拓扑三本，都是由该领域大数学家编著的“数学百科全书”的分册。对从事这方面研究的数学家了解该领域的前沿与全貌很有帮助。按照学科的特点，基础数学类的书以“经典”为主，应用和计算数学类的书以“前沿”为主。这些书的作者多数是国际知名的大数学家，例如《拓扑学》一书的作者诺维科夫是俄罗斯科学院的院士，曾获“菲尔兹奖”和“沃尔夫数学奖”。这些大数学家的著作无疑将会对我国的科研人员起到非常好的指导作用。

当然，23 本书只能涵盖数学的一部分，所以，这项工作还应该继续做下去。更进一步，有些读者面较广的好书还应该翻译成中文出版，使之有更大的读者群。

总之，我对科学出版社影印施普林格出版社的部分数学著作这一举措表示热烈的支持，并盼望这一工作取得更大的成绩。

王　元

2005 年 12 月 3 日

Basic Notions of Algebra

I.R. Shafarevich

Translated from the Russian
by M. Reid

Contents

Preface

This book aims to present a general survey of algebra, of its basic notions and main branches. Now what language should we choose for this? In reply to the question 'What does mathematics study?', it is hardly acceptable to answer 'structures' or 'sets with specified relations'; for among the myriad conceivable structures or sets with specified relations, only a very small discrete subset is of real interest to mathematicians, and the whole point of the question is to understand the special value of this infinitesimal fraction dotted among the amorphous masses. In the same way, the meaning of a mathematical notion is by no means confined to its formal definition; in fact, it may be rather better expressed by a (generally fairly small) sample of the basic examples, which serve the mathematician as the motivation and the substantive definition, and at the same time as the real meaning of the notion.

Perhaps the same kind of difficulty arises if we attempt to characterise in terms of general properties any phenomenon which has any degree of individuality. For example, it doesn't make sense to give a definition of the Germans or the French; one can only describe their history or their way of life. In the same way, it's not possible to give a definition of an individual human being; one can only either give his 'passport data', or attempt to describe his appearance and character, and relate a number of typical events from his biography. This is the path we attempt to follow in this book, applied to algebra. Thus the book accommodates the axiomatic and logical development of the subject together with more descriptive material: a careful treatment of the key examples and of points of contact between algebra and other branches of mathematics and the natural sciences. The choice of material here is of course strongly influenced by the author's personal opinions and tastes.

As readers, I have in mind students of mathematics in the first years of an undergraduate course, or theoretical physicists or mathematicians from outside algebra wanting to get an impression of the spirit of algebra and its place in mathematics. Those parts of the book devoted to the systematic treatment of notions and results of algebra make very limited demands on the reader: we presuppose only that the reader knows calculus, analytic geometry and linear algebra in the form taught in many high schools and colleges. The extent of the prerequisites required in our treatment of examples is harder to state; an acquaintance with projective space, topological spaces, differentiable and complex analytic manifolds and the basic theory of functions of a complex variable is desirable, but the reader should bear in mind that difficulties arising in the treatment of some specific example are likely to be purely local in nature, and not to affect the understanding of the rest of the book.

This book makes no pretence to teach algebra: it is merely an attempt to talk about it. I have attempted to compensate at least to some extent for this by giving a detailed bibliography; in the comments preceding this, the reader can find references to books from which he can study the questions raised in this book, and also some other areas of algebra which lack of space has not allowed us to treat.

A preliminary version of this book has been read by F.A. Bogomolov, R.V. Gamkrelidze, S.P. Dëmushkin, A.I. Kostrikin, Yu.I. Manin, V.V. Nikulin, A.N. Parshin, M.K. Polyvanov, V.L. Popov, A.B. Roiter and A.N. Tyurin; I am grateful to them for their comments and suggestions which have been incorporated in the book.

I am extremely grateful to N.I. Shafarevich for her enormous help with the manuscript and for many valuable comments.

Moscow, 1984 I.R. Shafarevich

I have taken the opportunity in the English translation to correct a number of errors and inaccuracies which remained undetected in the original; I am very grateful to E.B. Vinberg, A.M. Volkhonskii and D. Zagier for pointing these out. I am especially grateful to the translator M. Reid for innumerable improvements of the text.

Moscow, 1987 I.R. Shafarevich

§ 1. What is Algebra?

What is algebra? Is it a branch of mathematics, a method or a frame of mind? Such questions do not of course admit either short or unambiguous answers. One can attempt a description of the place occupied by algebra in mathematics by drawing attention to the process for which Hermann Weyl coined the unpronounceable word 'coordinatisation' (see [H. Weyl **109** (1939), Chap. I, § 4]). An individual might find his way about the world relying exclusively on his sense organs, sight, feeling, on his experience of manipulating objects in the world outside and on the intuition resulting from this. However, there is another possible approach: by means of *measurements*, subjective impressions can be transformed into objective marks, into numbers, which are then capable of being preserved indefinitely, of being communicated to other individuals who have not experienced the same impressions, and most importantly, which can be operated on to provide new information concerning the objects of the measurement.

The oldest example is the idea of *counting* (coordinatisation) and *calculation* (operation), which allow us to draw conclusions on the number of objects without handling them all at once. Attempts to 'measure' or to 'express as a number' a variety of objects gave rise to fractions and negative numbers in addition to the whole numbers. The attempt to express the diagonal of a square of side 1 as a number led to a famous crisis of the mathematics of early antiquity and to the construction of irrational numbers.

Measurement determines the points of a line by real numbers, and much more widely, expresses many physical quantities as numbers. To Galileo is due the most extreme statement in his time of the idea of coordinatisation: 'Measure everything that is measurable, and make measurable everything that is not yet so'. The success of this idea, starting from the time of Galileo, was brilliant. The creation of analytic geometry allowed us to represent points of the plane by pairs of numbers, and points of space by triples, and by means of operations with numbers, led to the discovery of ever new geometric facts. However, the success of analytic geometry is mainly based on the fact that it reduces to numbers not only points, but also curves, surfaces and so on. For example, a curve in the plane is given by an equation $F(x, y) = 0$; in the case of a line, F is a linear polynomial, and is determined by its 3 coefficients: the coefficients of x and y and the constant term. In the case of a conic section we have a curve of degree 2, determined by its 6 coefficients. If F is a polynomial of degree n then it is easy to see that it has $\frac{1}{2}(n + 1)(n + 2)$ coefficients; the corresponding curve is determined by these coefficients in the same way that a point is given by its coordinates.

In order to express as numbers the roots of an equation, the complex numbers were introduced, and this takes a step into a completely new branch of mathematics, which includes elliptic functions and Riemann surfaces.

For a long time it might have seemed that the path indicated by Galileo consisted of measuring 'everything' in terms of a known and undisputed collection of numbers, and that the problem consists just of creating more and more subtle methods of measurements, such as Cartesian coordinates or new physical instruments. Admittedly, from time to time the numbers considered as known (or simply called numbers) turned out to be inadequate: this led to a 'crisis', which had to be resolved by extending the notion of number, creating a new form of numbers, which themselves soon came to be considered as the unique possibility. In any case, as a rule, at any given moment the notion of number was considered to be completely clear, and the development moved only in the direction of extending it:

$$\text{'1, 2, many'} \Rightarrow \text{natural numbers} \Rightarrow \text{integers}$$
$$\Rightarrow \text{rationals} \Rightarrow \text{reals} \Rightarrow \text{complex numbers}.$$

But matrixes, for example, form a completely independent world of 'number-like objects', which cannot be included in this chain. Simultaneously with them, quaternions were discovered, and then other 'hypercomplex systems' (now called algebras). Infinitesimal transformations led to differential operators, for which the natural operation turns out to be something completely new, the Poisson bracket. Finite fields turned up in algebra, and p-adic numbers in number theory. Gradually, it became clear that the attempt to find a unified all-embracing concept of number is absolutely hopeless. In this situation the principle declared by Galileo could be accused of intolerance; for the requirement to 'make measurable *everything* which is not yet so' clearly discriminates against anything which stubbornly refuses to be measurable, excluding it from the sphere of interest of science, and possibly even of reason (and thus becomes a *secondary quality* or *secunda causa* in the terminology of Galileo). Even if, more modestly, the polemic term 'everything' is restricted to objects of physics and mathematics, more and more of these turned up which could not be 'measured' in terms of 'ordinary numbers'.

The principle of coordinatisation can nevertheless be preserved, provided we admit that the set of 'number-like objects' by means of which coordinatisation is achieved can be just as diverse as the world of physical and mathematical objects they coordinatise. The objects which serve as 'coordinates' should satisfy only certain conditions of a very general character.

They must be individually distinguishable. For example, whereas all points of a line have identical properties (the line is homogeneous), and a point can only be fixed by putting a finger on it, numbers are all individual: 3, $7/2$, $\sqrt{2}$, π and so on. (The same principle is applied when newborn puppies, indistinguishable to the owner, have different coloured ribbons tied round their necks to distinguish them.)

They should be sufficiently abstract to reflect properties common to a wide circle of phenomenons.

Certain fundamental aspects of the situations under study should be reflected in *operations* that can be carried out on the objects being coordinatised: addition, multiplication, comparison of magnitudes, differentiation, forming Poisson brackets and so on.

We can now formulate the point we are making in more detail, as follows:

Thesis. *Anything which is the object of mathematical study (curves and surfaces, maps, symmetries, crystals, quantum mechanical quantities and so on) can be 'coordinatised' or 'measured'. However, for such a coordinatisation the 'ordinary' numbers are by no means adequate.*

Conversely, when we meet a new type of object, we are forced to construct (or to discover) new types of 'quantities' to coordinatise them. The construction and the study of the quantities arising in this way is what characterises the place of algebra in mathematics (of course, very approximately).

From this point of view, the development of any branch of algebra consists of two stages. The first of these is the birth of the new type of algebraic objects out of some problem of coordinatisation. The second is their subsequent career, that is, the systematic development of the theory of this class of objects; this is sometimes closely related, and sometimes almost completely unrelated to the area in connection with which the objects arose. In what follows we will try not to lose sight of these two stages. But since algebra courses are often exclusively concerned with the second stage, we will maintain the balance by paying a little more attention to the first.

We conclude this section with two examples of coordinatisation which are somewhat less standard than those considered up to now.

Example 1. The Dictionary of Quantum Mechanics. In quantum mechanics, the basic physical notions are 'coordinatised' by mathematical objects, as follows.

Physical notion	Mathematical notion
State of a physical system	Line φ in an ∞-dimensional complex Hilbert space
Scalar physical quantity	Self-adjoint operator
Simultaneously measurable quantities	Commuting operators
Quantity taking a precise value λ in a state φ	Operator having φ as eigenvector with eigenvalue λ
Set of values of quantities obtainable by measurement	Spectrum of an operator
Probability of transition from state φ to state ψ	$\lvert(\varphi, \psi)\rvert$, where $\lvert\varphi\rvert = \lvert\psi\rvert = 1$

Example 2. Finite Models for Systems of Incidence and Parallelism Axioms. We start with a small digression. In the axiomatic construction of geometry, we often consider not the whole set of axioms, but just some part of them; to be

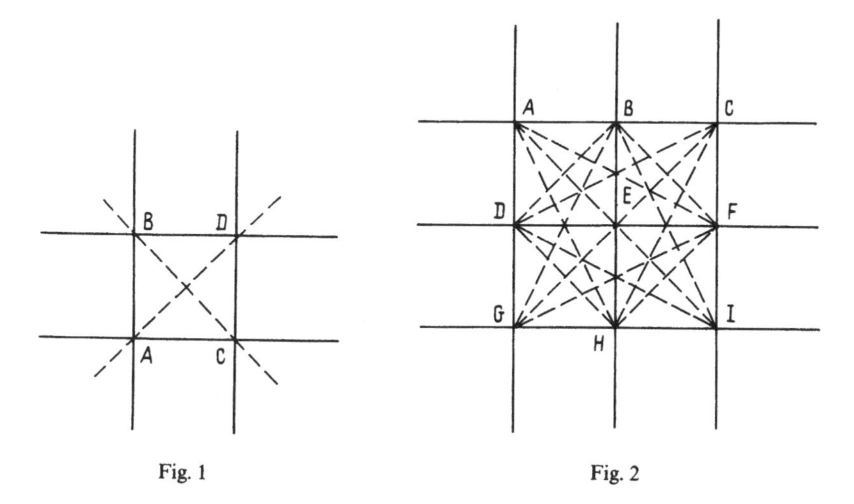

Fig. 1

Fig. 2

concrete we only discuss plane geometry here. The question then arises as to what realisations of the chosen set of axioms are possible: do there exists other systems of objects, apart from 'ordinary' plane geometry, for which the set of axioms is satisfied? We consider now a very natural set of axioms of 'incidence and parallelism'.

(a) Through any two distinct points there is one and only one line.

(b) Given any line and a point not on it, there exists one and only one other line through the point and not intersecting the line (that is, parallel to it).

(c) There exist three points not on any line.

It turns out that this set of axioms admits many realisations, including some which, in stark contrast to our intuition, have only a finite number of points and lines. Two such realisations are depicted in Figures 1 and 2. The model of Figure 1 has 4 points A, B, C, D and 6 lines AB, CD; AD, BC; AC, BD. That of Figure 2 has 9 points, A, B, C, D, E, F, G, H, I and 12 lines ABC, DEF, GHI; ADG, BEH, CFI; AEI, BFG, CDH; CEG, BDI, AFH. The reader can easily verify that axioms (a), (b), (c) are satisfied; in our list of lines, the families of parallel lines are separated by semicolons.

We return to our main theme, and attempt to 'coordinatise' the model of axioms (a), (b), (c) just constructed. For the first of these we use the following construction: write $\mathbb{0}$ and $\mathbb{1}$ for the property of an integer being even or odd respectively; then define operations on the symbols $\mathbb{0}$ and $\mathbb{1}$ by analogy with the way in which the corresponding properties of integers behave under addition and multiplication. For example, since the sum of an even and an odd integer is odd, we write $\mathbb{0} + \mathbb{1} = \mathbb{1}$, and so on. The result can be expressed in the 'addition and multiplication tables' of Figures 3 and 4.

The pair of quantities $\mathbb{0}$ and $\mathbb{1}$ with the operations defined on them as above serve us in coordinatising the 'geometry' of Figure 1. For this, we give points coordinates (X, Y) as follows:

$+$	$\mathbb{0}$	$\mathbb{1}$
$\mathbb{0}$	$\mathbb{0}$	$\mathbb{1}$
$\mathbb{1}$	$\mathbb{1}$	$\mathbb{0}$

Fig. 3

$\times$	$\mathbb{0}$	$\mathbb{1}$
$\mathbb{0}$	$\mathbb{0}$	$\mathbb{0}$
$\mathbb{1}$	$\mathbb{0}$	$\mathbb{1}$

Fig. 4

$$A = (\mathbb{0}, \mathbb{0}), \quad B = (\mathbb{0}, \mathbb{1}), \quad C = (\mathbb{1}, \mathbb{0}), \quad D = (\mathbb{1}, \mathbb{1}).$$

It is easy to check that the lines of the geometry are then defined by the linear equations:

$$AB\colon \mathbb{1}X = \mathbb{0}; \quad CD\colon \mathbb{1}X = \mathbb{1}; \quad AD\colon \mathbb{1}X + \mathbb{1}Y = \mathbb{0};$$

$$BC\colon \mathbb{1}X + \mathbb{1}Y = \mathbb{1}; \quad AC\colon \mathbb{1}Y = \mathbb{0}; \quad BD\colon \mathbb{1}Y = \mathbb{1}.$$

In fact these are the only 6 nontrivial linear equations which can be formed using the two quantities $\mathbb{0}$ and $\mathbb{1}$.

The construction for the geometry of Figure 2 is similar, but slightly more complicated: suppose that we divide up all integers into 3 sets U, V and W as follows:

$$U = \text{integers divisible by 3,}$$

$$V = \text{integers with remainder 1 on dividing by 3,}$$

$$W = \text{integers with remainder 2 on dividing by 3.}$$

The operations on the symbols U, V, W is defined as in the first example; for example, a number in V plus a number in W always gives a number in U, and so we set $V + W = U$; similarly, the product of two numbers in W is always a number in V, so we set $W \cdot W = V$. The reader can easily write out the corresponding addition and multiplication tables.

It is then easy to check that the geometry of Figure 2 is coordinatised by our quantities U, V, W as follows: the points are

$$A = (U, U), \quad B = (U, V), \quad C = (U, W), \quad D = (V, U) \quad E = (V, V),$$

$$F = (V, W), \quad G = (W, U), \quad H = (W, V), \quad I = (W, W);$$

and the lines are again given by all possible linear equations which can be written out using the three symbols U, V, W; for example, AFH is given by $VX + VY = U$, and DCH by $VX + WY = V$.

Thus we have constructed finite number systems in order to coordinatise finite geometries. We will return to the discussion of these constructions later.

Already these few examples give an initial impression of what kind of objects can be used in one or other version of 'coordinatisation'. First of all, the collection of objects to be used must be rigorously delineated; in other words, we must indicate a set (or perhaps several sets) of which these objects can be elements. Secondly, we must be able to operate on the objects, that is, we must define *operations*, which from one or more elements of the set (or sets) allow us to construct new elements. For the moment, no further restrictions on the nature of the sets to be used are imposed; in the same way, an operation may be a completely arbitrary rule taking a set of k elements into a new element. All the same, these operations will usually preserve some similarities with operations on numbers. In particular, in all the situations we will discuss, $k = 1$ or 2. The basic examples of operations, with which all subsequent constructions should be compared, will be: the operation $a \mapsto -a$ taking any number to its negative; the operation $b \mapsto b^{-1}$ taking any nonzero number b to its inverse (for each of these $k = 1$); and the operations $(a, b) \mapsto a + b$ and ab of addition and multiplication (for each of these $k = 2$).

§2. Fields

We start by describing one type of 'sets with operations' as described in §1 which corresponds most closely to our intuition of numbers.

A *field* is a set K on which two operations are defined, each taking two elements of K into a third; these operations are called *addition* and *multiplication*, and the result of applying them to elements a and b is denoted by $a + b$ and ab. The operations are required to satisfy the following conditions:

Addition:

Commutativity: $a + b = b + a$;

Associativity: $a + (b + c) = (a + b) + c$;

Existence of zero: there exists an element $0 \in K$ such that $a + 0 = a$ for every a (it can be shown that this element is unique);

Existence of negative: for any a there exists an element $(-a)$ such that $a + (-a) = 0$ (it can be shown that this element is unique).

Multiplication:

Commutativity: $ab = ba$;

Associativity: $a(bc) = (ab)c$;

Existence of unity: there exists an element $1 \in K$ such that $a1 = a$ for every a (it can be shown that this element is unique);

Existence of inverse: for any $a \neq 0$ there exists an element a^{-1} such that $aa^{-1} = 1$ (it can be shown that for given a, this element is unique).

Addition and multiplication:

Distributivity: $a(b + c) = ab + ac$.

Finally, we assume that a field does not consist only of the element 0, or equivalently, that $0 \neq 1$.

These conditions taken as a whole, are called the *field axioms*. The ordinary identities of algebra, such as

$$(a + b)^2 = a^2 + 2ab + b^2$$

or

$$a^{-1} - (a + 1)^{-1} = a^{-1}(a + 1)^{-1}$$

follow from the field axioms. We only have to bear in mind that for a natural number n, the expression na means $a + a + \cdots + a$ (n times), rather than the product of a with the number n (which may not be in K).

Working over an arbitrary field K (that is, assuming that all coordinates, coefficients, and so on appearing in the argument belong to K) provides the most natural context for constructing those parts of linear algebra and analytic geometry not involving lengths, polynomial algebras, rational fractions, and so on.

Basic examples of fields are the field of rational numbers, denoted by $\mathbb{Q}$, the field of real numbers $\mathbb{R}$ and the field of complex numbers $\mathbb{C}$.

If the elements of a field K are contained among the elements of a field L and the operations in K and L agree, then we say that K is a *subfield* of L, and L an *extension of* K, and we write $K \subset L$. For example, $\mathbb{Q} \subset \mathbb{R} \subset \mathbb{C}$.

Example 1. In §1, in connection with the 'geometry' of Figure 1, we defined operations of addition and multiplication on the set $\{\mathbb{0}, \mathbb{1}\}$. It is easy to check that this is a field, in which $\mathbb{0}$ is the zero element and $\mathbb{1}$ the unity. If we write 0 for $\mathbb{0}$ and 1 for $\mathbb{1}$, we see that the multiplication table of Figure 4 is just the rule for multiplying 0 and 1 in $\mathbb{Q}$, and the addition table of Figure 3 differs in that $1 + 1 = 0$. The field constructed in this way consisting of $\mathbb{0}$ and $\mathbb{1}$ is denoted by $\mathbb{F}_2$. Similarly, the elements U, V, W considered in connection with the geometry of Figure 2 also form a field, in which $U = 0$, $V = 1$ and $W = -1$. We thus obtain examples of fields with a finite number (2 or 3) of elements. Fields having only finitely many elements (that is, finite fields) are very interesting objects with many applications. A finite field can be specified by writing out the addition and multiplication tables of its elements, as we did in Figures 3–4. In §1 we met such fields in connection with the question of the realisation of a certain set of axioms of geometry in a finite set of objects; but they arise just as naturally in algebra as realising the field axioms in a finite set of objects. A field consisting of q elements is denoted by $\mathbb{F}_q$.

Example 2. An algebraic expression obtained from an unknown x and arbitrary elements of a field K using the addition, multiplication and division operations, can be written in the form

$$\frac{a_0 + a_1 x + \cdots + a_n x^n}{b_0 + b_1 x + \cdots + b_m x^m}, \tag{1}$$

where $a_i, b_i \in K$ and not all $b_i = 0$. An expression of this form is called a *rational fraction*, or a *rational function* of x. We can now consider it as a function, taking any x in K (or any x in L, for some field L containing K) into the given expression, provided only that the denominator is not zero. All rational functions form a field, called the *rational function field*; it is denoted by $K(x)$. We will discuss certain difficulties connected with this definition in §3. The elements of K are contained among the rational functions as 'constant' functions, so that $K(x)$ is an extension of K.

In a similar way we define the field $K(x, y)$ of rational functions in two variables, or in any number of variables.

An *isomorphism* of two fields K' and K'' is a 1-to-1 correspondence $a' \leftrightarrow a''$ between their elements such that $a' \leftrightarrow a''$ and $b' \leftrightarrow b''$ implies that $a' + b' \leftrightarrow a'' + b''$ and $a'b \leftrightarrow a''b''$; we say that two fields are *isomorphic* if there exists an isomorphism between them. If L' and L'' are isomorphic fields, both of which are extensions of the same field K, and if the isomorphism between them takes each element of K into itself, then we say that it is an isomorphism over K, and that L' and L'' are isomorphic over K. An isomorphism of fields K' and K'' is denoted by $K' \cong K''$. If L' and L'' are finite fields, then to say that they are isomorphic means that their addition and multiplication tables are the same; that is, they differ only in the notation for the elements of L' and L''. The notion of isomorphism for arbitrary fields is similar in meaning.

For example, suppose we take some line a and mark a point O and a 'unit interval' OE on it; then we can in a geometric way define addition and multiplication on the directed intervals (or vectors) contained in a. Their construction is given in Figures 5-6. In Figure 5, b is an arbitrary line parallel to a and U an arbitrary point on it, $OU \parallel AV$ and $VC \parallel UB$; then $OC = OA + OB$. In Figure 6, b is an arbitrary line passing through O, and $EU \parallel BV$ and $VC \parallel UA$; then $OC = OA \cdot OB$.

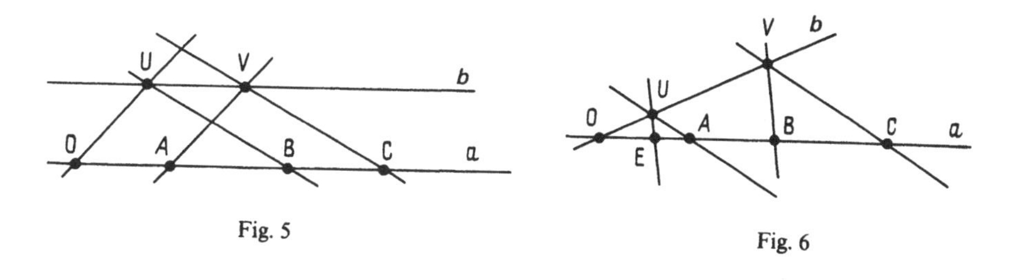

Fig. 5

Fig. 6

With this definition of the operations, intervals of the line form a field P; to verify all the axioms is a sequence of nontrivial geometric problems. Taking each interval into a real number, for example an infinite decimal fraction (this is again a process of measurement!), we obtain an isomorphism between P and the real number field $\mathbb{R}$.

Example 3. We return now to the plane curve given by $F(x, y) = 0$, where F is a polynomial; let C denote the curve itself. Taking C into the set of coefficients of F is one very primitive method of 'coordinatising' C. We now describe another method, which is much more precise and effective.

It is not hard to show that any nonconstant polynomial $F(x, y)$ can be factorised as a product of a number of nonconstant polynomials, each of which cannot be factorised any further. If $F = F_1 \cdot F_2 \ldots F_k$ is such a factorisation then our curve with equation $F = 0$ is the union of k curves with equations $F_1 = 0$, $F_2 = 0, \ldots, F_k = 0$ respectively. We say that a polynomial which does not factorise as a product of nonconstant polynomials is *irreducible*. From now on we will assume that F is irreducible.

Consider an arbitrary rational function $\varphi(x, y)$ in two variables; φ is represented as a ratio of two polynomials:

$$\varphi(x, y) = \frac{P(x, y)}{Q(x, y)}, \tag{2}$$

and we suppose that the denominator Q is not divisible by F. Consider φ as a function on points of C only; it is undefined on points (x, y) where both $Q(x, y) = 0$ and $F(x, y) = 0$. It can be proved that under the assumptions we have made there are only finitely many such points. In order that our treatment from now on has some content, we assume that the curve C has infinitely many points (that is, we exclude curves such as $x^2 + y^2 = -1$, $x^4 + y^4 = 0$ and so on; if we also consider points with complex coordinates, then the assumption is not necessary). Then $\varphi(x, y)$ defines a function on the set of points of C (for short, we say on C), possibly undefined at a finite number of points—in the same way that the rational function (1) is undefined at the finite number of values of x where the denominator of (1) vanishes. Functions obtained in this way are called *rational functions* on C. It can be proved that all the rational functions on a curve C form a field (for example, one proves that a function φ defines a nonzero function on C only if $P(x, y)$ is not divisible by $F(x, y)$, and then the function $\varphi^{-1} = \frac{Q(x, y)}{P(x, y)}$ satisfies the condition required for φ, that the denominator is not divisible by F; this proves the existence of the inverse). The field of rational functions on C is denoted by $\mathbb{R}(C)$; it is an extension of the real number field $\mathbb{R}$. Considering points with coordinates in an arbitrary field K, it is easy to replace $\mathbb{R}$ by K in this construction.

Assigning to a curve C the field $K(C)$ is a much more precise method of 'coordinatising' C than the coefficients of its equation. First of all, passing from a coordinate system (x, y) to another system (x', y'), the equation of a curve changes, but the field $K(C)$ is replaced by an isomorphic field, as one sees easily. Another important point is that an isomorphism of fields $K(C)$ and $K(C')$ establishes important relations between curves C and C'.

Suppose as a first example that C is the x-axis. Then since the equation of C is $y = 0$, restricting a function φ to C we must set $y = 0$ in (2), and we get a rational function of x:

$$\varphi(x, 0) = \frac{P(x, 0)}{Q(x, 0)}.$$

Thus in this case, the field $K(C)$ is isomorphic to the rational function field $K(x)$. Obviously, the same thing holds if C is an arbitrary line.

We proceed to the case of a curve C of degree 2. Let us prove that in this case also the field $K(C)$ is isomorphic to the field of rational functions of one variable $K(t)$. For this, choose an arbitrary point (x_0, y_0) on C and take t to be the slope of the line joining it to a point (x, y) with variable coordinates (Figure 7).

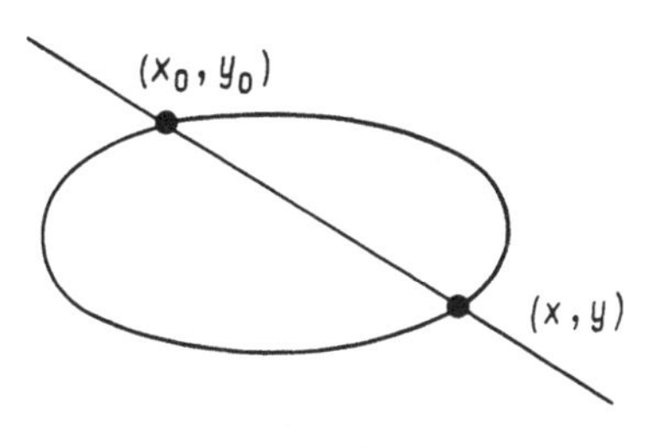

Fig. 7

In other words, set $t = \dfrac{y - y_0}{x - x_0}$, as a function on C. We now prove that x and y, as functions on C, are rational functions of t. For this, recall that $y - y_0 = t(x - x_0)$, and if $F(x, y) = 0$ is the equation of C, then on C we have

$$F(x, y_0 + t(x - x_0)) = 0. \tag{3}$$

In other words, the relation (3) is satisfied in $K(C)$. Since C is a curve of degree 2, this is a quadratic equation for x: $a(t)x^2 + b(t)x + c(t) = 0$ (whose coefficients involve t). However, one root of this equation is known, namely $x = x_0$; this simply reflects the fact that (x_0, y_0) is a point of C. The second root is then obtained from the condition that the sum of the roots equals $-\dfrac{b(t)}{a(t)}$. We get an expression $x = f(t)$ as a rational function of t, and a similar expression $y = g(t)$; of course, $F(f(t), g(t)) = 0$. Thus taking $x \leftrightarrow f(t)$, $y \leftrightarrow g(t)$ and $\varphi(x, y) \leftrightarrow \varphi(f(t), g(t))$, we obtain an isomorphism of $K(C)$ and $K(t)$ over K.

The geometric meaning of the isomorphism obtained in this way is that points of C can be parametrised by rational functions: $x = f(t)$, $y = g(t)$. If C has the equation $y^2 = ax^2 + bx + c$ then on C we have $y = \sqrt{ax^2 + bx + c}$, and another form of the result we have obtained is that both x and $\sqrt{ax^2 + bx + c}$ can be expressed as rational functions of some third function t. This expression is useful, for example, in evaluating indeterminate integrals: it shows that any integral

$$\int \varphi(x, \sqrt{ax^2 + bx + c})\, dx,$$

where φ is a rational function, reduces by substitutions to integrals of a rational function of t, and can hence be expressed in terms of elementary functions. In analysis our substitutions are called *Euler substitutions.* We mention two further applications.

(a) The *field of trigonometric functions* is defined as the field of all rational functions of $\sin \varphi$ and $\cos \varphi$. Since $\sin^2 \varphi + \cos^2 \varphi = 1$, this field is isomorphic to $\mathbb{R}(C)$, where C is the circle with equation $x^2 + y^2 = 1$. We know that $\mathbb{R}(C)$ is isomorphic to $\mathbb{R}(t)$. This explains why each trigonometric equation can be reduced to an algebraic equation.

(b) In the case of the circle $x^2 + y^2 = 1$, if we set $x_0 = 0$, $y_0 = -1$, our construction gives the formulas

$$x = \frac{2t}{1 + t^2}, \qquad y = \frac{1 - t^2}{1 + t^2}. \tag{4}$$

A problem of number theory which goes back to antiquity is the question of finding integers a, b, c for which $a^2 + b^2 = c^2$. Setting $\frac{a}{c} = x$, $\frac{b}{c} = y$, $t = \frac{p}{q}$ and reducing formula (4) to common denominators, we get the well-known expression

$$a = 2pq, \quad b = q^2 - p^2, \quad c = q^2 + p^2.$$

Already for the curve C with equation $y^2 = x^3 + 1$ the field $K(C)$ is not isomorphic to the field of rational functions. This is closely related to the fact that an elliptic integral, for example $\int \frac{dx}{\sqrt{x^3 + 1}}$ cannot be expressed in terms of elementary functions.

Of course, the field $K(C)$ also plays an important role in the study of other curves. It can also be defined for surfaces, given by $F(x, y, z) = 0$, where F is a polynomial, and if we consider spaces of higher dimensions, for an even wider class of geometric objects, algebraic varieties, defined in an n-dimensional space by an arbitrary system of equations $F_1 = 0, \ldots, F_m = 0$, where the F_i are polynomials in n variables.

In conclusion, we give examples of fields which arise in analysis.

Example 4. All meromorphic functions on some connected domain of the plane of one complex variable (or on an arbitrary connected complex manifold) form a field.

Example 5. Consider the set of all Laurent series $\sum_{n=-k}^{\infty} a_n z^n$ which are convergent in an annulus $0 < |z| < R$ (where different series may have different annuluses of convergence). With the usual definition of operations on series, these form a field, the *field of Laurent series*. If we use the same rules to compute the coefficients, we can define the sum and product of two Laurent series, even if these are nowhere convergent. We thus obtain the *field of formal Laurent series*. One can go further, and consider this construction in the case that the coefficients a_n belong to an arbitrary field K. The resulting field is called the field of formal Laurent series with coefficients in K, and is denoted by $K((z))$.

§3. Commutative Rings

The simplest possible example of 'coordinatisation' is counting, and it leads (once 0 and negative numbers have been introduced) to the integers, which do not form a field. Operations of addition and multiplication are defined on the set of all integers (positive, zero or negative), and these satisfy all the field axioms but one, namely the existence of an inverse element a^{-1} for every $a \neq 0$ (since, for example, $\frac{1}{2}$ is already not an integer).

A set having two operations, called addition and multiplication, satisfying all the field axioms except possibly for the requirement of existence of an inverse element a^{-1} for every $a \neq 0$ is called a *commutative ring*; it is convenient not to exclude the ring consisting just of the single element 0 from the definition.

The field axioms, with the axiom of the existence of an inverse and the condition $0 \neq 1$ omitted will from now on be referred to as the *commutative ring axioms*.

By analogy with fields, we define the notions of a *subring* $A \subset B$ of a ring, and *isomorphism* of two rings A' and A''; in the case that $A \subset A'$ and $A \subset A''$ we also have the notion of an isomorphism of A' and A'' over A; an isomorphism of rings is again written $A' \cong A''$.

Example 1. The Ring of Integers. This is denoted by $\mathbb{Z}$; obviously $\mathbb{Z} \subset \mathbb{Q}$.

Example 2. An example which is just a fundamental is the polynomial ring $A[x]$ with coefficients in a ring A. In view of its fundamental role, we spend some time on the detailed definition of $A[x]$. First we characterise it by certain properties.

We say that a commutative ring B is a *polynomial ring* over a commutative ring A if $B \supset A$ and B contains an element x with the property that every element of B can be uniquely written in the form

$$a_0 + a_1 x + \cdots + a_n x^n \qquad \text{with } a_i \in A$$

for some $n \geqslant 0$. If B' is another such ring, with x' the corresponding element, the correspondence

$$a_0 + a_1x + \cdots + a_nx^n \leftrightarrow a_0 + a_1x' + \cdots + a_n(x')^n$$

defines an isomorphism of B and B' over A, as one sees easily. Thus the polynomial ring is uniquely defined, in a reasonable sense.

However, this does not solve the problem as to its existence. In most cases the 'functional' point of view is sufficient: we consider the functions f of A into itself of the form

$$f(c) = a_0 + a_1c + \cdots + a_nc^n \qquad \text{for } c \in A. \tag{1}$$

Operations on functions are defined as usual: $(f + g)(c) = f(c) + g(c)$ and $(fg)(c) = f(c)g(c)$. Taking an element $a \in A$ into the constant function $f(c) = a$, we can view A as a subring of the ring of functions. If we let x denote the function $x(c) = c$ then the function (1) is of the form

$$f = a_0 + a_1x + \cdots + a_nx^n. \tag{2}$$

However, in some cases (for example if the number of elements of A is finite, and n is greater than this number), the expression (2) for f may not be unique. Thus in the field $\mathbb{F}_2$ of §2, Example 1, the functions x and x^2 are the same. For this reason we give an alternative construction.

We could define polynomials as 'expressions' $a_0 + a_1x + \cdots + a_nx^n$, with $+$ and x^i thought of as conventional signs or place-markers, serving to denote the sequence $(a_0, \ldots, a_n)$ of elements of a field K. After this, sum and product are given by formulas

$$\sum_k a_kx^k + \sum_k b_kx^k = \sum_k (a_k + b_k)x^k,$$

$$\left(\sum_k a_kx^k\right)\left(\sum_l b_lx^l\right) = \sum_m c_mx^m \qquad \text{where } c_m = \sum_{k+l=m} a_kb_l.$$

Rather more concretely, the same idea can be formulated as follows. We consider the set of all infinite sequences $(a_0, a_1, \ldots, a_n, \ldots)$ of elements of a ring A, every sequence consisting of zeros from some term onwards (this term may be different for different sequences). First we define addition of sequences by

$$(a_0, a_1, \ldots, a_n, \ldots) + (b_0, b_1, \ldots, b_n, \ldots) = (a_0 + b_0, a_1 + b_1, \ldots, a_n + b_n, \ldots).$$

All the ring axioms concerning addition are satisfied. Now for multiplication we define first just the multiplication of sequences by elements of A:

$$a(a_0, a_1, \ldots, a_n, \ldots) = (aa_0, aa_1, \ldots, aa_n, \ldots).$$

We write $E_k = (0, \ldots, 1, 0, \ldots)$ for the sequence consisting of 1 in the kth place and 0 everywhere else. Then it is easy to see that

$$(a_0, a_1, \ldots, a_n, \ldots) = \sum_{k \geqslant 0} a_kE_k. \tag{3}$$

Here the right-hand side is a finite sum in view of the condition imposed on sequences. Now define multiplication by

$$\left(\sum_k a_k E_k\right)\left(\sum_l b_l E_l\right) = \sum_{k,l} a_k b_l E_{k+l} \tag{4}$$

(on the right-hand side we must gather together all the terms for k and l with $k + l = n$ as the coefficient of E_n). It follows from (4) that E_0 is the unit element of the ring, and $E_k = E_1^k$. Setting $E_1 = x$ we can write the sequence (3) in the form $\sum a_k x^k$. Obviously this expression for the sequence is unique. It is easy to check that the multiplication (4) satisfies the axioms of a commutative ring, so that the ring we have constructed is the polynomial ring $A[x]$.

The polynomial ring $A[x, y]$ is defined as $A[x][y]$, or by generalising the above construction. In a similar way one defines the polynomial ring $A[x_1, \ldots, x_n]$ in any number of variables.

Example 3. All linear differential operators with constant (real) coefficients can be written as polynomials in the operators $\dfrac{\partial}{\partial x_1}, \ldots, \dfrac{\partial}{\partial x_n}$. Hence they form a ring

$$\mathbb{R}\left[\frac{\partial}{\partial x_1}, \ldots, \frac{\partial}{\partial x_n}\right].$$

Sending $\dfrac{\partial}{\partial x_i}$ to t_i defines an isomorphism

$$\mathbb{R}\left[\frac{\partial}{\partial x_1}, \ldots, \frac{\partial}{\partial x_n}\right] \cong \mathbb{R}[t_1, \ldots, t_n].$$

If $A = K$ is a field then the polynomial ring $K[x]$ is a subring of the rational function field $K(x)$, in the same way that the ring of integers $\mathbb{Z}$ is a subring of the rational field $\mathbb{Q}$. A ring which is a subring of a field has an important property: the relation $ab = 0$ is only possible in it if either $a = 0$ or $b = 0$; indeed, it follows easily from the commutative ring axioms that $a \cdot 0 = 0$ for any a. Hence if $ab = 0$ in a field and $a \neq 0$, multiplying on the left by a^{-1} gives $b = 0$. Obviously the same thing holds for a ring contained in a field.

A commutative ring with the properties that for any of its elements a, b the product $ab = 0$ only if $a = 0$ or $b = 0$, and that $0 \neq 1$, is called an *integral ring* or an *integral domain*. Thus a subring of any field is an integral domain.

Theorem I. *For any integral domain A, there exists a field K containing A as a subring, and such that every element of K can be written in the form ab^{-1} with $a, b \in A$ and $b \neq 0$. A field K with this property is called the* field of fractions *of A; it is uniquely defined up to isomorphism.*

For example, the field of fractions of $\mathbb{Z}$ is $\mathbb{Q}$, that of the polynomial ring $K[x]$ is the field of rational functions $K(x)$, and that of $K[x_1, \ldots, x_n]$ is $K(x_1, \ldots, x_n)$. Quite generally, fields of fractions give an effective method of constructing new fields.

Example 4. If A and B are two rings, their *direct sum* is the ring consisting of pairs (a, b) with $a \in A$ and $b \in B$, with addition and multiplication give by

$$(a_1, b_1) + (a_2, b_2) = (a_1 + a_2, b_1 + b_2),$$
$$(a_1, b_1)(a_2, b_2) = (a_1 a_2, b_1 b_2).$$

Direct sum is denoted by $A \oplus B$. The direct sum of any number of rings is defined in a similar way.

A direct sum is not an integral domain: $(a, 0)(0, b) = (0, 0)$, which is the zero element of $A \oplus B$.

The most important example of commutative rings, which includes non-integral rings, is given by rings of functions. Properly speaking, the direct sum $A \oplus \cdots \oplus A$ of n copies of A can be viewed as the ring of function on a set of n elements (such as $\{1, 2, \ldots, n\}$) with values in A: the element $(a_1, \ldots, a_n) \in A \oplus \cdots \oplus A$ can be identified with the function f given by $f(i) = a_i$. Addition and multiplication of functions are given as usual by operating on their values.

Example 5. The set of all continuous functions (to be definite, real-valued functions) on the interval $[0, 1]$ forms a commutative ring $\mathscr{C}$ under the usual definition of addition and multiplication of functions. This is not an integral domain: if f and g are the functions depicted in Figures 8 and 9, then obviously $fg = 0$. In the definition, we could replace real-valued functions by complex-valued ones, and the interval by an arbitrary topogical space. Rings of this form occuring in analysis are usually considered together with a topology on their set of elements, or a norm defining a topology. For example, in our case it is standard to consider the norm

$$\|f\| = \sup_{0 \leqslant x \leqslant 1} |f(x)|.$$

Examples analogous to those of Figures 8 and 9 can also be constructed in the ring of C^∞ functions on the interval.

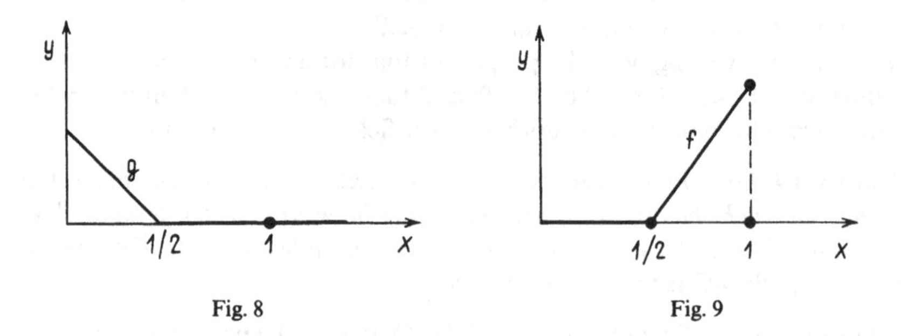

Fig. 8 Fig. 9

Example 6. The ring of functions of 1 complex variable holomorphic at the origin is an integral domain, and its field of fractions is the field of Laurent series

(§2, Example 5). Similarly to §2, Example 5 we can define the *ring of formal power series* $\sum_{n=0}^{\infty} a_n t^n$ with coefficients a_n in any field K. This can also be constructed as in Example 2, if we just omit the condition that the sequences $(a_0, a_1, \ldots, a_n, \ldots)$ are 0 from some point onwards. This is also an integral domain, and its field of fractions is the *field of formal Laurent series* $K((t))$. The ring of formal power series is denoted by $K[[t]]$.

Example 7. The ring $\mathcal{O}_n$ of functions in n complex variables holomorphic at the origin, that is of functions that can be represented as power series

$$\sum a_{i_1 \ldots i_n} z_1^{i_1} \ldots z_n^{i_n},$$

convergent in some neighbourhood of the origin. By analogy with Example 6 we can define the *rings of formal power series* $\mathbb{C}[[z_1, \ldots, z_n]]$ with complex coefficients, and $K[[z_1, \ldots, z_n]]$ with coefficients in any field K.

Example 8. We return to the curve C defined in the plane by the equation $F(x, y) = 0$, where F is a polynomial with coefficients in a field K, as considered in §2. With each polynomial $P(x, y)$ we associate the function on the set of points of C defined by restricting P to C. Functions of this form are *polynomial functions* on C. Obviously they form a commutative ring, which we denote by $K[C]$. If F is a product of factors then the ring $K[C]$ may not be an integral domain. For example if $F = xy$ then C is the union of the coordinate axes; then x is zero on the y-axis, and y on the x-axis, so that their product is zero on the whole curve C. However, if F is an irreducible polynomial then $K[C]$ is an integral domain. In this case the field of fractions of $K[C]$ is the rational function field $K(C)$ of C; the ring $k[C]$ is called the *coordinate ring* of C.

Taking an algebraic curve C into the ring $K[C]$ is also an example of 'coordinatisation', and in fact is more precise than taking C to $K(C)$, since $K[C]$ determines $K(C)$ (as its field of fractions), whereas there exist curves C and C' for which the fields $K(C)$ and $K(C')$ are isomorphic, but the rings $K[C]$ and $K[C']$ are not.

Needless to say, we could replace the algebraic curve given by $F(x, y) = 0$ by an algebraic surface given by $F(x, y, z) = 0$, and quite generally by an algebraic variety.

Example 9. Consider an arbitrary set M, and the commutative ring A consisting of all functions on M with values in the finite field with two elements $\mathbb{F}_2$ (§2, Example 1). Thus A consists of all maps from M to $\mathbb{F}_2$. Since $\mathbb{F}_2$ has only two elements 0 and 1, a function with values in $\mathbb{F}_2$ is uniquely determined by the subset $U \subset M$ of elements on which it takes the value 1 (on the remainder it takes the value 0). Conversely, any subset $U \subset M$ determines a function φ_U with $\varphi_U(m) = 1$ if $m \in U$ and $\varphi_U(m) = 0$ if $m \notin U$. It is easy to see which operations on subsets correspond to the addition and multiplication of functions:

$$\varphi_U \cdot \varphi_V = \varphi_{U \cap V} \quad \text{and} \quad \varphi_U + \varphi_V = \varphi_{U \Delta V},$$

where $U \Delta V$ is the symmetric difference, $U \Delta V = (U \cup V) \setminus (U \cap V)$. Thus our ring can be described as being made up of subsets $U \subset V$ with the operations of symmetric difference and intersection as sum and product. This ring was introduced by Boole as a formal notation for assertions in logic. Since $x^2 = x$ for every element of $\mathbb{F}_2$, this relations holds for any function with values in $\mathbb{F}_2$, that is, it holds in A. A ring for which every element x satisfies $x^2 = x$ is a *Boolean ring*.

More general examples of Boolean rings can be constructed quite similarly, by taking not all subsets of M, but only some system S of subsets containing together with U and V the subsets $U \cap V$ and $U \cup V$, and together with U its complement. For example, we could consider a topological space having the property that every open set is also closed (such a space is called 0-*dimensional*), and let S be the set of open subsets of M. It can be proved that every Boolean ring can be obtained in this way. In the following section § 4 we will indicate the principle on which the proof of this is based.

The qualitatively new phenomenon that occurs on passing from fields to arbitrary commutative rings is the appearance of a nontrivial theory of divisibility. An element a of a ring A is *divisible* by an element b if there exists c in A such that $a = bc$. A field is precisely a ring in which the divisibility theory is trivial: any element is divisible by any nonzero element, since $a = b(ab^{-1})$. The classical example of divisibility theory is the theory of divisibility in the ring $\mathbb{Z}$: this was constructed already in antiquity. The basic theorem of this theory is the fact that any integer can be uniquely expressed as a product of prime factors. The proof of this theorem, as is well known, is based on division with remainder (or the Euclidean algorithm).

Let A be an arbitrary integral domain. We say that an element $a \in A$ is *invertible* or is a *unit* of A if it has an inverse in A; in $\mathbb{Z}$ the units are ± 1, in $K[x]$ the nonzero constants $c \in K$, and in $K[\![x]\!]$ the series $\sum_{i=0}^{\infty} a_n x^n$ with $a_0 \neq 0$. Any element of A is divisible by a unit. An element a is said to be *prime* if its only factorisations are of the form $a = c(c^{-1}a)$ where c is a unit. If an integral domain A has the property that every nonzero element can be written as a product of primes, and this factorisation is unique up to renumbering the prime factors and multiplication by units, we say that A is a *unique factorisation domain* (*UFD*) or a *factorial ring*. Thus $\mathbb{Z}$ is a UFD, and so is $K[x]$ (the proof uses division with remainder for polynomials). It can be proved that if A is a UFD then so is $A[x]$; hence $A[x_1, \ldots, x_n]$ is also a UFD. The prime elements of a polynomial ring are called *irreducible polynomials*. In $\mathbb{C}[x]$ only the linear polynomials are irreducible, and in $\mathbb{R}[x]$ only linear polynomials and quadratic polynomials having no real roots. In $\mathbb{Q}[x]$ there are irreducible polynomials of any degree, for example the polynomial $x^n - p$ where p is any prime number.

Important examples of UFDs are the ring $\mathcal{O}_n$ of functions in n complex variables holomorphic at the origin, and the formal power series ring $K[\![t_1, \ldots, t_n]\!]$ (Example 7). The proof that these are UFDs is based on the Weierstrass preparation theorem, which reduces the problem to functions (or formal power series) which are polynomials in one of the variables. After this, one applies the fact that $A[t]$ is a UFD (provided A is) and an induction.

Example 10. The Gaussian Integers. It is easy to see that the complex numbers of the form $m + ni$, where m and n are integers, form a ring. This is a UFD, as can also be proved using division with remainder (but the quantity that decreases on taking the remainder is $m^2 + n^2$). Since in this ring

$$m^2 + n^2 = (m + ni)(m - ni),$$

divisibility in it can be used as the basis of the solution of the problem of representing integers as the sum of two squares.

Example 11. Let ε be a (complex) root of the equation $\varepsilon^2 + \varepsilon + 1 = 0$. Complex numbers of the form $m + n\varepsilon$, where m and n are integers, also form a ring, which is also a UFD. In this ring the expression $m^3 + n^3$ factorises as a product:

$$m^3 + n^3 = (m + n)(m + n\varepsilon)(m + n\bar{\varepsilon}),$$

where $\bar{\varepsilon} = \varepsilon^2 = -(1 + \varepsilon)$ is the complex conjugate of ε. Because of this, divisibility theory in this ring serves as the basis of the proof of Fermat's Last Theorem for cubes. The 18th century mathematicians Lagrange and Euler were amazed to find that the proof of a theorem of number theory (the theory of the ring $\mathbb{Z}$) can be based on introducing other numbers (elements of other rings).

Example 12. We give an example of an integral domain which is not a UFD; this is the ring consisting of all complex numbers of the form $m + n\sqrt{-5}$ where $m, n \in \mathbb{Z}$. Here is an example of two different factorisations into irreducible factors:

$$3^2 = (2 + \sqrt{-5})(2 - \sqrt{-5})$$

We need only check that 3, $2 + \sqrt{-5}$ and $2 - \sqrt{-5}$ are irreducible elements. For this, we write $N(\alpha)$ for the square of the absolute value of α; if $\alpha = n + m\sqrt{-5}$ then $N(\alpha) = (n + m\sqrt{-5})(n - m\sqrt{-5}) = n^2 + 5m^2$, which is a positive integer. Moreover, it follows from the properties of absolute value that $N(\alpha\beta) = N(\alpha)N(\beta)$. If, say, $2 + \sqrt{-5}$ is reducible, for example $2 + \sqrt{-5} = \alpha\beta$, then $N(2 + \sqrt{-5}) = N(\alpha)N(\beta)$. But $N(2 + \sqrt{-5}) = 9$, and hence there are only three possibilities: $(N(\alpha), N(\beta)) = (3, 3)$ or $(1, 9)$ or $(9, 1)$. The first of these is impossible, since 3 cannot be written in the form $n^2 + 5m^2$ with n, m integers. In the second $\beta = \pm 1$ and in the third $\alpha = \pm 1$, so α or β is a unit. This proves that $2 + \sqrt{-5}$ is irreducible.

To say that a ring is not a UFD does not mean to say that it does not have an interesting theory of divisibility. On the contrary, in this case the theory of divisibility becomes especially interesting. We will discuss this in more detail in the following section § 4.

§4. Homomorphisms and Ideals

A further difference of principle between arbitrary commutative rings and fields is the existence of nontrivial homomorphisms. A *homomorphism* of a ring A to a ring B is a map $f: A \to B$ such that

$$f(a_1 + a_2) = f(a_1) + f(a_2), \quad f(a_1 a_2) = f(a_1) \cdot f(a_2) \quad \text{and} \quad f(1_A) = 1_B$$

(we write 1_A and 1_B for the identity elements of A of B). An *isomorphism* is a homomorphism having an inverse.

If a ring has a topology, then usually only continuous homomorphisms are of interest.

Typical examples of homomorphisms arise if the rings A and B are realised as rings of functions on sets X and Y (for example, continuous, differentiable or analytic functions, or polynomial functions on an algebraic curve C). A map $\varphi: Y \to X$ transforms a function F on X into the function φ^*F on Y defined by the condition

$$(\varphi^*F)(y) = F(\varphi(y)).$$

If φ satisfies the natural conditions for the theory under consideration (that is, if φ is a continuous, differentiable or analytic map, or is given by polynomial expressions) then φ^* defines a homomorphism of A to B. The simplest particular case is when φ is an embedding, that is Y is a subset of X. Then φ^* is simply the restriction to Y of functions defined on X.

Example 1. If C is a curve, defined by the equation $F(x, y) = 0$ where $F \in K[x, y]$ is an irreducible polynomial, then restriction to C defines a homomorphism $K[x, y] \to K[C]$.

The case which arises most often is when Y is one point of a set X, that is $Y = \{x_0\}$ with $x_0 \in X$; then we are just evaluating a function, taking it into its value at x_0.

Example 2. If $x_0 \in C$ then taking each function of $K[C]$ into its value at x_0 defines a homomorphism $K[C] \to K$.

Example 3. If $\mathscr{C}$ is the ring of continuous functions on $[0, 1]$ and $x_0 \in [0, 1]$ then taking a function $\varphi \in \mathscr{C}$ into its value $\varphi(x_0)$ is a homomorphism $\mathscr{C} \to \mathbb{R}$. If A is the ring of functions which are holomorphic in a neighbourhood of 0, then taking $\varphi \in A$ into its value $\varphi(0)$ is a homomorphism $A \to \mathbb{C}$.

Interpreting the evaluation of a function at a point as a homomorphism has led to a general point of view on the theory of rings, according to which a commutative ring can very often be interpreted as a ring of functions on a set, the 'points' of which correspond to homomorphisms of the original ring into fields. The original example is the ring $K[C]$, where C is an algebraic variety, and from it, the geometric intuition spreads out to more general rings. Thus the concept that 'every geometric object can be coordinatised by some ring of functions on it' is complemented by another, that 'any ring coordinatises some geometric object'.

We have already run into these two points of view, the algebraic and functional, in the definition of the polynomial ring in §3. The relation between the two will gradually become deeper and clearer in what follows.

Example 4. Consider the ring A of functions which are holomorphic in the disc $|z| < 1$ and continuous for $|z| \leqslant 1$. In the same way as before, any point z_0 with $|z_0| \leqslant 1$ defines a homomorphism $A \to \mathbb{C}$, taking a function $\varphi \in A$ into $\varphi(z_0)$. It can be proved that all homomorphisms $A \to \mathbb{C}$ over $\mathbb{C}$ are provided in this way. Consider the boundary values of functions in A; these are continuous functions on the circle $|z| = 1$, whose Fourier coefficients with negative index are all zero, that is, with Fourier expansions of the form $\sum_{n \geqslant 0} c_n e^{2\pi i n \varphi}$. Since a function $f \in A$ is determined by its boundary values, A is isomorphic to the ring of continuous functions on the circle with Fourier series of the indicated type. However, in this interpretation, only the homomorphisms of A corresponding to points of the boundary circle $|z| = 1$ are immediately visible. Thus considering the set of all homomorphisms sometimes helps to reestablish the set on which the elements of the ring should naturally be viewed as functions.

In the ring of functions which are holomorphic and bounded for $|z| \leqslant 1$, by no means all homomorphisms are given in terms of points z_0 with $|z_0| < 1$. The study of these is related to delicate questions of the theory of analytic functions.

For a Boolean ring (see §3, Example 9), it is easy to see that the image of a homomorphism $\varphi: A \to F$ in a field F is a field with two elements. Hence, conversely, any element $a \in A$ sends a homomorphism φ to the element $\varphi(a) \in \mathbb{F}_2$. This is the idea of the proof of the main theorem on Boolean rings: for M one takes the set of all homomorphisms $A \to \mathbb{F}_2$, and A is interpreted as a ring of functions on M with values in $\mathbb{F}_2$.

Example 5. Let $\mathscr{X}$ be a compact subset of the space $\mathbb{C}^n$ of n complex variables, and A the ring of functions which are uniform limits of polynomials on $\mathscr{X}$. The homomorphisms $A \to \mathbb{C}$ over $\mathbb{C}$ are not exhausted by those corresponding to points $z \in \mathscr{X}$; they are in 1-to-1 correspondence with points of the so-called *polynomial convex hull* of $\mathscr{X}$, that is with the points $z \in \mathbb{C}^n$ such that $|f(z)| \leqslant \operatorname{Sup}_{\mathscr{X}} |f|$ for every polynomial f.

Example 6. Suppose we assign to an integer the symbol $\mathbb{0}$ if it is even and $\mathbb{1}$ if it is odd. We get a homomorphism $\mathbb{Z} \to \mathbb{F}_2$ of the ring of integers to the field with 2 elements $\mathbb{F}_2$ (addition and multiplication tables of which were given in §1, Figures 3 and 4). Properly speaking, the operations on $\mathbb{0}$ and $\mathbb{1}$ were defined in order that this map should be a homomorphism.

Let $f: A \to B$ be a homomorphism of commutative rings. The set of elements $f(a)$ with $a \in A$ forms a subring of B, as one sees from the definition of homomorphism; this is called the *image* of f, and is denoted by $\operatorname{Im} f$ or $f(A)$. The set of elements $a \in A$ for which $f(a) = 0$ is called the *kernel* of f, and denoted by $\operatorname{Ker} f$. If $B = \operatorname{Im} f$ then we say that B is a *homomorphic image* of A.

If $\operatorname{Ker} f = 0$ then f is an isomorphism from A to the subring $f(A)$ of B; for if $f(a) = f(b)$ then it follows from the definition of homomorphism that $f(a-b) = 0$, that is, $a - b \in \operatorname{Ker} f = 0$ and so $a = b$. Thus f is a 1-to-1 correspondence from A to $f(A)$, and hence an isomorphism. This fact draws our attention to the importance of the kernels of homomorphisms.

It follows at once from the definitions that if $a_1, a_2 \in \operatorname{Ker} f$ then $a_1 + a_2 \in \operatorname{Ker} f$, and if $a \in \operatorname{Ker} f$ then $ax \in \operatorname{Ker} f$ for any $x \in A$. We say that a nonempty subset I of a ring A is an *ideal* if it satisfies these two properties, that is,

$$a_1, a_2 \in I \Rightarrow a_1 + a_2 \in I, \quad \text{and} \quad a \in I \Rightarrow ax \in I \quad \text{for any } x \in A.$$

Thus the kernel of any homomorphism is an ideal. A universal method of constructing ideals is as follows. For an arbitrary set $\{a_\lambda\}$ of elements of A, consider the set I of elements which can be represented in the form $\sum x_\lambda a_\lambda$ for some $x_\lambda \in A$ (we assume that only a finite number of nonzero terms appears in each sum). Then I is an ideal; it is called the *ideal generated* by $\{a_\lambda\}$. Most commonly the set $\{a_\lambda\}$ is finite. An ideal $I = (a)$ generated by a single element is called a *principal ideal*. If a divides b then $(b) \subset (a)$.

A field K has only two ideals, (0) and $(1) = K$. For if $I \subset K$ is an ideal of K and $0 \neq a \in I$ then $I \ni aa^{-1}b = b$ for any $b \in K$, and hence $I = K$ (this is another way of saying that the theory of divisibility is trivial in a field). It follows from this that any homomorphism $K \to B$ from a field is an isomorphism with some subfield of B.

Conversely, if a commutative ring A does not have any ideals other than (0) and (1), and $0 \neq 1$ then A is a field. Indeed, then for any element $a \neq 0$ we must have $(a) = A$, and in particular $1 \in (a)$, so that $1 = ab$ for some $b \in A$, and a has an inverse.

In the ring of integers $\mathbb{Z}$, any ideal I is principal: it is easy to see that if $I \neq (0)$ then $I = (n)$, where n is the smallest positive integer contained in I. The same is true of the ring $K[x]$; here any ideal I is of the form $I = (f(x))$, where $f(x)$ is a polynomial of smallest degree contained in I. In the ring $K[x, y]$, it is easy to see that the ideal I of polynomials without constant term is not principal; it is of the form (x, y). An integral domain in which every ideal is principal is called a *principal ideal domain* (PID).

It is not by chance that the rings $\mathbb{Z}$ and $K[x]$ are unique factorisation domains: one can prove that any PID is a UFD. But the example of $K[x, y]$ shows that there are more UFDs than PIDs. In exactly the same way, the ring $\mathcal{O}_n$ of functions of $n > 1$ complex variables which are holomorphic at the origin (§3, Example 7) is a UFD but not a PID. The study of ideals in this ring plays an important role in the study of *local analytic varieties*, defined in a neighbourhood of the origin by equations $f_1 = 0, \ldots, f_m = 0$ (with $f_i \in \mathcal{O}_n$). The representation of such varieties as a union of irreducibles, the notion of their dimension, and so on, are based on properties of these ideals.

Example 7. In the ring $\mathscr{C}$ of continuous functions on the interval, taking a function φ to its value $\varphi(x_0)$ at x_0 is a homomorphism with kernel the ideal $I_{x_0} = \{\varphi \in \mathscr{C} | \varphi(x_0) = 0\}$. It is easy to see that I_{x_0} is not principal: any function which tends to 0 substantially slower than a given function $\varphi(x)$ as $x \to x_0$ (for example $\sqrt{|\varphi(x)|}$ is not contained in the ideal $(\varphi(x))$. One can prove in a similar way that I_{x_0} is not even generated by any finite number $\varphi_1, \ldots, \varphi_m \in I_{x_0}$ of functions in it.

Another example of a similar nature can be obtained in the ring $\mathscr{E}$ of germs of C^∞ functions at 0 on the line (by definition two functions defined the same germ at 0 if they are equal in some neighbourhood of 0). The ideal M_n of germs of functions which vanish at 0 together with all of their derivatives of order $\leqslant n$ is principal, equal to (x^{n+1}), but the ideal M_∞ of germs of functions all of whose derivatives vanish at 0 (such as e^{-1/x^2}) is not generated by any finite system of functions, as can be proved. In any case, the extent to which these examples carry conviction should not be exaggerated: it is more natural to use the topology of the ring $\mathscr{C}$ of continuous functions, and consider ideals *topologically generated* by functions $\varphi_1, \ldots, \varphi_m$, that is, the closure of the ideal $(\varphi_1, \ldots, \varphi_m)$. In this topological sense, any ideal of $\mathscr{C}$ is generated by one function. The same considerations apply to the ring $\mathscr{E}$, but its topology is defined in a more complicated way, and, for example, the fact that the ideal M_∞ is not generated by any finite system of functions then contains more genuine information.

Let I and J be two ideals of a ring A. The ideal generated by the set of all products ij with $i \in I$ and $j \in J$ is called the *product* of I and J and denoted by IJ. Multiplication of principal ideals agrees with that of elements: if $I = (a)$ and $J = (b)$ then $IJ = (ab)$. By analogy with the question of the unique factorisation of elements into prime factors, we can pose the question of factorising ideals of a ring as a product of ideals which cannot be factorised any further. Of course, both of these properties hold in a principal ideal domain. But there exist important types of ring which are not factorial, but in which the ideals have unique factorisation into products of irreducible factors.

Example 8. Consider the ring of numbers of the form $m + n\sqrt{-5}$ with $m, n \in \mathbb{Z}$, given in §3, Example 12 as an example of a nonfactorial ring. The factorisation

$$3^2 = (2 + \sqrt{-5})(2 - \sqrt{-5}) \tag{1}$$

which we gave in §3 is not a factorisation into prime factors if we replace the numbers by the corresponding principal ideals. It is not hard to see that

$$(2 + \sqrt{-5}) = (2 + \sqrt{-5}, 3)^2, \quad (2 - \sqrt{-5}) = (2 - \sqrt{-5}, 3)^2$$

and

$$(3) = (2 + \sqrt{-5}, 3)(2 - \sqrt{-5}, 3),$$

so that (1) is just the product $(2 + \sqrt{-5}, 3)^2(2 - \sqrt{-5}, 3)^2$ in which the factors are grouped in different ways. The possibility of an analogous factorisation is the basis of the arithmetic of algebraic numbers. This is the historical explanation of the term 'ideal': the prime ideals into which an irreducible number factorises (for example 3 or $2 + \sqrt{-5}$) were first considered as 'ideal prime factors'.

The numbers 3 and $2 + \sqrt{-5}$ do not have common factors other than ± 1, since they are irreducible. But the ideal $(3, 2 + \sqrt{-5})$ is their greatest common divisor (more precisely, it is the g.c.d. of the ideals (3), $(2 + \sqrt{-5})$). Similarly to the fact that the greatest common divisor of integers a and b can be expressed as $au + bv$, the ideal $(3, 2 + \sqrt{-5})$ consists of all numbers of the form $3\alpha + (2 + \sqrt{-5})\beta$.

The notion of ideal is especially important because of the fact that the relation between homomorphisms and ideals is reversible: every ideal is the kernel of some homomorphism. In order to construct from an ideal I of a ring A the ring B which A will map to under this homomorphism, we introduce the following definitions.

Elements a_1 and a_2 of a ring A are said to be *congruent* modulo an ideal I of A (or congruent mod I) if $a_1 - a_2 \in I$. This is written as follows:

$$a_1 \equiv a_2 \bmod I.$$

If $A = \mathbb{Z}$ and $I = (n)$ then we obtain the classical notion of congruence in number theory: a_1 and a_2 are congruent mod n if they have the same remainder on dividing by n.

Congruence modulo an ideal is an equivalence relation, and it decomposes A as a union of disjoint classes of elements congruent to one another mod I. These classes are also called *residue classes* modulo I.

Let Γ_1 and Γ_2 be two residue classes mod I. It is easy to see that however we choose elements $a_1 \in \Gamma_1$ and $a_2 \in \Gamma_2$, the sum $a_1 + a_2$ will belong to the same residue class Γ. This class is called the sum of Γ_1 and Γ_2. In a similar way we define the product of residue classes. It is not hard to see that the set of all residue classes modulo an ideal I with the above definition of addition and multiplication forms a commutative ring; this is called the *residue class ring* or the *quotient ring* of A modulo I, and denoted by A/I.

For example if $A = \mathbb{Z}$ and $I = (2)$ then I has 2 residue classes, the even and odd numbers; and the ring $\mathbb{Z}/(2)$ coincides with the field $\mathbb{F}_2$.

It is easy to see that taking an element $a \in A$ into its residue class mod I is a homomorphism $f: A \to A/I$, with kernel I. This is called the *canonical homomorphism* of a quotient ring.

Canonical homomorphisms of rings to their quotient rings give a more explicit description of arbitrary homomorphisms. Namely, the following assertion is easy to verify:

Theorem I. *For any ring homomorphism $\varphi: A \to B$, the image ring* $\operatorname{Im} \varphi$ *is isomorphic to the quotient ring $A/\operatorname{Ker} \varphi$, and the isomorphism σ between them can be chosen so as to take the canonical homomorphism $\psi: A \to A/\operatorname{Ker} \varphi$ into $\varphi: A \to \operatorname{Im} \varphi$.*

More precisely, for any $a \in A$, the isomorphism σ takes $\psi(a)$ into $\varphi(a)$ (recall that $\sigma(\psi(a)) \in \operatorname{Im} \varphi \subset B$, so that $\sigma(\psi(a))$ and $\varphi(a)$ are both elements of B).

This result is most often applied in the case $\operatorname{Im} \varphi = B$. In this case the assertion is the following.

II. Homomorphisms Theorem. *A homomorphic image is isomorphic to the quotient ring modulo the kernel of the homomorphism.*

Under the canonical homomorphism f, the inverse image $f^{-1}(J)$ of any ideal $J \subset A/I$ is an ideal of A containing I, and the image $f(I')$ of any ideal I' containing I is an ideal of A/I. This establishes a 1-to-1 correspondence between ideals of the quotient ring A/I and ideals of A containing I.

In particular, as we know, A/I is a field if and only if it has exactly two ideals, (0) and (1), and this means that I is not contained in any bigger ideal other that A itself. Such an I is called a *maximal ideal*. It can be proved (using Zorn's lemma from set theory) that any ideal $I \neq A$ is contained in at least one maximal ideal.

Together with the construction of fields of fractions, considering quotient rings modulo maximal ideals is the most important method of constructing fields. We now show how to use this to obtain a series of new examples of fields.

Example 9. In $\mathbb{Z}$, maximal ideals are obviously of the form (p), where p is a prime number. Thus $\mathbb{Z}/(p)$ is a field; it has p elements, and is denoted by $\mathbb{F}_p$. Up to now we have only constructed fields $\mathbb{F}_2$ and $\mathbb{F}_3$ with 2 or 3 elements. If n is not prime, then the ring $\mathbb{Z}/(n)$ is not a field, and as one sees easily, is not even an integral domain.

Example 10. Consider now the polynomial ring $K[x]$; its maximal ideals are of the form $(\varphi(x))$ with $\varphi(x)$ an irreducible polynomial. In this case, the quotient ring $L = K[x]/(\varphi(x))$ is a field. Write α for the image of x under the homomorphism $K[x] \to L = K[x]/(\varphi(x))$. Then for tautological reasons, $\varphi(\alpha) = 0$, so that the polynomial φ has a root in L. Write n for the degree of φ. Using division with remainder, we can represent any polynomial $u(x) \in K[x]$ in a unique way in the form $u(x) = \varphi(x)\psi(x) + v(x)$, where v is a polynomial of degree less than n. It follows from this that any element of L can be uniquely expressed in the form

$$a_0 + a_1\alpha + a_2\alpha^2 + \cdots + a_{n-1}\alpha^{n-1}, \tag{2}$$

where $a_0, \ldots, a_{n-1}$ are arbitrary elements of K.

If $K = \mathbb{R}$ and $\varphi(x) = x^2 + 1$ then we construct in this way the field $\mathbb{C}$ of complex numbers; here i is the image of x in $\mathbb{R}[x]/(x^2 + 1)$, and $a + bi$ is the image of $a + bx$.

The above construction gives an extension field L/K in which a given polynomial $\varphi(t)$ has a root. Iterating this process, it can be proved that for any field K, there exists an extension Σ/K such that any polynomial $\varphi \in \Sigma[t]$ has a root in Σ. A field having this property is said to be *algebraically closed*. For example, $\mathbb{C}$ is algebraically closed.

Let K be a field with p elements. If φ is an irreducible polynomial of degree n over K then the expression (2) shows that L has p^n elements. Based on these ideas, one can prove the following results, which together describe all finite fields.

III. Theorem on Finite Fields.

(i) *The number of elements of a finite field is of the form p^n, where p is the characteristic.*

(ii) *For each p and n there exists a field $\mathbb{F}_q$ with $q = p^n$ elements.*

(iii) *Two finite fields with the same number of elements are isomorphic.*

Finite fields have very many applications. One of them, which specifically uses the fact that they are finite, relates to the theory of error-correcting codes. By definition, a *code* consists of a finite set E (an 'alphabet') and a subset U of the set E^n of all possible sequences $(a_1, \dots, a_n)$ with $a_i \in E$. This subset is to be chosen in such a way that any two sequences in U should differ at a sufficiently large number of places. Then when we transmit a 'message' $(u_1, \dots, u_n) \in U$, we can still reconstruct the original message even if a small number of the u_i are corrupted. A wealth of material for making such choices is provided by taking E to be some finite field $\mathbb{F}_q$, and U to be a subspace of the vector space $\mathbb{F}_q^n$. Furthermore, the greatest success has been achieved by taking $\mathbb{F}_q^n$ and U to be finite-dimensional subspaces of the field $\mathbb{F}_q(t)$ or even of $\mathbb{F}_q(C)$, where C is an algebraic curve, and determining the choice of these subspaces by means of certain geometric conditions (such as considering functions with specified zeros and poles). Thus coding theory has turned out to be related to very delicate questions of algebraic geometry over finite fields.

Considering already the simplest ring $\mathbb{Z}/(n)$ leads to interesting conclusions. Let K be an arbitrary field, with identity element 1. Consider the map f from $\mathbb{Z}$ to K defined by

$$f(n) = n \cdot 1, \quad \text{that is } f(n) = \begin{cases} 1 + \cdots + 1 \quad (n \text{ times}) & \text{if } n > 0 \\ 0 & \text{if } n = 0 \\ -(1 + \cdots + 1) \quad (-n \text{ times}) & \text{if } n < 0. \end{cases}$$

It is easy to see that f is a homomorphism. Two cases are possible, either $\operatorname{Ker} f = 0$ or $\operatorname{Ker} f \neq 0$.

In the first case $f(\mathbb{Z})$ is a subring of K isomorphic to $\mathbb{Z}$. Since K is a field, it must also contain the ratio of elements of this ring, which one easily checks form a subfield $K_0 \subset K$. It follows from the uniqueness of fields of fractions that K_0 is isomorphic to $\mathbb{Q}$, that is, K contains a subfield isomorphic to $\mathbb{Q}$.

In the second case, suppose that $\operatorname{Ker} f = (n)$. Obviously, n must be a prime number, since otherwise $f(\mathbb{Z}) \cong \mathbb{Z}/n$ would not be integral. But then $f(\mathbb{Z}) = \mathbb{Z}/(p) = \mathbb{F}_p$ is a field with p elements.

Thus we have seen that any field K contains either the field $\mathbb{Q}$ of rational numbers, or a field $\mathbb{F}_p$ with some prime number of elements. These fields are called the *prime fields*; any field is an extension of one of these. If K contains a field with p elements, then $px = 0$ for every $x \in K$. In this case, p is called the *characteristic* of K, and we say that K is a field of *finite characteristic*, and write $\operatorname{char} K = p$. If K contains $\mathbb{Q}$ then $nx = 0$ only if $n = 0$ or $x = 0$; in this case, we say that K has *characteristic* 0, and write $\operatorname{char} K = 0$ (or sometimes $\operatorname{char} K = \infty$).

The fields $\mathbb{Q}$, $\mathbb{R}$, $\mathbb{C}$, $\mathbb{Q}(x)$, $\mathbb{R}(x)$, $\mathbb{C}(x)$ are of characteristic 0. The field $\mathbb{F}_p$ with p elements has characteristic p, as have $\mathbb{F}_p(x)$, $\mathbb{F}_p(x, y)$ and so on.

A ring A/I can be embedded in a field if and only if it is an integral domain. This means that $I \neq A$ and if $a, b \in A$ and $ab \in I$ then either $a \in I$ or $b \in I$. We say that an ideal is *prime* if it satisfies this condition. For example, the principal ideal $I = (F(x, y)) \subset K[x, y]$ is prime if F is an irreducible polynomial: the ring $K[x, y]/I = K[C]$ (where C is the algebraic curve with equation $F(x, y) = 0$) can be embedded in the field $K(C)$. We can say that a prime ideal is the kernel of a homomorphism $\varphi \colon A \to K$, where K is a field (but possibly $\varphi(A) \neq K$).

It can be shown that the ideals of Example 8 which are irreducible (in the sense that they do not decompose as a product of factors) are exactly the prime ideals in the sense of the above definition.

At the beginning of this section we discussed the point of view that any ring can be thought of as a ring of functions on some space X. The 'points' of the space correspond to homomorphisms of the ring into fields. Hence we can interpret them as maximal ideals (or in another version, prime ideals) of the ring. If M is an ideal 'specifying a point $x \in X$' and $a \in A$, then the 'value' $a(x)$ of a at x is the residue class $a + M$ in A/M. The resulting geometric intuition might at first seem to be rather fanciful. For example, in $\mathbb{Z}$, maximal ideals correspond to prime numbers, and the value at each 'point' (p) is an element of the field $\mathbb{F}_p$ corresponding to p (thus we should think of $1984 = 2^6 \cdot 31$ as a function on the set of primes[1], which vanishes at (2) and (31); we can even say that it has a zero of multiplicity 6 at (2) and of multiplicity 1 at (31)). However, this is nothing more than a logical extension of the analogy between the ring of integers $\mathbb{Z}$ and the polynomial ring $K[t]$, under which prime numbers $p \in \mathbb{Z}$ correspond to irreducible polynomials $P(t) \in K[t]$. Continuing the analogy, the equation $a_0(t) + a_1(t)x + \cdots + a_n(t)x^n = 0$ with $a_i(t) \in K[t]$ defining an algebraic function $x(t)$ should be considered as analogous to the defining equation $a_0 + a_1 x + \cdots + a_n x^n = 0$ with $a_i \in \mathbb{Z}$ of an algebraic number. In fact, in the study of algebraic numbers, it has turned out to be possible to apply the intuition of the theory of algebraic functions, and even of the Riemann surfaces associated with them. Several of the most beautiful achievements of number theory can be attributed to the systematic development of this point of view.

Another version of the same ideas plays an important role in considering maps $\varphi \colon Y \to X$ (for example, analytic maps between complex analytic manifolds). If A is the ring of analytic functions on X and B that on Y, then as we said at the beginning of this section, a map φ determines a homomorphism $\varphi^* \colon A \to B$. Let

---

[1] This example was chosen because of the year the book was written, and has nothing to do with the fiction of George Orwell (translator's footnote).

$Z \subset X$ be a submanifold and $I \subset A$ the ideal of functions vanishing on Z. If $I = (f_1, \ldots, f_r)$, this means that Z is defined by the equations $f_1 = 0, \ldots, f_r = 0$. The inverse image $\varphi^{-1}(Z)$ of Z in Y is defined by the equations $\varphi^* f_1 = 0, \ldots, \varphi^* f_r = 0$, and it is natural to associate with it the ring $B/(\varphi^* f_1, \ldots, \varphi^* f_r) = B/(\varphi^* I)B$. Suppose for example that φ is the map of a line Y to a line X given by $x = y^2$. If Z is the point $x = \alpha \neq 0$ then $\varphi^{-1}(Z)$ consists of two points $y = \pm\sqrt{\alpha}$, and

$$B/(\varphi^* I)B \cong \mathbb{C}[y]/(y^2 - \alpha) \cong \mathbb{C}[y]/(y - \sqrt{\alpha}) \oplus \mathbb{C}[y]/(y + \sqrt{\alpha}) \cong \mathbb{C} \oplus \mathbb{C};$$

that is, it is in fact the ring of functions on a pair of points. But if Z is the point $x = 0$ then $\varphi^{-1}(Z)$ is the single point $y = 0$, and $B/(\varphi^* I)B \cong \mathbb{C}[y]/y^2$. This ring consists of elements of the form $\alpha + \beta\varepsilon$, with $\alpha, \beta \in \mathbb{C}$, and ε the image of y, with $\varepsilon^2 = 0$; it can be interpreted as the 'ring of functions on a double point', and it gives much more precise information on the behaviour of the map $x = y^2$ in a neighbourhood of $x = 0$ than just the set-theoretic inverse image of this point. In the same way, the study of singularities of analytic maps leads to considering much more complicated commutative rings as invariants of these singularities.

Example 11. Let $K_1, K_2, \ldots, K_n, \ldots$ be an infinite sequence of fields. Consider all possible infinite sequences $(a_1, a_2, \ldots, a_n, \ldots)$ with $a_i \in K_i$, and define operations on them by

$$(a_1, a_2, \ldots, a_n, \ldots) + (b_1, b_2, \ldots, b_n, \ldots) = (a_1 + b_1, a_2 + b_2, \ldots, a_n + b_n, \ldots)$$

and

$$(a_1, a_2, \ldots, a_n, \ldots)(b_1, b_2, \ldots, b_n, \ldots) = (a_1 b_1, a_2 b_2, \ldots, a_n b_n, \ldots).$$

We thus obtain a commutative ring called the *product* of the fields K_i, and denoted $\prod K_i$.

Certain homomorphisms of the ring $\prod K_i$ into fields (and hence, its maximal ideals) are immediately visible: we take the sequence $(a_1, a_2, \ldots, a_n, \ldots)$ into its nth component a_n (for fixed n). But there are also less trivial homomorphisms. In fact, consider all the sequences with only finitely many nonzero components a_i; these form an ideal I. Every ideal is contained in a maximal ideal, so let $\mathcal{M}$ be some maximal ideal of $\prod K_i$ containing I. This is distinct from the kernels of the above trivial homomorphisms, since these do not contain I. The quotient ring $\prod K_i/\mathcal{M}$ is a field, and is called an *ultraproduct* of the fields K_i. We obtain an interesting 'mixture' of the fields K_i; for example, if all the K_i have different finite characteristics, then their ultraproduct is of characteristic 0. This is one method of passing from fields of finite characteristic to fields of characteristic 0, and using it allows us to prove certain hard theorems of number theory.

If all the fields K_i coincide with the field $\mathbb{R}$ of real numbers, then their ultraproduct has applications in analysis. It lies at the basis of so-called *non-standard analysis*, which allows us, for example, to avoid hard estimates and verifications of convergence in certain questions of the theory of differential equations.

From the point of view of mathematical logic, ultraproducts are interesting in that any 'elementary' statement which is true in all the fields K_i remains true in their ultraproduct.

Example 12. Consider differential operators of the form $\mathscr{D} = \sum_{i=0}^{k} f_i(z)\frac{d^i}{dz^i}$, where the $f_i(z)$ are Laurent series (either convergent or formal). Multiplication of such operators is not necessarily commutative; but for certain pairs of operators $\mathscr{D}$ and Δ it may nevertheless happen that $\mathscr{D}\Delta = \Delta\mathscr{D}$; for example, if

$$\mathscr{D} = \frac{d^2}{dz^2} - 2z^{-2} \quad \text{and} \quad \Delta = \frac{d^3}{dz^3} - 3z^{-2}\frac{d}{dz} + 3z^{-3}.$$

Then the set of all polynomial $P(\mathscr{D}, \Delta)$ in $\mathscr{D}$ and Δ with constant coefficients is a commutative ring, denoted by $R_{\mathscr{D},\Delta}$. Now something quite unexpected happens: if $\mathscr{D}\Delta = \Delta\mathscr{D}$ then there exists a nonzero polynomial $F(x, y)$ with constant coefficients such that $F(\mathscr{D}, \Delta) = 0$, that is, $\mathscr{D}$ and Δ satisfy a polynomial relation. For example, if

$$\mathscr{D} = \frac{d^2}{dz^2} - 2z^{-2} \quad \text{and} \quad \Delta = \frac{d^3}{dz^3} - 3z^{-2}\frac{d}{dz} + 3z^{-3},$$

then $F = \mathscr{D}^3 - \Delta^2$; we can assume that F is irreducible. Then the ring $R_{\mathscr{D},\Delta}$ is isomorphic to $\mathbb{C}[x, y]/(F(x, y))$, or in other words, to the ring $\mathbb{C}[C]$ where C is an irreducible curve with equation $F(x, y) = 0$. If the operators $\mathscr{D}$ and Δ have a common eigenfunction f, then this function will also be an eigenfunction for all operators of $R_{\mathscr{D},\Delta}$. Taking any operator into its eigenvalue on the eigenfunction f is a homomorphism $R_{\mathscr{D},\Delta} \to \mathbb{C}$. In view of the isomorphism $R_{\mathscr{D},\Delta} \cong \mathbb{C}[C]$, this homomorphism defines a point of C. It can be shown that every point of the curve corresponds to a common eigenfunction of the operators $\mathscr{D}$ and Δ. The relation between commuting differential operators and algebraic curves just described has in recent times allowed a significant clarification of the structure of commuting rings of operators.

§5. Modules

Consider some domain V in space and the vector fields defined on it. These can be added and multiplied by numbers, carrying out these operations on vectors applied to one point. Thus all vector fields form an infinite-dimensional vector space. But in addition to this, they can be multiplied by functions. This operation is very useful, since every vector field can be written in the form

$$A\frac{\partial}{\partial x} + B\frac{\partial}{\partial y} + C\frac{\partial}{\partial z},$$

where A, B and C are functions; hence it is natural to consider the set of vector fields as being 3-dimensional over the ring of functions. We thus arrive at the notion of a *module* over a ring (in this section, we only deal in commutative rings). This differs from a vector space only in that for a module, multiplication of its elements by ring elements is defined, rather than by field elements as for a vector space. The remaining axioms, both those for the addition of elements, and for multiplication by ring elements, are exactly as before, and we will not repeat them.

Example 1. A ring is a module over itself; this is an analogue of a 1-dimensional vector space.

Example 2. Differential forms of a given degree on a differentiable (or real or complex analytic) manifold form a module over the ring of differentiable (or real or complex analytic) functions on the manifold. The same holds for vector fields, and quite generally for tensor fields of a fixed type. (We will discuss the definition of all these notions in more detail later in § 5 and in § 7).

Example 3. If φ is a linear transformation of a vector space L over a field K, then we can make L into a module over the ring $K[t]$ by setting

$$f(t)x = (f(\varphi))(x) \quad \text{for} \quad f(t) \in K[t] \text{ and } x \in L.$$

Example 4. The ring of linear differential operators with constant coefficients (§ 3, Example 3) acts on the space of functions (C^∞, of compact support, exponentially decaying, polynomial), and makes each of these spaces into a module over this ring. Since this ring is isomorphic to the polynomial ring $\mathbb{R}[t_1, \ldots, t_n]$ (§ 3, Example 3), each of the indicated spaces is a module over the polynomial ring. Of course, the same remains true if we replace the field $\mathbb{R}$ by $\mathbb{C}$.

Example 5. Let M and N be modules over a ring A. Consider the module consisting of pairs (m, n) for $m \in M$, $n \in N$, with addition and multiplication by elements of A given by

$$(m, n) + (m_1, n_1) = (m + m_1, n + n_1) \quad \text{and} \quad a(m, n) = (am, an).$$

This module is called the *direct sum* of M and N and is denoted by $M \oplus N$. The direct sum of any number of modules can be defined in the same way. The sum of n copies of the module A (Example 1) is denoted by A^n and is called the *free module* of rank n. This is the most direct generalisation of an n-dimensional vector space; elements of A^n are n-tuples of the form

$$m = (a_1, \ldots, a_n) \quad \text{with} \quad a_i \in A.$$

If $e_i = (0, \ldots, 1, \ldots, 0)$ with 1 in the ith place then $m = \sum a_i e_i$, and this representation is unique.

It is sometimes also useful to consider algebraic analogues of infinite-dimensional vector spaces, the direct sum of a family Σ of modules isomorphic to A. Elements of this sum are specified as sequences $\{a_\sigma\}_{\sigma\in\Sigma}$ with $a_\sigma \in A$ as σ runs through Σ, and $a_\sigma \neq 0$ for only a finite number of $\sigma \in \Sigma$. With the elements e_σ defined as before, every element of the direct sum has a unique representation as a finite sum $\sum a_\sigma e_\sigma$. The module we have constructed is a *free module*, and the $\{e_\sigma\}$ a *basis* or a *free family of generators* of it.

Example 6. In a module M over the ring $\mathbb{Z}$ the multiplication by a number $n \in \mathbb{Z}$ is already determined once the addition is defined:

$$\text{if } n > 0 \quad \text{then} \quad nx = x + \cdots + x \quad (n \text{ times})$$

and if $n = -m$ with $m > 0$ then $nx = -(mx)$. Thus M is just an Abelian group[2] written additively.

We omit the definitions of *isomorphism* and *submodule*, which repeat word for word the definition of isomorphism and subspace for vector spaces. An isomorphism of modules M and N is written $M \cong N$.

Example 7. Any differential r-form on n-dimensional Euclidean space $\mathbb{R}^n$ can be uniquely written in the form

$$\sum_{i_1<\cdots<i_r} a_{i_1\ldots i_r}\, dx_{i_1} \wedge \cdots \wedge dx_{i_r},$$

where $a_{i_1\ldots i_r}$ belongs to the ring A of functions on $\mathbb{R}^n$ (differentiable, real analytic or complex analytic, see Example 2). Hence the module of differential forms is isomorphic to $A^{\binom{n}{r}}$, where $\binom{n}{r}$ is the binomial coefficient.

Example 8. Consider the polynomial ring $\mathbb{C}[x_1,\ldots,x_n]$ as a module M over itself (Example 1); on the other hand, consider it as a module over the ring of differential operators with constant coefficients (Example 4). Since this ring is isomorphic to the polynomial ring, we get a new module N over $\mathbb{C}[x_1,\ldots,x_n]$. These modules are not isomorphic; in fact for any $m' \in N$ there exists a non-zero element $a \in \mathbb{C}[x_1,\ldots,x_n]$ such that $am' = 0$ (take a to be any differential operator of sufficiently high order). But since $\mathbb{C}[x_1,\ldots,x_n]$ is an integral domain, it follows that in M, $am = 0$ implies that $a = 0$ or $m = 0$.

In a series of cases, Fourier transform establishes an isomorphism of modules M and N over the ring $A = \mathbb{C}[x_1,\ldots,x_n]$, where M and N are modules consisting of functions, and A acts on M by multiplication, and on N via the isomorphism

$$\mathbb{C}[t_1,\ldots,t_n] \cong \mathbb{C}\left[\frac{\partial}{\partial x_1},\ldots,\frac{\partial}{\partial x_n}\right].$$

For example, this is the case if $M = N$ is the space of C^∞ functions $F(x_1,\ldots,x_n)$ for which

[2] We assume that the reader knows the definition of a group and of an Abelian group; these will be repeated in § 12.

$$\left| x_1^{\alpha_1} \dots x_n^{\alpha_n} \frac{\partial^{\beta_1+\dots+\beta_n}}{\partial x_1^{\beta_1} \dots \partial x_n^{\beta_n}} F \right|$$

is bounded for all $\alpha \geqslant 0$, $\beta_i \geqslant 0$.

Recalling the definition of §4, we can now say that an ideal of a ring A is a submodule of A, if A is considered as a module over itself (as in Example 1). Ideals which are distinct as subsets of A can be isomorphic as A-modules. For example, an ideal I of an integral domain A is isomorphic to A as an A-module if and only if it is principal (because if $I = (i)$ then $a \mapsto ai$ is the required homomorphism; conversely, if $\varphi: A \to I$ is an isomorphism of A-modules, and 1 is the identity element of A then $\varphi(1) = i \in I$ implies that $\varphi(a) = \varphi(a1) = a\varphi(1) = ai$, that is $I = (i)$). Hence the set of ideals of a ring which are non-isomorphic as modules is a measure of its failure to be a principal ideal domain. For example, in the ring $A_d = \mathbb{Z} + \mathbb{Z}\sqrt{d}$ consisting of numbers of the form $a + b\sqrt{d}$ with $a, b \in \mathbb{Z}$ (where d is some integer), there are only a finite number of non-isomorphic ideals. This number is called the *class number* of A_d and is a basic arithmetic invariant.

Example 9. Let $\{m_\alpha\}$ be a set of elements of a module M over a ring A. Consider all possible linear combinations $\sum a_i m_{\alpha_i}$ with coefficients $a_i \in A$ (even if the set $\{m_\alpha\}$ is infinite, each linear combination only involves finitely many terms). These form a submodule of the module M, called the submodule *generated* by the $\{m_\alpha\}$. In particular, if $M = A$ as a module over itself, we arrive back at the notion of the ideal generated by elements $\{m_\alpha\}$ which we have already met. If the system $\{m_\alpha\}$ generates the whole of M, it is called a *system of generators* of M.

The notion of a linear map of one vector space to another carries over word-for-word to modules; in this case such a map is called an *A-linear map*, or a *homomorphism*. Exactly as for the case of an ideal in a ring, for a submodule $N \subset M$ we can define its cosets $m + N$, the *quotient module* M/N and the *canonical homomorphism* $M \to M/N$. The notions of image and kernel, and the relation between homomorphisms and submodules formulated in §4 for the case of rings and ideals also carry over.

These notions allow us to define certain important constructions. By definition, we know how to add elements of a module M and multiply them by elements of A, but we don't know how to multiply two elements together. However, in some situations there arises an operation of multiplying elements of a module M by elements of a module N, and getting a value in some third module L For example, if M consists of vector fields $\sum f_i \dfrac{\partial}{\partial x_i}$ and N of differential 1-forms $\sum p_i\, dx_i$ then the product $\sum f_i p_i$ is defined, and belongs to the ring of functions (and is independent of the choice of coordinates $x_1, \dots, x_n$). In a similar way, one can define (independently of the choice of coordinates) a product of a vector field by a differential r-form, the result of which is a differential $(r - 1)$-form.

We define a *multiplication* defined on two modules M and N and with values in a third module L to be a map which takes a pair of elements $x \in M$, $y \in N$ into an element $xy \in L$, having the following bilinearity properties:

$$(x_1 + x_2)y = x_1 y + x_2 y \quad \text{for} \quad x_1, x_2 \in M \text{ and } y \in N;$$

$$x(y_1 + y_2) = xy_1 + xy_2 \quad \text{for} \quad x \in M \text{ and } y_1, y_2 \in N;$$

$$(ax)y = x(ay) = a(xy) \quad \text{for} \quad x \in M, y \in N \text{ and } a \in A.$$

If a multiplication xy is defined on two modules M and N with values in L, and if $\varphi\colon L \to L'$ is a homomorphism, then $\varphi(xy)$ defines a multiplication with values in L'. It turns out that all possible multiplications on given modules M and N can be obtained in this way from a single 'universal' one. This has values in a module which we denote by $M \otimes_A N$, and the product of elements x and y is also denoted by $x \otimes y$. The universality consists of the fact that for any multiplication xy defined on M and N with values in L, there exists a unique homomorphism

$$\varphi\colon M \otimes_A N \to L \quad \text{for which} \quad xy = \varphi(x \otimes y).$$

It is easy to show that if a module and a product with this universality property exist, then they are defined uniquely up to isomorphism. The construction of the module $M \otimes_A N$ and the multiplication $x \otimes y$ is as follows: suppose that M has a finite set of generators $x_1, \ldots, x_m$ and N a set $y_1, \ldots, y_n$. We consider symbols (x_i, y_j), and the free module $S = A^{mn}$ with these as generators. In S, consider the elements

$$\sum_i a_i(x_i, y_j) \quad \text{for which} \quad \sum a_i x_i = 0 \text{ in } M,$$

and the elements

$$\sum_j b_j(x_i, y_j) \quad \text{for which} \quad \sum b_j x_j = 0 \text{ in } N,$$

and consider the submodule S_0 generated by these elements. We set

$$M \otimes_A N = S/S_0$$

and if $x = \sum a_i x_i$ and $y = \sum b_j y_j$ then

$$x \otimes y = \sum_{i,j} a_i b_j (x_i \otimes y_j),$$

where $x_i \otimes y_j$ denotes the image in S/S_0 of (x_i, y_j) under the canonical homomorphism $S \to S/S_0$. It is easy to check that $x \otimes y$ does not depend on the choice of the expressions of x and y in terms of generators, and that in this way we actually get a universal object. More intrinsically, and without requiring that M and N have finite systems of generators, we could construct the module $M \otimes_A N$ by taking as generators of S all possible pairs (x, y) with $x \in M$ and $y \in N$, and S_0 to be the submodule generated by the elements

$$(x_1 + x_2, y) - (x_1, y) - (x_2, y), \quad (x, y_1 + y_2) - (x, y_1) - (x, y_2),$$

$$a(x, y) - (x, ay), \quad a(x, y) - (ax, y).$$

This way, we have to use a free module S on an infinite set of generators, even if we are dealing with modules M and N having finitely many generators. However, there is nothing arbitrary in the construction related to the choice of systems of generators.

The module $M \otimes_A N$ defined in this way is called the *tensor product* of the modules M and N, and $x \otimes y$ the tensor product of the elements x and y. If M and N are finite-dimensional vector spaces over a field K, then $M \otimes_K N$ is also a vector space, and

$$\dim(M \otimes_K N) = \dim M \cdot \dim N.$$

If M is a module over the ring $\mathbb{Z}$, then $M \otimes_{\mathbb{Z}} \mathbb{Q}$ is a vector space over $\mathbb{Q}$; for example if $M \cong \mathbb{Z}^n$ then $M \otimes_{\mathbb{Z}} \mathbb{Q} \cong \mathbb{Q}^n$. But if $M \cong \mathbb{Z}/(n)$ then $M \otimes_{\mathbb{Z}} \mathbb{Q} = 0$, that is, M is killed off on passing to $M \otimes_{\mathbb{Z}} \mathbb{Q}$; although any element $m \in M$ corresponds to $m \otimes 1$ in $M \otimes_{\mathbb{Z}} \mathbb{Q}$, this is 0, as one checks easily from the bilinearity conditions. In a similar way, from a module M over an integral domain A we can get a vector space $M \otimes_A K$ over its field of fractions K. In exactly the same way, a vector space E over a field K defines a vector space $E \otimes_K L$ over any extension L of K. When $K = \mathbb{R}$ and $L = \mathbb{C}$ this is the operation of *complexification* which is very useful in linear algebra (for example, in the study of linear transformations).

If M_i is a vector space of functions $f(x_i)$ of a variable x_i (for example, the polynomials $f(x_i)$ of degree $< k_i$), then $M_1 \otimes \cdots \otimes M_n$ consists of linear combinations of functions

$$f_1(x_1) \dots f_n(x_n) \quad \text{with} \quad f_i \in M_i$$

in the space of functions of $x_1, \dots, x_n$. In particular, the 'degenerate kernels' of the theory of integral equations are of this form. It is natural quite generally to try to interpret spaces of functions (of one kind or another) $K(x, y)$ of variables x, y as tensor products of spaces of functions of x and of y. This is how the analogues of the notion of tensor products arise in the framework of Banach and topological vector spaces. The classical functions $K(x, y)$ arise as kernels of integral operators

$$f \mapsto \int K(x, y) f(y)\, dy.$$

In the general case the elements of tensor products are also used for specifying operators of Fredholm type. A similar role is played by tensor products in quantum mechanics. If spaces M_1 and M_2 are state spaces of quantum-mechanical systems S_1 and S_2 then $M_1 \otimes M_2$ describe the state of the system composed of S_1 and S_2.

Example 10. The module $M \otimes_A \cdots \otimes_A M$ (r factors) is denoted by $T^r(M)$. If M is a finite-dimensional vector space over K, then $T^r(M)$ is the space of *contravariant tensors* of degree r.

Example 11. The quotient module of $M \otimes_A M$ by the submodule generated by the elements $x \otimes y - y \otimes x$ for $x, y \in M$ is called the *symmetric square* of M, and is denoted by S^2M; it is universal for commutative multiplication xy of $x, y \in M$. In a similar way we can define the rth *symmetric power* S^rM; this is the quotient module of $T^r(M)$ by the submodule generated by all possible elements

$$x_1 \otimes \cdots \otimes x_i \otimes x_{i+1} \otimes \cdots \otimes x_r - x_1 \otimes \cdots \otimes x_{i+1} \otimes x_i \otimes \cdots \otimes x_r$$

for $i = 1, \ldots, r-1$, where $x_i \in M$. For example, if M is the module of linear forms in variables $t_1, \ldots, t_n$ with coefficients in the field K, then S^rM consists of all forms (that is, homogeneous polynomials) of degree r in $t_1, \ldots, t_n$.

Obviously, a product of r elements $x_1, \ldots, x_r \in M$ with values in S^rM is always defined, and does not depend on the order of the factors: just consider the image of $x_1 \otimes \cdots \otimes x_r$ under the canonical homomorphism $T^r(M) \to S^rM$. These products generate S^rM.

Example 12. The rth *exterior power* of a module M is the quotient module of $T^r(M)$ by the submodule generated by expressions $x_1 \otimes \cdots \otimes x_r$ in which two factors coincide, say $x_i = x_j$. The exterior power is denoted by $\bigwedge^r M$. For example, the module of differential r-forms on a differential manifold is isomorphic to $\bigwedge^r M$, where M is the module of differential 1-forms. By analogy with the case of the symmetric power, the multiplication of r elements $x_1, \ldots, x_r$ of M with values in $\bigwedge^r M$ is defined; it is denoted by $x_1 \wedge \cdots \wedge x_r$, and is called their *exterior product*. By definition, $x_1 \wedge \cdots \wedge x_r = 0$ if $x_i = x_j$. It follows easily from this that $x_1 \wedge \cdots \wedge x_i \wedge x_{i+1} \wedge \cdots \wedge x_r = -x_1 \wedge \cdots \wedge x_{i+1} \wedge x_i \wedge \cdots \wedge x_r$. If M has a finite number of generators $x_1, \ldots, x_n$ then the products

$$x_{i_1} \wedge \cdots \wedge x_{i_r} \quad \text{for} \quad 1 \leqslant i_1 < i_2 < \cdots < i_r \leqslant n$$

are generators for $\bigwedge^r M$. In particular, $\bigwedge^r M = 0$ for $r > n$. If M is an n-dimensional vector space over a field K, then $\dim \bigwedge^r M = \binom{n}{r}$ for $r \leqslant n$.

Example 13. If M is a module over a ring A then the set M^* of all homomorphisms of M to A is a module, if we define operations by

$$(f + g)(m) = f(m) + g(m) \quad \text{for} \quad f, g \in M^* \text{ and } m \in M;$$

$$(af)(m) = af(m), \quad \text{for} \quad f \in M^*, a \in A \text{ and } m \in M.$$

This module is called the *dual* module of M. If M is a vector space over a field K, then M^* is the dual vector space. The space of differential 1-forms on a differentiable manifold (as a module over the ring of differentiable functions) is the dual of the module of vector fields.

The elements of the space $T^r(M^*)$ are called *covariant tensors*; the elements of $T^p(M) \otimes T^q(M^*)$ are called *tensors of type* (p, q).

If M is the space of tensors of type (p, q) over a vector space and N the space of tensors (p', q'), then $M \otimes N$ is the space of tensors of type $(p + p', q + q')$, and $\otimes$ is the operation of multiplying tensors.

In conclusion, we attempt to extend to modules the 'functional' intuition which we discussed in §4 as applied to rings. We start with an example.

Let X be a differentiable manifold, A the ring of differentiable functions on it, and M the A-module of vector fields on X. At a given point $x \in X$, every vector field τ takes a value $\tau(x)$, that is, there is defined a map $M \to T_x$ where T_x is the tangent space to X at x. This map can be described in algebraic terms, by defining multiplication of constants $\alpha \in \mathbb{R}$ by a function $f \in A$ by $f \cdot \alpha = f(x)\alpha$. Then $\mathbb{R}$ will be a module over A, and $T_x \cong M \otimes_A \mathbb{R}$, and our map takes τ into the element $\tau \otimes 1$. In this form, we can construct this map for an arbitrary module M over an arbitrary ring A. Let $\varphi\colon A \to K$ be a homomorphism of A into a field with $\varphi(A) = K$ and kernel the maximal ideal $\mathfrak{m}$; then K is a module over A if we set $a\alpha = \varphi(a)\alpha$ for $a \in A$ and $\alpha \in K$. Hence there is defined a vector space $M_{\mathfrak{m}} = M \otimes_A K$ over K, the 'value of M at the point $\mathfrak{m}$'. For example, if $A = K[C]$, where C is an algebraic curve (or any algebraic variety), then as we saw in §4, any point $c \in C$ defines a homomorphism $\varphi_c\colon A \to K$, where $\varphi_c(f) = f(c)$, and the maximal ideal $\mathfrak{m}_c$ consisting of functions $f \in A$ with $f(c) = 0$.

Thus each module M over $K[C]$ defines a family of vector spaces M_x 'parametrised by' the variety C, and in an entirely similar way, a module M over an arbitrary ring defines a family of vector spaces $M \otimes_A (A/\mathfrak{m})$ over the various residue fields $A/\mathfrak{m}$, 'parametrised by' the set of maximal ideals $\mathfrak{m}$ of A.

The geometrical analogue of this situation is the following: a *family of vector spaces* over a topological space X is a topological space $\mathscr{E}$ with a continuous map

$$f\colon \mathscr{E} \to X,$$

in which every fibre $f^{-1}(x)$ is given a vector space structure (over $\mathbb{R}$ or $\mathbb{C}$), compatible with the topology of $\mathscr{E}$ in the natural sense. A *homomorphism* between families $f\colon \mathscr{E} \to X$ and $g\colon \mathscr{F} \to X$ is a continuous map

$$\varphi\colon \mathscr{E} \to \mathscr{F},$$

taking each fibre $f^{-1}(x)$ into the fibre $g^{-1}(x)$, and inducing a linear map between them. A family $\mathscr{E}$ of vector spaces defines a module $M_{\mathscr{E}}$ over the ring $A(X)$ of continuous functions on X. If the family $\mathscr{E}$ is a generalisation of a vector space, then an element of $M_{\mathscr{E}}$ is a generalisation of a vector: it is a choice of a vector in each fibre $f^{-1}(x)$ for $x \in X$. More precisely, elements of $M_{\mathscr{E}}$, called *sections*, are defined as continuous maps

$$s\colon X \to \mathscr{E},$$

for which the point $s(x)$ belongs to the fibre $f^{-1}(x)$, for all $x \in X$ (that is, $fs(x) = x$). The operations

$$(s_1 + s_2)(x) = s_1(x) + s_2(x) \quad \text{for} \quad s_1, s_2 \in M_{\mathscr{E}} \text{ and } x \in X;$$

$$(\varphi s)(x) = \varphi(x)s(x), \quad \text{for} \quad \varphi \in A(X), x \in X \text{ and } s \in M_{\mathscr{E}},$$

make $M_{\mathscr{E}}$ into a module over $A(X)$.

§6. Algebraic Aspects of Dimension

The basic invariant of a vector space is its dimension, and in this context the class of finite-dimensional vector spaces is distinguished. For modules, which are a direct generalisation of vector spaces, there are analogous notions, which play the same fundamental role. On the other hand, we have considered algebraic curves, surfaces, and so on, and have 'coordinatised' each such object C by assigning to it the coordinate ring $K[C]$ or the rational function field $K(C)$. The intuitive notion of dimension (1 for an algebraic curve, 2 for a surface, and so on) is reflected in algebraic properties of the ring $K[C]$ or of the field $K(C)$, and these properties are meaningful and important for more general types of rings and fields. As one might expect, the situation becomes more complicated in comparison with the simplest examples: we will see that there exist various ways of expressing the 'dimension' of rings or modules as a number, and various analogues of finite dimensionality.

The dimension of a vector space can be defined from various different starting points: firstly, as the maximal number of linearly independent vectors; secondly, as the number of vectors in a basis (and here we need to prove that all bases of the same vector space consist of the same number of vectors); finally, one can make use of the fact that if the dimension is already defined, then an n-dimensional space L contains an $(n-1)$-dimensional subspace L_1, and L_1 an $(n-2)$-dimensional subspace L_2, and so on. We thus get a chain

$$L \supsetneqq L_1 \supsetneqq L_2 \supsetneqq \cdots \supsetneqq L_n = 0.$$

Hence the dimension can be defined as the greatest length of such a chain. Each of these definitions applies to modules, but here we already get different properties, which provide different numerical characteristics of modules; they also lead to different analogues of finite dimensionality for modules. We will consider all three of these approaches. For the first of these we assume that A is an integral domain.

Elements $m_1, \ldots, m_k$ of a module M over a ring A are *linearly dependent* if there exist elements $a_1, \ldots, a_k \in A$, not all zero, with

$$a_1 m_1 + \cdots + a_k m_k = 0;$$

otherwise they are *linearly independent*. The maximal number of linearly independent elements of a module M is called its *rank*, rank M; if this is finite, then M is a module of *finite rank*. The ring A itself is of rank 1 as an A-module, and the free module A^n has rank n in the new definition.

Despite the apparent similarity, the notion of rank is in substance very far from the dimension of a vector space. Even if the rank n is finite and $m_1, \ldots, m_n$ is a maximal set of linearly independent elements of a module, then it is quite false that every element m can be expressed in terms of them: in a linear dependence relation $am + a_1 m_1 + \cdots + a_k m_k = 0$, we cannot in general divide through by a. Thus we do not get the same kind of canonical description of all elements of the module as that provided by the basis of a vector space. Moreover, one might think that modules of rank 0, being analogues of 0-dimensional vector spaces, should be in some way quite trivial, whereas they can be arbitrarily complicated. Indeed, a single element $m \in M$ is linearly dependent if there exists a nonzero element $a \in A$ such that $am = 0$; in this case we say that m is a *torsion element*. A module has rank 0 if it consists entirely of torsion elements; it is then called a *torsion module*. For example, any finite Abelian group considered as a $\mathbb{Z}$-module is a torsion module. A vector space L with a linear transformation φ considered as a module over the polynomial ring $K[x]$ (§ 5, Example 3) is also a torsion module: there exists a polynomial $f(x) \neq 0$ such that $f(\varphi) = 0$, (that is $(f(\varphi))(x) = 0$ or $f \cdot x = 0$) for every $x \in L$. The polynomial ring $\mathbb{R}[x_1, \dots, x_n]$ as a module over the ring of differential operators $\mathbb{R}\left[\frac{\partial}{\partial x_1}, \dots, \frac{\partial}{\partial x_n}\right]$ (§ 5, Example 4) is another example of a torsion module. All of these modules have rank 0, although, for example, it is intuitively hard to accept the last example as being even finite-dimensional.

A better approximation to an intuitive notion of finite dimensionality is provided by the definition of finite dimensionality of a vector space in terms of the existence of a basis.

A module M having a finite set of generators is said to be *finitely generated*, or a module *of finite type*. Thus M contains a finite system $m_1, \dots, m_k$ of elements such that any element is a linear combination of these, although in contrast to vector spaces, we cannot require that this representation is unique.

A ring as a module over itself, and more generally a free module of finite rank, is of finite type, as is a finite Abelian group as a $\mathbb{Z}$-module and a vector space with a given linear transformation as a $K[x]$-module. The polynomial ring $\mathbb{R}[x_1, \dots, x_n]$ is not of finite type as a module over the ring of differential operators $\mathbb{R}\left[\frac{\partial}{\partial x_1}, \dots, \frac{\partial}{\partial x_n}\right]$: starting from a finite number of polynomials $F_1, \dots, F_k$, it is not possible to get polynomials of higher degree by applying differentiations.

A homomorphic image of a module of finite type has the same property: the image of a system of generators is a system of generators. In particular, homomorphic images of the free module A^n are all of finite type and are generated by at most n elements. The converse is also true. If M has generators $m_1, \dots, m_k$ then taking a k-tuple $(a_1, \dots, a_k) \in A^k$ (by definition A^k consists of such k-tuples) into the element $a_1 m_1 + \dots + a_k m_k$ is a homomorphism with image M. This proves the following:

Theorem I. *Any module of finite type is a homomorphic image of a free module of finite type* A^n.

In particular, a module with a single generator is a homomorphic image of the ring A itself, that is (by the homomorphisms theorem) is of the form A/I, where I is an ideal of A; if $I = 0$ then M is isomorphic to A. A module of this form is called a *cyclic module*. We can think of these as analogues of 1-dimensional vector spaces.

In some cases, modules of finite type are rather close to finite-dimensional vector spaces. For example, if A is an integral domain in which all the ideals are principal (that is, a PID), then we have the following result.

II. Theorem on Modules over a Principal Ideal Domain. *A module of finite type over a PID is isomorphic to a direct sum of a finite number of cyclic modules. A cyclic module is either isomorphic to A or decomposes further as a direct sum of cyclic modules of the form $A/(\pi^k)$ where π is a prime element. The representation of a module as a direct sum of such modules is unique.*

If a module M is a torsion module then there are no summands isomorphic to A. This happens for example if $A = \mathbb{Z}$ and M is a finite Abelian group. In this case the theorem we have stated gives a classification of finite Abelian groups. The same holds if $A = \mathbb{C}[x]$, and $M = L$ is a finite-dimensional vector space over $\mathbb{C}$ with a given linear transformation (§5, Example 3). In this case it is easy to see that our theorem gives the reduction of a linear transformation to Jordan normal form.

One proof of Theorem II is based on a representation of M in the form

$$M = A^n/N \quad \text{with} \quad N \subset A^n$$

(by Theorem I). It is easy to prove that N is also a module of finite type. If

$$A^n = Ae_1 \oplus \cdots \oplus Ae_n \quad \text{and} \quad N = (u_1, \ldots, u_m), \quad \text{then } u_i = \sum c_{ij}e_j,$$

and the representation $M = A^n/N$ shows that M is 'defined by the system of linear equations'

$$\sum_{j=1}^{n} c_{ij}e_j = 0 \quad \text{for} \quad i = 1, \ldots, m.$$

We now apply to this system the idea of Gauss' method from the classical theory of systems of linear equations.

Main Lemma. *Over a PID, any matrix can be reduced to diagonal form by multiplying on either side by unimodular matrixes.*

If the analogue of the Euclidean algorithm holds in the ring then multiplication on either side by unimodular matrixes can be performed by the well-known elementary transformations (row and column operations): interchanging two rows, adding a multiple of one row to another, and similar operations on columns. Applied to the matrix (c_{ij}), row and column operations correspond to the simplest possible transformations of the systems of generators $e_1, \ldots, e_n$ and relations $u_1, \ldots, u_m$. In this case, the analogy with Gauss' method is particularly obvious.

The main lemma allows us to find systems of generators for which the matrix (c_{ij}) is diagonal. If

$$(c_{ij}) = \begin{bmatrix} a_1 & & & & & \\ & \ddots & & & \text{\Large 0} & \\ & & a_r & & & \\ & & & 0 & & \\ & & & & \ddots & \\ \text{\Large 0} & & & & & 0 \end{bmatrix}, \quad \text{with } a_1 \neq 0, \ldots, a_r \neq 0$$

then $M = A^n/N \cong A/(a_1) \oplus \cdots \oplus A/(a_r) \oplus A^{n-r}$. From this it is not hard to get to the assertion of Theorem II.

In particular if $A = \mathbb{Z}$, Theorem II describes the structure of Abelian groups with a finite number of generators. Such groups arise, for example, in topology as the homology or cohomology groups of a finite complex (see § 21 for these).

However, one property, which intuitively is closely related to finite dimensionality, does not hold in general for a module of finite type: a submodule may no longer be of finite type. This can fail even in the simplest case: a submodule of a ring A, that is, an ideal, is not always of finite type. For example in the ring $\mathscr{E}$ of germs of C^∞ functions at $0 \in \mathbb{R}$, the ideal of functions vanishing at 0 together with all derivatives does not have a finite number of generators (§ 4, Example 7). In the same way, in the polynomial ring in an infinite number of generators x_1, $x_2, \ldots, x_n, \ldots$ (each polynomial depends of course only on finitely many of them) the polynomials with no constant term form an ideal which does not have a finite number of generators. Thus it is natural to strengthen the finite dimensionality condition, by considering modules all of whose submodules are of finite type. We say that a module with this property is *Noetherian*. This notion can be related to the so far unused characterisation of the dimension of a vector space in terms of chains of subspaces. Namely, the Noetherian condition is equivalent to the following property of a module (called the *ascending chain condition* or *a.c.c.*): any sequence of submodules

$$M_1 \subsetneqq M_2 \subsetneqq \cdots \subsetneqq M_k \subsetneqq \cdots,$$

is finite. The verification of this equivalence is almost obvious.

These ideas can also be applied to the classification of rings from the point of view of analogues of finite dimensionality. It is natural to consider rings over which any module of finite type is Noetherian; a ring with this property is a *Noetherian ring*. For this, it is necessary first of all that the ring should be Noetherian as a module over itself, that is, that every ideal should have a finite system of generators. But it is not hard to check that this is also sufficient: if all ideals of a ring A have a finite basis then the free modules A^n are also Noetherian, and hence also their homomorphic images, that is, all modules of finite type.

How wide is the notion of a Noetherian ring? Obviously any ring all of whose ideals are principal is Noetherian. Another fundamental fact is the following theorem:

III. The Hilbert Basis Theorem. *For a Noetherian ring A the polynomial ring $A[x]$ is again Noetherian.*

The proof is based on considering the ideals $J_n \subset A$ (for $n = 1, 2, \ldots$), consisting of elements which are coefficients of leading terms of polynomials of degree n contained in a given ideal $I \subset A[x]$, and then making repeated use of the Noetherian property of A. It follows from the Hilbert basis theorem that the polynomial ring $A[x_1, \ldots, x_n]$ in any number of variables is Noetherian if A is. In particular, the ring $K[x_1, \ldots, x_n]$ is Noetherian. It was for this purpose that Hilbert proved this theorem; he formulated it in the following explicit form.

Theorem. *Given any set $\{F_\alpha\}$ of polynomials in $K[x_1, \ldots, x_n]$, there exists a finite subset $F_{\alpha_1}, \ldots, F_{\alpha_m}$ such that any polynomial F_α can be expressed as a linear combination*

$$P_1 F_{\alpha_1} + \cdots + P_m F_{\alpha_m} \quad \textit{with} \quad P_1, \ldots, P_m \in K[x_1, \ldots, x_n].$$

But we can go even further. Obviously, if A is Noetherian then the same is true of any homomorphic image B of A. We say that a ring R containing a subring A is *finitely generated* over A, or is a ring *of finite type* over A if there exists a finite system of elements $r_1, \ldots, r_n$ of R such that all elements of R can be expressed in terms of them as polynomials with coefficients in A; the elements $r_1, \ldots, r_n$ are called *generators* of R over A. Consider the polynomial ring $A[x_1, \ldots, x_n]$ and the map

$$F(x_1, \ldots, x_n) \mapsto F(r_1, \ldots, r_n).$$

This is a homomorphism, and its image is R. Thus we have the result:

Theorem IV. *Any ring of finite type over a ring A is a homomorphic image of the polynomial ring $A[x_1, \ldots, x_n]$. From the above it then follows that a ring of finite type over a Noetherian ring is Noetherian.*

For example, the coordinate ring $K[C]$ of an algebraic curve C (or surface, or an algebraic variety) is Noetherian. If C is given by an equation $F(x, y) = 0$ then x and y are generators of $K[C]$ over K.

Other examples of Noetherian rings which are important in applications are the rings $\mathcal{O}_n$ of functions of n complex variables which are holomorphic at the origin, and the formal power series ring $K[\![t_1, \ldots, t_n]\!]$.

Noetherian rings are the most natural candidates for the role of finite-dimensional rings. A notion of dimension can also be defined for these, but this would require a rather more precise treatment.

While the condition that a ring should be a ring of finite type over some simple ring (for example, over a field) is a concrete, effective form of a finite dimensionality condition, the Noetherian condition is more intrinsic, although a weaker assertion. In one important case these notions coincide.

A ring A is *graded* if it has specified subgroups A_n (that is, submodules of A as a $\mathbb{Z}$-module) for $n = 0, 1, \ldots$, such that for $x \in A_n$ and $y \in A_m$ we have $xy \in A_{n+m}$, and any element $x \in A$ can be uniquely represented in the form

$$x = x_0 + x_1 + \cdots + x_k \quad \text{with} \quad x_i \in A_i. \tag{1}$$

We say that elements $x \in A_n$ are *homogeneous*, and the representation (1) is the decomposition of x into homogeneous components. The subset A_0 is obviously a subring of A.

For example, the ring $K[x_1, \ldots, x_m]$ is graded, with A_n the space of homomogeneous polynomials of degree n in $x_1, \ldots, x_m$, and $A_0 = K$.

One checks easily the following result:

Theorem V. *Let A be a graded ring; then A is Noetherian if and only if A_0 is Noetherian and A is a ring of finite type over A_0.*

Proof. Obviously, the set of elements $x \in A$ for which $x_0 = 0$ in (1) is an ideal I_0. It turns out that for the truth of the assertion in the theorem, it is sufficient for just this single ideal to be finitely generated. Indeed, we take a set of generators of I_0, represent each generator in the form (1), and consider all the homogeneous terms x_i appearing in this way. We get a set of homogeneous elements $x_1, \ldots, x_N$ (with $x_i \in A_{n_i}$) which again obviously generate I_0. These elements $x_1, \ldots, x_N$ are generators for A over A_0. Indeed, it is enough to prove that any element $x \in A_n$ with $n > 0$ can be expressed as a polynomial in $x_1, \ldots, x_N$ with coefficients in A. By assumption $I_0 = (x_1, \ldots, x_N)$, and in particular

$$x = a_1 x_1 + \cdots + a_N x_N \quad \text{with} \quad a_i \in A.$$

Considering the decomposition of the elements a_i into homogeneous components, and noting that on the left-hand side $x \in A_n$, we can assume that $a_i \in A_{m_i}$ and $x_i \in A_{n_i}$ with $n_i + m_i = n$. For $n_i = n$ the component $a_i x_i$ is expressed in terms of x_i with coefficient $a_i \in A_0$ as required, whereas for $n_i < n$ we can apply to a_i the same argument as for x. After a finite number of steps we get the required expression for x.

For fields, the intuitive notion of finite-dimensionality is realised by analogy with rings. We say that a field L is an *extension of finite type* of a subfield K if there exists a finite number of elements $\alpha_1, \ldots, \alpha_n \in L$ such that all the remaining elements of L can be represented as rational functions of $\alpha_1, \ldots, \alpha_n$ with coefficients in K. In this case we write $L = K(\alpha_1, \ldots, \alpha_n)$, and say that L is the extension of K generated by $\alpha_1, \ldots, \alpha_n$. For example, the field of rational functions $K(x_1, \ldots, x_n)$ is an extension of K of finite type. The complex number field is an extension of finite type of the real number field: complex numbers can be represented as extremely simple rational functions $a + bi$ of the single element i. Any finite field $\mathbb{F}_q$ is a extension of finite type of its prime subfield: we could take $\alpha_1, \ldots, \alpha_n$ to be, for example, all the elements of $\mathbb{F}_q$. If C is an irreducible algebraic curve, given by an equation $F(x, y) = 0$ then $K(C)$ is an extension of finite type of K, since all the functions in $K(C)$ are rational functions of the coordinates x and y. The same holds if C is an algebraic surface, and so on.

These examples make it plausible that for extensions of finite type there exists an analogue of the notion of dimension, corresponding to the intuitive notion of dimension for algebraic curves, surfaces, and any algebraic varieties.

A system of elements $\alpha_1, \ldots, \alpha_n$ of a field L is said to be *algebraically dependent* over a subfield K of L if there exists an irreducible polynomial $F \in K[x_1, \ldots, x_n]$, not identically zero, such that

$$F(\alpha_1, \ldots, \alpha_n) = 0.$$

If α_n actually occurs in this relation, we say that the element α_n is *algebraically dependent* on $\alpha_1, \ldots, \alpha_{n-1}$. Certain very simple properties of algebraic dependence are just the same as the well-known properties of linear dependence. For example if an element α is algebraically dependent on $\alpha_1, \ldots, \alpha_n$ and each of the α_i is algebraically dependent on elements $\beta_1, \ldots, \beta_m$, then α is algebraically dependent on $\beta_1, \ldots, \beta_m$. From this, repeating formally the well-known arguments for the case of linear dependence, we can prove that in an extension of finite type there exists an upper bound for the number of algebraically independent elements. The maximal number of algebraically independent elements of an extension of finite type L/K is called the *transcendence degree* of the extension, and is denoted by $\operatorname{tr\,deg} L/K$.

If the transcendence degree of an extension L/K is n, then L contains a set of n algebraically independent elements such that any other element is algebraically dependent on them; conversely, if n elements with this property exist, then the transcendence degree equals n.

For example, the transcendence degree of the rational function field $K(x_1, \ldots, x_n)$ as an extension of K is n. Let C be an irreducible algebraic curve, defined by an equation $F(x, y) = 0$. If for example y actually occurs in the equation F then in the field $K(C)$, the element x is algebraically independent and y is algebraically dependent on x, and hence so are all other elements of $K(C)$. Hence the transcendence degree of $K(C)/K$ is 1. In the same way, one proves that if C is an algebraic surface then the transcendence degree of the field $K(C)$ is 2. We thus arrive at a notion of dimension which really agrees with geometric intuition. The transcendence degree of the field $K(C)$, where C is an algebraic variety, is called the *dimension* of C, and is denoted by $\dim C$. It enjoys natural properties: for example,

$$\dim C_1 \leqslant \dim C_2 \quad \text{if} \quad C_1 \subset C_2.$$

Example 1. Let X be a compact complex analytic manifold of dimension n and $\mathcal{M}(X)$ the field of all meromorphic functions on X. It can be proved that

$$\operatorname{tr\,deg} \mathcal{M}(X)/\mathbb{C} \leqslant n.$$

If X is an algebraic variety over $\mathbb{C}$ then

$$\mathcal{M}(X) = \mathbb{C}(X) \quad \text{and} \quad \operatorname{tr\,deg} \mathcal{M}(X) = n.$$

Thus the number $\operatorname{tr deg} \mathcal{M}(X)/\mathbb{C}$ is a measure of how close the complex manifold X is to being an algebraic variety; all possible values from 0 to n occur already in the particular case of complex toruses (see § 15).

What does an extension L/K of finite type and of transcendence degree 0 look like? To say that the transcendence degree is 0 means that any element $\alpha \in L$ satisfies an equation $F(\alpha) = 0$, where F is a polynomial. Such an element α is said to be *algebraic* over K. Since L/K is an extension of finite type,

$$L = K(\alpha_1, \ldots, \alpha_n) \quad \text{for certain} \quad \alpha_1, \ldots, \alpha_n \in L.$$

Thus L/K can be obtained as a composite of extensions of the form $K(\alpha)/K$, where α is an algebraic element. Conversely, a composite of such extensions always has transcendence degree 0.

Suppose that $L = K(\alpha)$ where α is an algebraic element over K. Among all polynomials $F(x) \in K[x]$ for which $F(\alpha) = 0$ (these exist, since α is algebraic over K), there exists one of smallest degree; all others are divisible by this one: for otherwise, by division with remainder, we would arrive at a polynomial of smaller degree with the same property. This polynomial of smallest degree P is uniquely determined up to a constant multiple. It is called the *minimal polynomial* of α. Obviously, P is irreducible over K. Knowing the minimal polynomial P we can specify all the elements of the field $L = K(\alpha)$ in a very explicit form. For this, consider the homomorphism

$$\varphi\colon K[x] \to L$$

which takes a polynomial $F \in K[x]$ into the element $F(\alpha) \in L$. The kernel of φ is the principal ideal (P), as one sees easily. Hence its image is isomorphic to $K[x]/(P)$ (by the homomorphisms theorem). It is not hard to show that its image is the whole of L; for this we should note that $\operatorname{Im} \varphi$ is a field and contains α. Hence L is isomorphic to $K[x]/(P)$. If the degree of P is n then, as we saw in §4, Formula (2), every element of the field $L \cong K[x]/(P)$ can be expressed in the form

$$\xi = a_0 + a_1\alpha + \cdots + a_{n-1}\alpha^{n-1} \quad \text{with} \quad a_i \in K, \tag{2}$$

and the expression is unique. The classic example of this situation is $K = \mathbb{R}$, $L = \mathbb{C} = \mathbb{R}[i]$, $P(x) = x^2 + 1$: every complex number can be represented as $a + bi$ with $a, b \in \mathbb{R}$.

The representation (2) for elements of the field $L = K(\alpha)$ leads to an important corollary. Suppose we forget about the multiplication in L and keep only addition and multiplication by elements of K. Then (2) shows that the vector space L is finite-dimensional over K and the elements $1, \alpha, \ldots, \alpha^{n-1}$ form a basis of it. An extension L/K is *finite* if L is finite-dimensional as a vector space over K. Its dimension is called the *degree* of the extension L/K, and is denoted by $[L:K]$. In the previous example $[L:K] = n$; in particular $[\mathbb{C}:\mathbb{R}] = 2$.

For example, if $\mathbb{F}_q$ is a finite field and p the characteristic of $\mathbb{F}_q$, then $\mathbb{F}_q$ contains the prime field with p elements $\mathbb{F}_p$. Obviously, $\mathbb{F}_q/\mathbb{F}_p$ is a finite extension. If $[\mathbb{F}_q:\mathbb{F}_p] = n$ then there exist n elements $\alpha_1, \ldots, \alpha_n \in \mathbb{F}_q$ such that any other element can be uniquely represented in the form

$$\alpha = a_1\alpha_1 + \cdots + a_n\alpha_n \quad \text{with} \quad a_i \in \mathbb{F}_p,$$

and it follows from this that the number of elements of a finite field $\mathbb{F}_q$ is equal to p^n, that is, it is always a power of p.

It is easy to prove that the condition that an extension be finite is transitive, that is, if L/K and Λ/L are finite extensions, then Λ/K is also finite, and

$$[\Lambda : K] = [\Lambda : L][L : K]. \tag{3}$$

It follows from the above that any extension of finite type and of transcendence degree 0 is finite. Conversely, if L/K is a finite extension and $[L : K] = n$ then for any $\alpha \in L$ the elements $1, \alpha, \ldots, \alpha^n$ must be linearly dependent over K (since there are $n + 1$ of them). It follows from this that α is algebraic, and hence L has transcendence degree 0. Thus we obtain another characterisation of extensions of finite type and of transcendence degree 0; these are the finite extensions. From what we have said above, any finite extension is obtained as a composite of extensions of the form $K(\alpha)$. But we have the following result:

VI. Primitive Element Theorem. *Suppose that K is a field of characteristic* 0, *and that $L = K(\alpha, \beta)$ is an extension generated by two algebraic elements α and β; then there exists an element $\gamma \in L$ such that $L = K(\gamma)$.*

Under this condition, any finite extension $L = K(\alpha_1, \ldots, \alpha_n)$ can be expressed in the form $L = K(\alpha)$, so that $L \cong K[x]/(P)$, and we have the representation (2) *of the elements of L.*

In fact the result holds under much wider assumptions, and in particular for finite fields.

If every polynomial has a root in a field K, that is, if K is *algebraically closed*, then all the irreducible polynomials are linear, and an extension of K cannot contain algebraic elements other than the elements of K. Hence K does not have any finite extensions other than K itself. This is the case for the complex number field $\mathbb{C}$. The real number field has only two finite extensions, $\mathbb{R}$ and $\mathbb{C}$. But the rational number field $\mathbb{Q}$ and the field $K(t)$ of rational functions (even for $K = \mathbb{C}$) have very many finite extensions. These are instruments for the study of algebraic numbers (in the case of $\mathbb{Q}$) and of algebraic functions (in the case $\mathbb{C}(t)$). It can be shown that any finite extension of $K(t)$ is of the form $K(C)$ where C is some algebraic curve, and a finite extension of the field $K(x_1, \ldots, x_n)$ is of the form $K(V)$, where V is an algebraic variety (of dimension n).

An extension $K(\alpha)$, where α is a root of an irreducible polynomial $P(x)$, is determined by this polynomial, and so the theory of finite extensions is a certain language (and also a 'philosophy') in the theory of polynomials in one variable. In one and the same extension L/K there exist many elements α for which $L = K(\alpha)$, and many polynomials $P(x)$ corresponding to these. The extension itself reflects those properties which all of these have in common. We have here another example of 'coordinatisation', analogous to assigning the function field $K(C)$ to an algebraic curve C. The construction of a field $K(\alpha)$ in the form

$K[x]/(P)$ is entirely parallel to the construction of the field $K(C)$ from the equation of the curve C.

The most elementary example illustrating applications of properties of extensions to concrete questions is the theory of ruler and compass constructions. Translating these constructions into the language of coordinates, it is easy to see that they lead either to addition, subtraction, multiplication and division operations on the numbers representing intervals already constructed; or to solving quadratic equations, the coefficients of which are numbers of this type (to find the points of intersection of a line and a circle, or of two circles). Hence if we let K denote the extension of $\mathbb{Q}$ generated by all the quantities given in the statement of the problem, and α the numerical value of the quantity we are looking for, then the problem of constructing this quantity by ruler and compass reduces to the question of whether α is contained in an extension L/K which can be represented as a chain

$$L/L_1, L_1/L_2, \ldots, L_{n-2}/L_{n-1}, L_{n-1}/L_n = K,$$

in which each extension is of the form $L_{i-1} = L_i(\beta)$, where β satisfies a quadratic equation. This condition is equivalent to $[L_{i-1} : L_i] = 2$. Applying the relation (3) we obtain that $[L : K] = 2^n$. If $\alpha \in L$ then $K(\alpha) \subset L$, and again it follows from (3) that the degree $[K(\alpha) : K]$ must be a power of 2. This is only a necessary condition; a sufficient condition for the solvability of a problem by ruler and compass can also be formulated in terms of the field $K(\alpha)$, but is slightly more complicated. However, already the necessary condition we have obtained proves, for example, that the problem of doubling the cube is not solvable by ruler and compass: it reduces to the construction of a root of the polynomial

$$x^3 - 2, \quad \text{and} \quad [\mathbb{Q}(\sqrt[3]{2}) : \mathbb{Q}] = 3.$$

In exactly the same way, the problem of trisecting an angle leads, for example, to the construction of $\alpha = \cos \varphi/3$, given that $a = \cos \varphi$ is known. This is related to the cubic equation

$$4\alpha^3 - 3\alpha - a = 0.$$

We should consider a as an independent variable, since φ is arbitrary. Hence K is the field of rational functions $\mathbb{Q}(a)$, and $[K(\alpha) : K] = 3$, and again the problem is not solvable by ruler and compass.

In the same way, the question of solving algebraic equations by radicals also leads to certain questions on the structure of finite extensions. We will deal with this in detail in § 18.A.

§ 7. The Algebraic View of Infinitesimal Notions

Considering quantities 'up to infinitesimals of order n' can be translated in algebraic terms quite conveniently, considering elements ε (of certain rings) satisfying $\varepsilon^n = 0$ as analogues of infinitesimals. Suppose, for example, that C is an algebraic curve, for simplicity considered over the complex field $\mathbb{C}$. We introduce the commutative ring

$$U = \{a + a_1\varepsilon | a, a_1 \in \mathbb{C}, \varepsilon^2 = 0\}.$$

This can be described more precisely as $\mathbb{C}[x]/(x^2)$, with ε the image of x under the canonical homomorphism $\mathbb{C}[x] \to U$. Consider homomorphisms $\varphi\colon \mathbb{C}[C] \to U$ over $\mathbb{C}$ (that is, such that $\mathbb{C} \subset \mathbb{C}[C]$ is mapped by the identity to $\mathbb{C} \subset U$). Such a φ is determined by the images $\varphi(x)$ and $\varphi(y)$ of the coordinates x and y, since the other elements of $\mathbb{C}[C]$ are polynomials $h(x, y)$ in x and y, and $\varphi(h(x, y)) = h(\varphi(x), \varphi(y))$. Also, if $F(x, y) = 0$ is the equation of C then the elements $\varphi(x)$ and $\varphi(y)$ of U must satisfy the same equation

$$F(\varphi(x), \varphi(y)) = 0. \tag{1}$$

We write $\varphi(x) = a + a_1\varepsilon$, and $\varphi(y) = b + b_1\varepsilon$. The ring U has a standard homomorphism $\psi\colon U \to \mathbb{C}$ given by $\psi(a + a_1\varepsilon) = a$. Applying this to the relation (1), we get $F(a, b) = 0$, that is, φ defines a point $(a, b) \in C$. However, knowing this point, we can reconstruct only the terms a and b in the expressions for $\varphi(x)$ and $\varphi(y)$. What is the meaning of the coefficients a_1 and b_1? We substitute the values for $\varphi(x)$ and $\varphi(y)$ in (1) and write $F(a + a_1\varepsilon, b + b_1\varepsilon)$ in the standard form $c + c_1\varepsilon$. Expanding F as a Taylor series and using the fact that $F(a, b) = 0$ and $\varepsilon^2 = 0$ we see that $F(a + a_1\varepsilon, b + b_1\varepsilon) = (a_1F'_x(a, b) + b_1F'_y(a, b))\varepsilon$, and condition (1) can be written

$$F(a, b) = 0 \quad \text{and} \quad a_1F'_x(a, b) + b_1F'_y(a, b) = 0.$$

This means that (a, b) is a point of C and (a_1, b_1) is a vector lying on the tangent line to C at (a, b). Here we assume that (a, b) is not a singular point of C, that is, the partial derivatives $F'_x(a, b)$ and $F'_y(a, b)$ do not both vanish. It is easy to see that our arguments give a description of all homomorphisms of $\mathbb{C}[C]$ to U: these correspond to pairs consisting of a point of C and a vector of the tangent line to the curve at this point. In a similar way, for the case of an algebraic surface we get a description of the tangent planes, and so on.

We formulate the previous arguments in a somewhat different way. We compose $\varphi\colon \mathbb{C}[C] \to U$ with the standard homomorphism $\psi\colon U \to \mathbb{C}$, to get the sequence

$$\mathbb{C}[C] \xrightarrow{\varphi} U \xrightarrow{\psi} \mathbb{C}.$$

As in §4, Example 2, the composite $\bar{\varphi} = \psi\varphi$ defines the point $x_0 \in C$, taking a function into its value at x_0. Hence the kernel is the maximal ideal $\mathfrak{M}_{x_0}$ of $\mathbb{C}[C]$, consisting of functions vanishing at x_0. If $x_0 = (a, b)$ then $x - a$ and $y - b$ belong to $\mathfrak{M}_{x_0}$. This corresponds to the fact that $\varphi(x - a)$ and $\varphi(y - b)$ are of the form $a_1\varepsilon$ and $b_1\varepsilon$, that is, they belong to the ideal $I = \operatorname{Ker}\psi$ of U. A vector of the tangent space at x_0 (in the present case, of the tangent line) is defined by the images $x - a$ and $y - b$ lying in this ideal, that is, by the restriction of φ to $\mathfrak{M}_{x_0}$.

Since $\varepsilon^2 = 0$, φ obviously vanishes on $\mathfrak{M}_{x_0}^2$. Hence φ defines a linear map of the space $\mathfrak{M}_{x_0}/\mathfrak{M}_{x_0}^2$ into $\mathbb{C}$, and precisely this linear function determines a vector of the tangent line at x_0. It is not hard to prove that any linear function $\mathfrak{M}_{x_0}/\mathfrak{M}_{x_0}^2 \to \mathbb{C}$ defines a tangent vector at x_0.

Theorem. *The tangent space at a point x_0 is the dual vector space to $\mathfrak{M}_{x_0}/\mathfrak{M}_{x_0}^2$, where $\mathfrak{M}_{x_0}$ is the maximal ideal corresponding to x_0.*

The same thing holds for an algebraic surface C with equation $F(x, y, z) = 0$: the tangent plane to C at a nonsingular point $x_0 = (a, b, c)$ (that is, a point at which the three derivatives

$$F'_x(a, b),\ F'_y(a, b) \text{ and } F'_z(a, b)$$

do not all vanish simultaneously) can be identified with the dual vector space to $\mathfrak{M}_{x_0}/\mathfrak{M}_{x_0}^2$. Later we will apply these arguments to an arbitrary algebraic variety, but for the moment we show that they also have applications outside the algebraic case.

Example 1. Let A be the ring of differentiable functions in a neighbourhood of a point O of an n-dimensional vector space E, and let $\mathfrak{M}$ be the ideal of functions vanishing at O. By Taylor's formula, $f \in \mathfrak{M}$ can be represented in the form $f \equiv l \bmod \mathfrak{M}^2$ where l is a linear function. Linear functions on E form the dual vector space E^*, and we again get an isomorphism $\mathfrak{M}/\mathfrak{M}^2 \cong E^*$. If $\zeta \in E$ then $l(\xi)$ can be interpreted as the partial derivative $l(\xi) = \dfrac{\partial f}{\partial \xi}(O)$.

A similar situation holds if A is the ring of differentiable functions on a differentiable manifold X and $\mathfrak{M}$ consists of the functions vanishing at $x_0 \in X$. Again we have $\mathfrak{M}/\mathfrak{M}^2 \cong T_{x_0}^*$, where T_{x_0} is the tangent space at x_0, and the isomorphism is given by

$$l(\xi) = \frac{\partial f}{\partial \xi}(x_0) \quad \text{for} \quad \xi \in T_{x_0} \text{ and } l = f + \mathfrak{M}^2. \tag{2}$$

The preceding argument presupposed that we already had a definition of the tangent space of a differentiable manifold, but the argument can be reversed and turned into the definition of the *tangent space*,

$$T_{x_0} = (\mathfrak{M}_{x_0}/\mathfrak{M}_{x_0}^2)^*. \tag{3}$$

Thus $\xi \in T_{x_0}$ is by definition a linear function l on $\mathfrak{M}_{x_0}$ which is zero on $\mathfrak{M}_{x_0}^2$. Setting l to be equal to zero by definition on constants, we get a function on the whole of A. It is easy to see that the conditions imposed on l can be written as

$$l(\alpha f + \beta g) = \alpha l(f) + \beta l(g) \quad \text{for} \quad \alpha, \beta \in \mathbb{R} \text{ and } f, g \in A$$

and (4)

$$l(fg) = l(f)g(x_0) + l(g)f(x_0).$$

In this form they axiomatise the intuitive notion of a tangent vector as 'that with respect to which a function can be differentiated' (as in (2)). The relation (3) or the equivalent conditions (4) gives perhaps the most intrinsic definition of the tangent space at a point of a differentiable manifold.

In this connection it is natural to consider the notion of a vector field on a differentiable manifold. By definition, a vector field θ assigns to any point $x \in X$ a vector $\theta(x) \in T_x$. For any function $f \in A$ and a point $x \in X$, the vector $\theta(x)$ defines a number $\theta(x)(f)$, that is, a function $g(x) = \theta(x)(f)$. We write $\mathscr{D}(f)$ for this operator. The relations (4) show that $\mathscr{D}$ satisfies the conditions

$$\begin{aligned} &\mathscr{D}(\alpha f + \beta g) = \alpha\mathscr{D}(f) + \beta\mathscr{D}(g), \\ \text{and} \quad &\mathscr{D}(fg) = f\mathscr{D}(g) + \mathscr{D}(f)g. \end{aligned} \tag{5}$$

An operator of this type is called a *first order linear differential operator*. It is easy to see that in a coordinate system $(x_1, \ldots, x_n)$ it can be written

$$\mathscr{D}(f) = \sum_1^n a_i \frac{\partial f}{\partial x_i}, \tag{6}$$

where $a_i = \mathscr{D}(x_i)$. Conversely, every operator $\mathscr{D}$ satisfying (5) defines a vector field θ for which

$$\theta(x)(f) = \mathscr{D}(f)(x).$$

For any arbitrary ring A a *derivation* of A is a map $\mathscr{D}\colon A \to A$ which satisfies

$$\mathscr{D}(a + b) = \mathscr{D}(a) + \mathscr{D}(b),$$
$$\mathscr{D}(ab) = a\mathscr{D}(b) + \mathscr{D}(a)b.$$

If $B \subset A$ is a subring, we say that $\mathscr{D}$ is a derivation of A over B if $\mathscr{D}(b) = 0$ for $b \in B$. Then $\mathscr{D}(ab) = \mathscr{D}(a)b$ for $a \in A$, $b \in B$. If we set

$$(\mathscr{D}_1 + \mathscr{D}_2)(a) = \mathscr{D}_1(a) + \mathscr{D}_2(a),$$
$$(c\mathscr{D})(a) = c\mathscr{D}(a) \quad \text{for } a, c \in A$$

then derivations of A over B form an A-module.

We can thus say that the module of vector fields on a differentiable manifold X is by definition the module of derivations over $\mathbb{R}$ of the ring of differentiable functions on X. Together with the assertions of §5, Examples 13 and 12, we now get an algebraic definition of all the basic notions: vector fields, differential 1-forms and r-forms on a manifold.

We now return to arbitrary commutative rings. In §4 we formulated a general conception according to which the elements of an arbitrary commutative ring A can be viewed as functions on a 'space', the points of which are maximal ideals (or in another version, prime ideals) of the ring, and the homomorphisms $A \to A/\mathfrak{M}$ define the value of a 'function' $a \in A$ at the 'point' corresponding to the maximal ideal $\mathfrak{M}$. Now we can make this connection deeper by assigning a tangent space to each point. For this, consider the maximal ideal $\mathfrak{M}$ defining a point, and the quotient $\mathfrak{M}/\mathfrak{M}^2$. Suppose that $k = A/\mathfrak{M}$ is the 'field of values' at the point corresponding to $\mathfrak{M}$. For elements $m \in \mathfrak{M}$ and $a \in A$ the residue class of $am \bmod \mathfrak{M}^2$ depends only on the residue class of $a \bmod \mathfrak{M}$, that is, on the element of k determined by a. This shows that $\mathfrak{M}/\mathfrak{M}^2$ is a vector space over k. The dual vector space, that is, the set of k-valued linear functions on $\mathfrak{M}/\mathfrak{M}^2$ is the analogue of the tangent space at the point corresponding to $\mathfrak{M}$.

This point of view is useful in the analysis of various geometric and algebraic situations. For example, if an irreducible algebraic curve C is given by an equation $F(x, y) = 0$, then for $(a, b) \in C$ the tangent space is given by the equation

$$F'_x(a, b)(x - a) + F'_y(a, b)(y - b) = 0.$$

This is 1-dimensional for all points (a, b), except for points at which $F'_x(a, b) = F'_y(a, b) = 0$. We say that a point of C is *singular* if both F'_x and F'_y vanish there, and *nonsingular* otherwise. It is easy to see that the number of singular points is finite. We see that the tangent space is 1-dimensional (that it, it has the same dimension as C) for nonsingular points, and has bigger dimension (namely 2) for singular points. A similar situation holds for more general algebraic varieties: the dimension of the tangent spaces is the same at all points, except at the points of a certain proper algebraic subvariety, at which it jumps up. This gives us, firstly a new characterisation of the dimension of an irreducible algebraic variety (as the dimension of the tangent spaces at all points except those of some proper subvariety); secondly, it distinguishes the singular points (the points of this proper subvariety); and thirdly, it gives an important invariant of a singular point (the jump in dimension of the tangent space). But perhaps most remarkable of all is that these notions are applicable to arbitrary rings, not necessarily geometric in origin, and allow us to use geometric intuition in their study. For example, the maximal ideals of the ring of integers $\mathbb{Z}$ are described by prime numbers, and for $\mathfrak{M} = (p)$ the vector space $\mathfrak{M}/\mathfrak{M}^2$ is 1-dimensional over $\mathbb{F}_p$, so that here singular points do not occur.

Example 2. Consider the ring A consisting of elements $a + b\sigma$ with $a, b \in \mathbb{Z}$, with operations defined on them as usual, together with the condition $\sigma^2 = 1$ (this ring turns up in connection with the arithmetical properties of representations of the group of order 2). Its maximal ideals can be described as follows. For any prime number $p \neq 2$ we have two maximal ideals

$$\mathfrak{M}_p = \{a + b\sigma | p \text{ divides } a + b\}$$

and

$$\mathfrak{M}'_p = \{a + b\sigma | p \text{ divides } a - b\}.$$

Obviously, $\mathfrak{M}_p = (p, 1 - \sigma)$ and $\mathfrak{M}'_p = (p, 1 + \sigma)$. For each of these, the space $\mathfrak{M}/\mathfrak{M}^2$ is 1-dimensional over $\mathbb{F}_p$. In addition, there exists a further maximal ideal

$$\mathfrak{M}_2 = \{a + b\sigma \mid a \text{ and } b \text{ have the same parity}\} = (2, 1 + \sigma).$$

It is easy to see that $\mathfrak{M}_2^2 = (4, 2 + 2\sigma)$ and that $\mathfrak{M}_2/\mathfrak{M}_2^2$ consists of 4 elements, the cosets of the elements 0, 2, $1 + \sigma$ and $3 + \sigma$. Thus this is a 2-dimensional vector space over $\mathbb{F}_2$. The ideal $\mathfrak{M}_2$ corresponds to the unique singular point.

All of our considerations so far have been connected with considering quantities 'up to infinitesimals of order 2', which for an arbitrary ring A and its maximal ideal $\mathfrak{M}$ reduces to considering the ring $A/\mathfrak{M}^2$. Of course, it is also possible to consider quantities 'up to infinitesimals of order r', which leads to the ring $A/\mathfrak{M}^r$. For example, if A is the polynomial ring $\mathbb{C}[x_1, \ldots, x_n]$ or the ring of analytic functions of variables $z, \ldots, z_n$ in a neighbourhood of the origin, or the ring of C^∞ complex-valued functions in n variables, and $\mathfrak{M}$ is the ideal of functions which vanish at the origin $O = (0, \ldots, 0)$ then $A/\mathfrak{M}^r$ is a finite-dimensional vector space over $\mathbb{C}$. It generalises the space $A/\mathfrak{M}^2$ we have already considered, and is called the *space of jets* of order $(r - 1)$.

Example 3. Differential Operators of Order >1. A *linear differential operator* of order $\leqslant r$ on a differentiable manifold X can be defined formally as an $\mathbb{R}$-linear map $\mathscr{D}: A \to A$ of the ring A of differentiable functions on X to itself such that for any function $g \in A$ the operator $\mathscr{D}_1(f) = \mathscr{D}(gf) - g\mathscr{D}(f)$ has order $\leqslant r - 1$. Formula (5) defining a first order operator shows that $\mathscr{D}(gf) - g\mathscr{D}(f)$ is the operator of multiplying by a function (namely, $\mathscr{D}(g)$); conversely if $\tilde{\mathscr{D}}(gf) - g\tilde{\mathscr{D}}(f)$ is multiplication by a function then it is easy to check that $\tilde{\mathscr{D}}(f) = \mathscr{D}(f) + \mathscr{D}(1)f$, where $\mathscr{D}$ is a first order operator.

From the definition it follows by induction that if $\mathscr{D}$ is a operator of order $\leqslant r$ then $\mathscr{D}(\mathfrak{M}_{x_0}^{r+1}) \subset \mathfrak{M}_{x_0}$, where $\mathfrak{M}_{x_0} \subset A$ is the maximal ideal corresponding to a point $x_0 \in X$. In coordinates this means that $\mathscr{D}(f)(x_0)$ depends only on the values at x_0 of the partial derivatives of f of order $\leqslant r$. In other words, we have

$$\mathscr{D}(f) = \sum_{i_1 + \cdots + i_n \leqslant r} a_{i_1 \ldots i_n}(x_1, \ldots, x_n) \frac{\partial^{i_1 + \cdots + i_n} f}{\partial x_1^{i_1} \ldots \partial x_n^{i_n}}, \quad \text{with } a_{i_1 \ldots i_n} \in A.$$

For any point $x_0 \in X$ the map $f(x) \mapsto \mathscr{D}(f)(x_0)$ defines a linear function l on the space of all jets of order r: $l \in (A/\mathfrak{M}^r)^*$, in exactly the same way that a first order linear differential operator defines a linear function on $\mathfrak{M}_{x_0}/\mathfrak{M}_{x_0}^2$.

However, the most precise apparatus for studying the ring A 'in a neighbourhood of a maximal ideal $\mathfrak{M}$' is obtained if we consider simultaneously all the rings $A/\mathfrak{M}^n$ for $n = 1, 2, 3, \ldots$ They can all be put together into one ring $\hat{A}$ called the *projective limit* of the $A/\mathfrak{M}^n$. For this we observe that there exists a canonical homomorphism $\varphi_n: A/\mathfrak{M}^{n+1} \to A/\mathfrak{M}^n$ with kernel $\mathfrak{M}^n/\mathfrak{M}^{n+1}$. The ring $\hat{A}$ is defined as the set of sequences of elements $\{\alpha_n \mid \alpha_n \in A/\mathfrak{M}^n\}$ which are compatible in the sense that $\varphi_n(\alpha_{n+1}) = \alpha_n$; the ring operations on sequences are defined element-by-element. Each element $a \in A$ defines such a sequence, by $\alpha_n = a + \mathfrak{M}^n$, and we thus get a homomorphism $\varphi: A \to \hat{A}$. The kernel of φ is the intersection of all the ideals $\mathfrak{M}^n$. In many interesting cases this intersection is 0, and hence A embeds in $\hat{A}$ as a subring.

Example 4. Let $A = K[x]$, and $\mathfrak{M} = (x)$. An element α_n of the ring $K[x]/(x^n)$ is uniquely determined by a polynomial

$$f_n = a_0 + a_1 x + \cdots + a_{n-1} x^{n-1},$$

and a sequence of elements $\{\alpha_n\}$ is compatible if the polynomial f_{n+1} representing α_{n+1} is obtained from f_n by adding in a term of degree n. The whole sequence thus defines an infinite (formal) power series. In other words, the ring $\hat{A}$ is isomorphic to the ring of formal power series $K[\![x]\!]$ of § 3, Example 6. The inclusion $\varphi\colon K[x] \to K[\![x]\!]$ extends to an inclusion of the fields of fractions $\varphi\colon K(x) \to K((x))$, where $K((x))$ is the field of formal Laurent power series (§ 2, Example 5). It is easy to see that this inclusion is the same thing as sending a rational function to its Laurent series at $x = 0$. In particular, if a function does not have a pole at $x = 0$ then it is sent to its Taylor series. For example, if $f(x) = 1/(1 - x)$ then $f(x) \equiv 1 + x + \cdots + x^{n-1} \bmod x^n$, or in other words, the function $f(x) - 1 - x - \cdots - x^{n-1}$ has denominator not divisible by x, and numerator divisible by x^n. This means that $f(x)$ is sent to the series $1 + x + x^2 + \cdots$

Example 5. Let $A = K[C]$ be the coordinate ring of an arbitrary algebraic variety C. If $\mathfrak{M}_c$ is the maximal ideal of A corresponding to a nonsingular point $c \in C$, then $\hat{A}$ is isomorphic to the ring $K[\![x_1,\ldots,x_n]\!]$ of formal power series, where n is the dimension of C (in any of the definitions of this notion discussed above). Moreover, the inclusion

$$K[C] \to K[\![x_1,\ldots,x_n]\!]$$

extends to those functions in $K(C)$ that are finite at c, that is, can be represented as P/Q where $P, Q \in K[C]$ and $Q(c) \neq 0$. This gives a representation of such functions as formal power series. If K is the complex or real number field $\mathbb{C}$ or $\mathbb{R}$ then it can be proved that the corresponding functions converge for sufficiently small values of $x_1, \ldots, x_n$. This is how one proves that an algebraic variety without singular points is also a topological, differentiable and analytic manifold.

Example 6. Let A be the ring of C^∞ functions in a neighbourhood of $x = 0$, and I the ideal of functions that vanish at $x = 0$. Then I^n is the ideal of functions that vanish at $x = 0$ together with all of their derivatives of order $< n$; A/I^n is the ring $\mathbb{R}[x]/(x^n)$, and the homomorphism $A \to \mathbb{R}[x]/(x^n)$ takes a function into its Taylor series. In this case $\bigcap I^n \neq 0$, since there exist nonzero C^∞ functions all of whose derivatives vanish at $x = 0$. The homomorphism $A \to \hat{A}$ takes each function to its formal Taylor series. Since by a theorem of E. Borel there exist C^∞ functions all of whose derivatives at $x = 0$ take preassigned values, $\hat{A} \cong \mathbb{R}[\![t]\!]$.

But the same ideas can also be applied to rings of a completely different nature.

Example 7. Suppose that $A = \mathbb{Z}$ is the ring of integers and $\mathfrak{M} = (p)$ for some prime number p. As $\hat{A}$ we get a ring $\mathbb{Z}_p$ called the *ring of p-adic integers*. By analogy with the case of the ring $K[x]$ considered above, one can see that an element of $\mathbb{Z}_p$ is given as a sequence $\{\alpha_n\}$ of integers of the form

$$\alpha_n = a_0 + a_1 p + \cdots + a_{n-1} p^{n-1},$$

where the a_i belong to the fixed system $0 \leqslant a_i < p$ of representatives of the classes of residues mod p, and α_{n+1} is obtained from α_n by adding on a term $a_n p^n$. This sequence can be written as a formal series

$$a_0 + a_1 p + a_2 p^2 + \cdots$$

The ring operations on these sequences are carried out in exactly the same way as the operations on integers written in base p; that is, if operating on the coefficient a_i we get a number $c > p$, we must divide c by p with remainder $c = c_0 + c_1 p$ and 'carry c_1 into the next place'. The ring $\mathbb{Z}_p$ is integral, and its field of fractions $\mathbb{Q}_p$ is the *field of p-adic numbers*. The inclusion $\mathbb{Z} \hookrightarrow \mathbb{Z}_p$ extends to an inclusion $\mathbb{Q} \hookrightarrow \mathbb{Q}_p$.

To get a more rounded view of the relation between the constructions described above, we return to the example of the ring $K[x]$ and the field $K(x)$. For a more precise numerical characterisation of the fact that a nonzero function $f \in K(x)$ vanishes to a given order at $x = 0$, we introduce the exponent $v(f)$, equal to n if f has a zero of order $n > 0$ at $x = 0$, or to $-n$ if f has a pole of order $n > 0$ at x. We fix a real number c with $0 < c < 1$ once and for all (for example, $c = \frac{1}{2}$), and set $\varphi(f) = c^{v(f)}$ for $f \neq 0$, and $\varphi(0) = 0$. Then $\varphi(f)$ is small if f vanishes to a high order at $x = 0$. The expression $\varphi(f)$ we have introduced has the formal properties of the absolute value of a rational, real or complex number: $\varphi(f) = 0$ if and only if $f = 0$, and

$$\varphi(fg) = \varphi(f)\varphi(g), \quad \varphi(f + g) \leqslant \varphi(f) + \varphi(g). \tag{7}$$

We say that a field L having a real-valued function φ with these three properties is a *normed field* and the function φ a *valuation*. The simplest example of a normed field is the rational number field $\mathbb{Q}$ with $\varphi(x) = |x|$. The procedure of constructing the reals starting from the rationals, by means of Cauchy series, can be taken over word-for-word to any normed field. We get a new normed field $\hat{L}$, into which L embeds as a subfield with the valuation preserved, such that the image of L is everywhere dense; and $\hat{L}$ is complete (in the sense of its valuation), that is, it satisfies the Cauchy convergence criterion; $\hat{L}$ is called the *completion* of L with respect to the valuation φ.

It is very easy to see that the construction of the field $K((x))$ and of the embedding $K(x) \to K((x))$ is an application of the general construction to the case of the valuation $\varphi(f) = c^{v(f)}$ introduced above. Now we can use the fact that the field $K(X)^\wedge = K((x))$ has a valuation extending the valuation φ of $K(x)$. It is easy to see what this is: if $f \in K((x))$ and

$$f = c_n x^n + c_{n+1} x^{n+1} + \cdots \quad \text{with } c_n \neq 0$$

then $\varphi(f) = c^n$, and $\varphi(0) = 0$. But in a normed field the convergence of series is meaningful, and it is easy to see that any formal Laurent series converges in this sense; in particular, $x^n \to 0$ as $n \to \infty$ in the sense of our theory. Taking a rational function f into its Laurent series now turns into an equality, in the sense that in $K(X)^\wedge = K((x))$, f is equal to the sum of the series converging to it.

In this connection it is interesting to determine quite generally which valuations can be defined on the field $K(x)$. We restrict ourselves to the case of the complex number field $K = \mathbb{C}$, and strengthen the notion of valuation by adding a further condition to (7):

$$\varphi(\alpha) = 1 \quad \text{if} \quad \alpha \in \mathbb{C} \quad \text{and} \quad \alpha \neq 0. \tag{8}$$

Obviously the valuation $\varphi(f) = c^{v(f)}$ we have constructed satisfies this extra condition. Of course, we could vary our construction, considering any point $x = \alpha$ in place of $x = 0$, that is, defining $v(f)$ as the order of the zero or pole of a function f at $x = \alpha$. The valuation so obtained is denoted by φ_α. We can consider another similar valuation by considering the order of zero or pole of a function at infinity; we denote this valuation by φ_∞. It can most simply be defined by $\varphi_\infty(f) = c^{m-n}$ if $f = \dfrac{P}{Q}$ and P, Q are polynomials of degree n, m respectively (and of course $\varphi(0) = 0$).

It is not hard to see that these valuations exhaust all the valuations of $\mathbb{C}(x)$.

Theorem I. *All valuations of* $\mathbb{C}(x)$ *(with the extra condition* (8)*) are given by the valuations* φ_α *for* $\alpha \in \mathbb{C}$, *and the valuation* φ_∞.

Thus the valuations of $\mathbb{C}(x)$ give us in a very natural way all the points of the line (including the point at infinity), or of the Riemann sphere, on which the rational functions are defined.

We now ask the same question for finite extensions of $\mathbb{C}(x)$. These are of the form $\mathbb{C}(C)$, where C is some irreducible curve. The answer turns out to be similar, but rather more delicate. Every nonsingular point c of the curve C corresponds to some valuation φ_c, characterised for example by the fact that $\varphi_c(f) < 1$ if and only if $f(c) = 0$. But there are a finite number of valuations to be added to these; firstly, the points at infinity of the curve C (which occur if we consider a curve in the projective plane). Secondly, singular points of C may correspond to several distinct valuations. The entire set of valuations is in 1-to-1 correspondence with the points of a certain nonsingular curve lying in projective space, and defining the same field $\mathbb{C}(C)$, the so-called *nonsingular projective model* of C. The points of this model are thus characterised in a very remarkable way quite intrinsically by the field $\mathbb{C}(C)$. Another way of stating the same description is that if the curve C is given by the equation $F(x, y) = 0$ then all valuations of $\mathbb{C}(C)$ are in 1-to-1 correspondence with the points of the Riemann surface of the function y as an analytic function of x. This can be considered as a purely algebraic description of the Riemann surface of an algebraic function.

Let $\xi = (a, b)$ be some point of an algebraic curve C with equation $F(x, y) = 0$, and φ one of the valuations corresponding to ξ. Then the completion of $\mathbb{C}(C)$ with respect to the valuation φ is again isomorphic to the field $\mathbb{C}((t))$ of formal

Laurent series. Suppose that under the inclusion $\mathbb{C}(C) \subset \mathbb{C}((t))$,

$$x - a = c_k t^k + c_{k+1} t^{k+1} + \cdots, \quad \text{with} \quad c_k \neq 0.$$

Then $x - a = t^k f(t)$ with $f(0) \neq 0$. Thus

$$x - a = \tau^k,$$

where

$$\tau = t f(t)^{1/k},$$

and $f(t)^{1/k}$ must be understood as a formal power series, which is meaningful in view of the condition $f(0) \neq 0$. It is easy to show that τ is a 'parameter' of the field $\mathbb{C}((t))$ as well as t; that is, all elements of $\mathbb{C}((t))$ can be represented as Laurent series in τ also, so that $\mathbb{C}((t)) = \mathbb{C}((\tau))$. In particular,

$$y = \sum d_i \tau^i = \sum d_i (x - a)^{i/k}.$$

This type of expansion of an algebraic function y as a fractional power series in $x - a$ is called a *Puiseux expansion.*

We now proceed to the rational number field $\mathbb{Q}$. Let p be a prime, and c a real number with $0 < c < 1$. We write $v(n)$ for the highest power of p which divides n, and for a rational number $a = \dfrac{n}{m}$ with $n, m \in \mathbb{Z}$, we set

$$\varphi_p(a) = c^{v(n) - v(m)}.$$

It is easy to check that φ_p is a valuation on the rational number field $\mathbb{Q}$. Considering the completion of $\mathbb{Q}$ in this valuation, we arrive at the p-adic number field $\mathbb{Q}_p$ which was introduced earlier. In it, the notion of convergence of series makes sense, and the formal power series which we used to specify p-adic numbers are convergent. For example, the equality

$$\frac{1}{1 - p} = 1 + p + p^2 + \cdots$$

has the meaning that the number on the left-hand side is the sum of the convergent series on the right.

By analogy with the field $\mathbb{C}(x)$ it is natural to ask: what are all the valuations of $\mathbb{Q}$?

II. Ostrowski's Theorem. *Every valuation of $\mathbb{Q}$ is either a p-adic valuation φ_p or a valuation of the form $\varphi(a) = |a|^c$, where c is a real number with* $0 < c < 1$.

The number c here is an inessential parameter, exactly as that occuring in the definition of a p-adic valuation or of the valuation φ_α of $\mathbb{C}(x)$: valuations obtained for different choices of c define the same notion of convergence and isomorphic completions. The completion with respect to the valuation $|\ |^c$ gives of course the real number field. Thus all the p-adic number fields $\mathbb{Q}_p$ and the real number field $\mathbb{R}$ play entirely similar roles. The comparison with the field $\mathbb{C}(x)$ shows that primes p (defining the fields $\mathbb{Q}_p$) are analogous to finite points $x = \alpha$, and the inclusions $\mathbb{Q} \to \mathbb{Q}_p$ are analogous to expansions in Laurent series at finite points; then the inclusion $\mathbb{Q} \to \mathbb{R}$ is an analogue of the Laurent expansion at infinity. This gives a unified point of view on two types of properties of integers (or rational numbers): divisibility, and size. For example, for $f \in \mathbb{Z}[x]$, the fact that the equation $f(x) = 0$ has a real root means that there exist rational numbers a_n for which $|f(a_n)|$ is arbitrarily small. In the same way, the fact that $f(x)$ is solvable in the p-adic field means that there exist rational numbers a_n for which $\varphi_p(f(a_n))$ is arbitrarily small, that is, that $f(a_n)$ is divisible by larger and larger powers of p. It can be shown that for a polynomial $f(x_1, \ldots, x_n)$ the solvability of the equation

$$f(x_1, \ldots, x_n) = 0$$

in $\mathbb{Q}_p$ is equivalent to the solvability of the congruence

$$f(x_1, \ldots, x_n) \equiv 0 \mod p^k$$

for any k. Since a congruence to any modulus reduces to congruences mod p^k, the solvability of the equation $f = 0$ in all the fields $\mathbb{Q}_p$ is equivalent to the solvability of the congruence

$$f \equiv 0 \mod N$$

for any modulus N. For example, the following assertion is a classical result of number theory.

III. Legendre's Theorem. *The equation*

$$ax^2 + by^2 = c \quad (\text{for } a, b, c \in \mathbb{Z} \text{ and } c > 0)$$

is solvable in rational numbers if and only if the following conditions hold:

(1) *either $a > 0$ or $b > 0$;*

(2) *the congruence $ax^2 + by^2 \equiv c \mod N$ is solvable for all N.*

By what we have said above, this means that the equation $ax^2 + by^2 = c$ is solvable in rationals if and only if it is solvable in each of the fields $\mathbb{Q}_p$ and $\mathbb{R}$.

This result can be generalised.

IV. Minkowski-Hasse Theorem. *The equation*

$$f(x_1, \ldots, x_n) = c,$$

where f is a quadratic form with rational coefficients, and $c \in \mathbb{Q}$, is solvable in $\mathbb{Q}$ if and only if it is solvable in all the fields $\mathbb{Q}_p$ and $\mathbb{R}$.

The p-adic number field reflects arithmetic properties of the rational numbers (divisibility by powers of p), but on the other hand, it has a number of properties in common with the field $\mathbb{R}$; in $\mathbb{Q}_p$ we can consider measures, integrals, analytic functions, interpolation and so on. This gives a powerful number-theoretic method (especially if all the fields $\mathbb{Q}_p$ and $\mathbb{R}$ are considered together), the use of which has led to a large number of deep arithmetic results.

In conclusion, we consider a finite extension field K of $\mathbb{Q}$; a field of this type is called an *algebraic number field*. What valuations are there on K? Every valuation induces a certain valuation of $\mathbb{Q}$, and one can prove that any valuation of $\mathbb{Q}$ is induced by a finite number of valuations of K. Those which induce the usual absolute value $|a|$ on $\mathbb{Q}$ are related to embeddings of K into the real field $\mathbb{R}$ or the complex field $\mathbb{C}$ and the function $|x|$ on these fields. We consider other valuations. In $\mathbb{Q}$ the subring $\mathbb{Z}$ is distinguished by the conditions $\varphi_p(a) \leqslant 1$ for all p. By analogy, consider the elements of K satisfying $\varphi(a) \leqslant 1$ for all the valuations of K inducing the valuations φ_p of $\mathbb{Q}$ for some prime p. One sees easily that these elements form a ring A, which plays the role of the *ring of integers* of K; the elements of A are called *algebraic integers*. (It can be proved that $\alpha \in K$ is an algebraic integer if and only if it satisfies a equation

$$\alpha^n + a_1\alpha^{n-1} + \cdots + a_n = 0 \quad \text{with} \quad a_1, \ldots, a_n \in \mathbb{Z};$$

this is often taken as the definition of an algebraic integer.) The field of fractions of A equals K. Obviously, $A \supset \mathbb{Z}$. It can be proved that A is a free module over $\mathbb{Z}$, of rank equal to the degree $[K : \mathbb{Q}]$ of the extension $K/\mathbb{Q}$. The ring A is in general not a unique factorisation domain, but the theorem on unique factorisation of ideals into a product of prime ideals holds in it. In particular, for any prime ideal $\mathfrak{p}$ and element $\alpha \in A$ there is a well-defined exponent $\nu(\alpha)$ which tells us what power of $\mathfrak{p}$ divides the principal ideal (α). We choose a real number c with $0 < c < 1$, and for any element $\xi \in K, \xi \neq 0$, write $\xi = \dfrac{\alpha}{\beta}$ with $\alpha, \beta \in A$ and set

$$\varphi_{\mathfrak{p}}(\xi) = c^{\nu(\alpha)-\nu(\beta)}.$$

Thus to each prime ideal $\mathfrak{p}$ of A we assign a valuation $\varphi_{\mathfrak{p}}$. It turns out that these exhaust all the valuations of K that induce one of the valuations φ_p on $\mathbb{Q}$. These facts make up the first steps in the arithmetic of algebraic number fields. Comparing them with the analogous facts which we have discussed above in connection with the fields $\mathbb{C}(C)$ for an algebraic number field C, we can observe a far-reaching parallelism between the arithmetic of algebraic number fields and the geometry of algebraic curves (or properties of the corresponding Riemann surfaces). This is a further realisation of the 'functional' point of view of numbers which we discussed in §4 (see the remark after Example 3).

§8. Noncommutative Rings

The set of linear transformations of a finite-dimensional vector space has two operations defined on it, addition and multiplication; writing out linear transformations in terms of matrixes, these operations can be transferred to matrixes as well. The existence of both these operations is extremely important and is constantly used. It is, for example, only because of this that we can define polynomials in a linear operator; and, among other uses, they are used in the study of the structure of a linear transformation, which depends in an essential way on the multiplicity of roots of its minimal polynomial. The same two operations, together with a passage to limits, make it possible to define analytic functions of a (real or complex) matrix. For example,

$$e^A = \sum_{n=0}^{\infty} \frac{A^n}{n!};$$

by writing out a system of n first order linear ordinary differential equations with constant coefficients in n unknowns in the form $\frac{dx}{dt} = Ax$, where x is the vector of unknown functions and A the matrix of coefficients, this allows us to write the solution in the form $x(t) = e^{At}x_0$, where x_0 is the vector of initial data.

The operations of addition and multiplication of linear transformations are subject to all the axioms of a commutative ring, except commutativity of multiplication. Omitting this requirement from the definition of a commutative ring, we also omit the adjective 'commutative' in the name of the new notion.

Thus, a *ring* is a set with operations of addition and multiplication, satisfying the conditions:

$$a + b = b + a,$$

$$a + (b + c) = (a + b) + c,$$

$$(ab)c = a(bc),$$

$$a(b + c) = ab + ac$$

$$(b + c)a = ba + ca.$$

There exists an element 0 such that $a + 0 = 0 + a = a$ for all a; for any a there exists an element $-a$ with the property $a + (-a) = 0$. There exists an element 1 such that $1 \cdot a = a \cdot 1 = a$ for all a.

We now give some examples of rings (noncommutative ones; we have already seen any number of commutative ones).

Example 1. The ring of linear transformations of a vector space L, and its natural generalisation, the ring of all homomorphisms of a module M to itself over a commutative ring A. Homomorphisms of a module to itself are called *endomorphisms*, and the ring defined above is denoted by $\mathrm{End}_A M$. If $A = K$ is a field, we get the ring of linear transformations of a vector space L, which we will also denote by $\mathrm{End}_K L$.

Example 2. The simplest infinite-dimensional analogue of the ring of linear transformations is the rings of bounded linear operators in a Banach space.

Example 3. The ring of linear differential operators in 1 or n variables, whose coefficients are polynomials, or analytic functions, or C^∞ functions, or formal power series (in the same number of variables, of course).

Before proceeding to consider further examples, we note those notions which we introduced for commutative rings, but which did not in fact use commutativity. These are: *isomorphism, homomorphism, kernel and image of a homomorphism, subring, graded ring.*

For example, choosing a basis in an n-dimensional vector space L over a field K determines an isomorphism of the ring $\mathrm{End}_K L$ with the ring of $n \times n$ matrixes, which we denote $M_n(K)$.

In a ring R the set of elements a commuting with all elements of R (that is, $ax = xa$ for all $x \in R$) forms a subring, called the *centre* of R, and denoted by $Z(R)$.

If the centre of a ring R contains a subring A then we say that R is an *A-algebra* or an *algebra over A*. Forgetting about multiplication in R and considering only multiplication of elements of R by elements of A turns R into an A-module. The notion of homomorphism of two algebras over a commutative ring A differs from an ordinary ring homomorphism in that we insist that each element of A is taken into itself, that is, that the homomorphism defines a homomorphism of the corresponding A-modules. The notion of subalgebra of an algebra R over A is defined in the same way: it should be a subring containing A.

If $A = K$ is a field and R is an algebra over K then the dimension of R as a vector space over K is the *rank* of the algebra R. We have already met this notion: a finite extension L/K is an extension which is an algebra of finite rank. An algebra of finite rank n over a field K has by definition a basis $e_1, \ldots, e_n$, and multiplication in the algebra is determined by the multiplication of elements of this basis. Since $e_i e_j$ is again an element of the algebra, it can be written in the form

$$e_i e_j = \sum c_{ijk} e_k \quad \text{with} \quad c_{ijk} \in K. \tag{1}$$

The elements c_{ijk} are called the *structure constants* of the algebra. They determine multiplication in the algebra:

$$(\sum a_i e_i)(\sum b_j e_j) = \sum a_i b_j c_{ijk} e_k.$$

The relations (1) are referred to as the multiplication table of the algebra. Of course, the structure constants cannot be given in an arbirary way: they have to satisfy the conditions that reflect the requirement that multiplication is associative and there exists a unit element.

For example, the matrix ring $M_n(K)$ is an algebra of rank n^2 over K. As a basis we can take the n^2 elements E_{ij}, where E_{ij} is the matrix with all entries equal to 0 except for the entry in the ith row and jth column, which is 1. Its structure constants are determined by

$$\begin{aligned} E_{ij}E_{kl} &= 0 \quad \text{if } j \neq k, \\ E_{ij}E_{jl} &= E_{il}. \end{aligned} \tag{2}$$

Now we can introduce some more examples of rings, given most simply as algebras over a field.

Example 4. Let G be a finite group (we assume that the reader is familiar with this notion, although in any case it is recalled in § 12). We construct an algebra over a field K whose basis elements e_g for $g \in G$ are indexed by elements of the group, and which multiply together as elements of G:

$$e_{g_1}e_{g_2} = e_{g_1g_2}.$$

The algebra so obtained is called the *group algebra* of G, and is denoted by $K[G]$. In the same way, we can define the group algebra $A[G]$ of a finite group G over a commutative ring A. Identifying elements $g \in G$ with the corresponding basis elements e_g, we can view the elements of $K[G]$ as sums $\sum_{g \in G} \alpha_g g$. The product $\left(\sum_{g \in G} \alpha_g g\right)\left(\sum_{h \in G} \beta_h h\right)$ can of course also be written in the same form $\sum_{g \in G} \gamma_g g$, where, as is easy to check,

$$\gamma_g = \sum_{u \in G} \alpha_u \beta_{u^{-1}g}. \tag{3}$$

An element $\sum \alpha_g g$ is determined by its coefficients, which we can view as functions on G, and write accordingly $\alpha(g)$. We then get an interpretation of $K[G]$ as the algebra of functions on G, with multiplication taking functions $\alpha(g)$, $\beta(g)$ into the function $\gamma(g)$ given as in (3) by

$$\gamma(g) = \sum_{u \in G} \alpha(u)\beta(u^{-1}g). \tag{4}$$

This notation is the starting point for generalisations to infinite groups. For example, if G is the unit circle $|z| = 1$, writing elements of G in terms of their argument φ, we see that a function on G is just a periodic function of φ with period 2π. By analogy with formula (4), the group algebra of our group is defined as the algebra of periodic functions $\alpha(\varphi)$ (for example continuous and absolutely integrable) with the multiplication law which takes $\alpha(\varphi)$, $\beta(\varphi)$ into the function

$$\gamma(f) = \frac{1}{2\pi}\int_0^{2\pi} \alpha(t)\beta(\varphi - t)\,dt.$$

In analysis, this operation is called the *convolution* of two functions.

This definition fails in one formal respect: the group algebra does not contain the identity element, which is the delta-function of the unit element. We can easily overcome this failure by adjoining a unit to R, that is, considering $\mathbb{C} \oplus R$ with multiplication $(\alpha + x)(\beta + y) = \alpha\beta + (\alpha y + \beta x + xy)$.

Another way of generalising the notion of group algebra to infinite groups is applicable to countable groups, and is related to considering series instead of functions: we consider infinite series (for example, absolutely convergent) of the form $\sum \alpha_g g$ with $\alpha_g \in \mathbb{C}$, and the multiplication law given by (3).

Example 5. The most famous example of a noncommutative ring is the *quaternion algebra* $\mathbb{H}$. This is an algebra of rank 4 over the field of real numbers $\mathbb{R}$ with basis 1, i, j, k, having the multiplication law

$$i^2 = j^2 = k^2 = -1, \quad ij = k, \ ji = -k, \ jk = i, \ kj = -i, \ ki = j, \ ik = -j,$$

that is, if we write i, j, k around the circle,

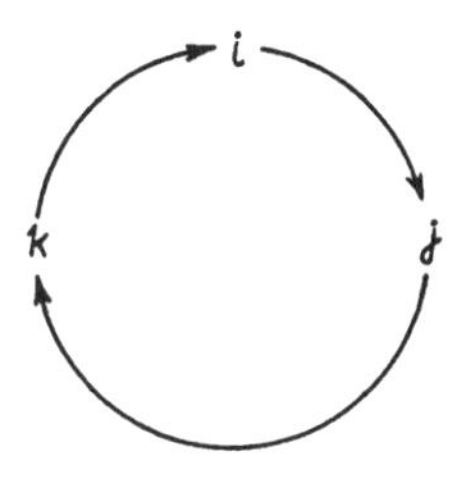

Fig. 10

then the product of two adjacent elements taken in clockwise order is equal to the third, and taken anticlockwise is equal to minus the third.

The *modulus* (or *absolute value*) of a quaternion $q = a + bi + cj + dk$ is the number $|q| = \sqrt{a^2 + b^2 + c^2 + d^2}$; the *conjugate* of q is the quaternion $\bar{q} = a - bi - cj - dk$. The relations

$$q\bar{q} = \bar{q}q = |q|^2 \quad \text{and} \quad \overline{q_1 q_2} = \bar{q}_2 \bar{q}_1 \tag{5}$$

are easy to check. It follows from these that if $q \neq 0$ then the quaternion $q^{-1} = \frac{1}{|q|^2}\bar{q}$ is an *inverse* of q, that is, $qq^{-1} = q^{-1}q = 1$. If $q = a + bi + cj + dk$ then a is called the *real part* of q and $bi + cj + dk$ the *imaginary part*; they are denoted by $\operatorname{Re} q$ and $\operatorname{Im} q$. If $a = 0$ then q is *purely imaginary*. In this case it corresponds to a 3-dimensional vector $x = (b, c, d)$. The product of two purely imaginary quaternions can expressed in terms of the two basic algebraic operations on 3-dimensional vectors, the scalar product (x, y) and the vector product $[x, y]$; in fact if purely imaginary quaternions p and q correspond to vectors x and y then $\operatorname{Re}(pq) = (x, y)$ and $\operatorname{Im}(pq)$ corresponds to the vector $[x, y]$.

From the equalities (5) it follows easily that $|q_1 q_2| = |q_1| \cdot |q_2|$ for two quaternions q_1 and q_2. This means that if a, b, c, d and a_1, b_1, c_1, d_1 are arbitrary numbers, then the product

$$(a^2 + b^2 + c^2 + d^2)(a_1^2 + b_1^2 + c_1^2 + d_1^2)$$

can be written in the form $a_2^2 + b_2^2 + c_2^2 + d_2^2$, where a_2, b_2, c_2, d_2 (which are the coefficients of 1, i, j, k in the quaternion $q_1 q_2$), can be expressed very simply in terms of a, b, c, d and a_1, b_1, c_1, d_1 (the reader can easily write out these expressions). The resulting identity was discovered by Euler long before Hamilton's introduction of quaternions; it is useful, for example, in the proof of Lagrange's famous theorem that any natural number n is equal to a sum of squares of four integers: using this identity, the problem reduces at once to the case of a prime number n.

Example 6. The quaternions contain the field $\mathbb{C}$ of complex numbers, as elements of the form $a + bi$. Any quaternion can be uniquely written in the form $z_1 + z_2 j$ with $z_1, z_2 \in \mathbb{C}$. This expression

$$\mathbb{H} = \mathbb{C} \oplus \mathbb{C}j \tag{6}$$

gives a convenient way of representing quaternions. When handling quaternions written in this form, we need only remember that $z \in \mathbb{C}$ and j do not commute. However, it is easy to check that their commutation is subject to the simple rule

$$jz = \bar{z}j. \tag{7}$$

The representation (6) has one important geometric application. Suppose we consider pairs $(q_1, q_2) \neq (0, 0)$ with $q_1, q_2 \in \mathbb{H}$ and identify pairs which are proportional 'on the left': $(q_1, q_2) \sim (qq_1, qq_2)$ for $q \neq 0$. We obtain the *quaternionic projective line* $\mathbb{P}^1(\mathbb{H})$. Just as the real and complex projective plane, it contains a finite part, the pairs (q_1, q_2) with $q_2 \neq 0$, which we can identify with $\mathbb{H}$ (by taking $q_2 = 1$), and $\mathbb{P}^1(\mathbb{H})$ is obtained from $\mathbb{H}$ by adding the point at infinity $(q_1, 0)$. This shows that, as a manifold, $\mathbb{P}^1(\mathbb{H})$ is diffeomorphic to the 4-dimensional sphere S^4. Representing $\mathbb{H}$ in the form (6) and setting $q_1 = z_1 + z_2 j$, $q_2 = z_3 + z_4 j$, we replace the pair (q_1, q_2) by the 4-tuple (z_1, z_2, z_3, z_4) in which not all z_i are zero. These 4-tuples, considered up to nonzero complex multiples, form the 3-dimensional complex projective space $\mathbb{P}^3(\mathbb{C})$. Both $\mathbb{P}^1(\mathbb{H})$ and $\mathbb{P}^3(\mathbb{C})$ are obtained from the same set of pairs (q_1, q_2), but by means of different identification processes, differing by the choice of proportionality factors: $q \in \mathbb{H}$ in the first case, and $q \in \mathbb{C}$ in the second. Since pairs identified in the second case are obviously also identified in the first, we get a map

$$\mathbb{P}^3(\mathbb{C}) \to S^4.$$

This is the *twistor space* over the sphere S^4, which is very important in geometry; its fibres form a certain 4-dimensional family of lines of $\mathbb{P}^3(\mathbb{C})$. It allows us to reduce many differential-geometric questions concerning the sphere S^4 to questions of complex analytic geometry of $\mathbb{P}^3(\mathbb{C})$.

Other applications of quaternions, to the study of the groups of orthogonal transformations of 3- and 4-dimensional space, will appear in § 15.

A ring in which any nonzero element a has an inverse a^{-1} (that is an element such that $aa^{-1} = a^{-1}a = 1$) is a *division algebra* or *skew field*. In fact it is enough to assume only the existence of a left inverse a^{-1}, such that $a^{-1}a = 1$ (or only a right inverse). If a' is a left inverse of a and a'' a left inverse of a' then, by associativity, $a''a'a$ is equal to both of a and a''. This gives $aa' = 1$, so that a' is also a right inverse. A field is a commutative division algebra, and the quaternions are the first example we have met of a noncommutative division algebra. It is easy to check that there is only one inverse in a division algebra for a given element. In a division algebra any equation $ax = b$ with $a \neq 0$ can be solved: $x = a^{-1}b$; similarly, for $ya = b$ with $a \neq 0$, $y = ba^{-1}$.

The standard notions of linear algebra over a field K carry over word-for-word to the case of vector spaces over an arbitrary division algebra D. We observe only the single distinction, which is significant, although formal: if a linear transformation φ of a n-dimensional vector space over a division algebra is given in a basis $e_1, \ldots, e_n$ by a matrix (a_{ij}) and ψ by a matrix (b_{kl}) then, as one can easily check, the transformation $\varphi\psi$ is given by the matrix c_{il}, where

$$c_{il} = \sum_k b_{kl} a_{ik}. \tag{8}$$

In other words, in the usual formula for multiplying matrixes, we must interchange the order of the terms. (This can already be observed in the example of 1-dimensional vector spaces!)

In this connection we introduce the following definition.

Rings R and R' are said to be *opposite* or *skew-isomorphic* if there exists a 1-to-1 correspondence $a \leftrightarrow a'$ between $a \in R$ and $a' \in R'$ with the properties that

$$a_1 \leftrightarrow a_1' \text{ and } a_2 \leftrightarrow a_2' \Rightarrow a_1 + a_2 \leftrightarrow a_1' + a_2' \text{ and } a_1 a_2 = \leftrightarrow a_2' a_1'.$$

A correspondence $a \leftrightarrow a'$ which establishes a skew-isomorphism of R with itself is called an *involution* of R. Examples are the correspondences $a \leftrightarrow a^*$ (where a^* is the transpose matrix) in the matrix ring $M_n(A)$ over a commutative ring A, $\sum \alpha_g g \leftrightarrow \sum \alpha_g g^{-1}$ in the group ring $A[G]$, and $q \leftrightarrow \bar{q}$ in the quaternion algebra $\mathbb{H}$.

For each ring R there exists an opposite ring R' skew-isomorphic to R. To get this, we simply take the set of elements of R with the same addition and define the product of two elements a and b to be ba instead of ab.

Now we can describe the result expressed in (8) above as:

Theorem I. *The ring of linear transformations of an n-dimensional vector space over a division algebra D is isomorphic to the matrix ring $M_n(D')$ over the opposite division algebra D'.*

With the exception of this alteration, the well-known results of linear algebra are preserved for vector spaces over division algebras. Going further, we can also define the projective space $\mathbb{P}^n(D)$ over D, and this will again have most of the properties we are familiar with.

Example 7. We consider the space $T^r(L)$ of contravariant tensors of degree r over an n-dimensional vector space L over a field K (see §5 for the definition of the module $T^n(M)$). The tensor product operation defines the product of tensors $\varphi \in T^r(L)$ and $\psi \in T^s(L)$ as a tensor $\varphi \otimes \psi \in T^{r+s}(L)$. To construct a ring by means of this operation, consider the direct sum $\bigoplus T^r(L)$ of all the spaces $T^r(L)$, consisting of sequences $(\varphi_0, \varphi_1, \ldots)$, with only a finite number of nonzero terms, and $\varphi_r \in T^r(L)$. We define the sum of sequences component-by-component, and the product of $(\varphi_0, \varphi_1, \ldots)$ and $(\psi_0, \psi_1, \ldots)$ as $(\xi_0, \xi_1, \ldots)$, where $\xi_p = \sum_{0 \leqslant r \leqslant p} \varphi_r \psi_{p-r}$. It follows from properties of multiplication of tensors that we get a ring in this way. It contains subspaces $T^r(L)$ for $r = 0, 1, \ldots$, and each element can be represented as a finite sum $\varphi_0 + \varphi_1 + \cdots + \varphi_k$ with $\varphi_r \in T^r(L)$. The elements $\varphi_0 \in T^0(L) = K$ are identified with elements of K, so that the ring constructed is a K-algebra. It is called the *tensor algebra* of the vector space L, and denoted by $T(L)$. The decomposition of $T(L)$ as the sum of the $T^r(L)$ makes $T(L)$ into a graded algebra.

Let us choose a basis $\xi_1, \xi_2, \ldots, \xi_n$ of $T^1(L) = L$. The well-known properties of tensor multiplication show that the products $\xi_{i_1} \cdot \ldots \cdot \xi_{i_m}$, where $(i_1, \ldots, i_m)$ is any collection of m indexes, each of which can take the values $1, \ldots, n$, form a basis of $T^m(L)$. Hence all such products (for all m) form an infinite basis of the tensor algebra over K. Thus any element of the tensor algebra can be written as a linear combination of products of the elements $\xi_1, \xi_2, \ldots, \xi_n$ and different products are linearly independent (the order of the factors is distinguished). In view of this, $T(L)$ is also called the *noncommuting polynomial algebra* in n variables $\xi_1, \ldots, \xi_n$. As such it is denoted by $K\langle \xi_1, \ldots, \xi_n \rangle$.

The characterisation of the algebra $T(L)$ indicated above has important applications. We say that elements $\{x_\alpha\}$ (finite or infinite in number) are *generators* of an algebra R over a commutative ring A if any element of R can be written as a linear combination with coefficients in A of certain products of them. Suppose that an algebra R has a finite number of generators $x_1, \ldots, x_n$ over a field K. Consider the map which takes any element $\alpha = \sum a_{i_1 \ldots i_m} \xi_{i_1} \cdot \ldots \cdot \xi_{i_m}$ of the algebra $K\langle \xi_1, \ldots, \xi_n \rangle$ into the element $\alpha' = \sum a_{i_1 \ldots i_m} x_{i_1} \cdot \ldots \cdot x_{i_m}$ of R. It is easy to see that we thus get a homomorphism $K\langle \xi_1, \ldots, \xi_n \rangle \to R$ whose image is the whole of R. Thus any algebra having a finite number of generators is a homomorphic image of a noncommuting polynomial algebra. In this sense, the noncommuting polynomial algebras play the same role in the theory of noncommutative algebras as the commutative polynomial algebras in commutative algebra, or free modules in the theory of modules.

We must again interrupt our survey of examples of noncommutative rings to get to know the simplest method of constructing them. As in the case of commutative rings, it is natural to pay attention to properties enjoyed by kernels of homomorphisms. Obviously, if $\varphi: R \to R'$ is a homomorphism, then $\operatorname{Ker} \varphi$ contains the sum $a + b$ of two elements $a, b \in \operatorname{Ker} \varphi$, and both of the products ax and xa of an element $a \in \operatorname{Ker} \varphi$ with any element $x \in R$. We have run up against the fact that the notion of ideal of a commutative ring can be generalised to the noncommutative case in the three ways (a), (b), (c) below. Consider a subset $I \subset R$ containing the sum $a + b$ of any two elements $a, b \in I$

(a) If the product xa is contained in I for any $a \in I$ and $x \in R$, then we say that I is a *left ideal*;

(b) if (under the same conditions) ax is contained in I then we say that I is a *right ideal*;

(c) if both conditions (a) and (b) hold, we say that I is a *two-sided ideal*. Thus the kernel of a homomorphism is a two-sided ideal.

We give examples of these notions. In the ring of linear transformations of a finite-dimensional vector space L over a division algebra D, a subspace $V \subset L$ defines a left ideal ${}_V I$, consisting of all transformations φ such that $\varphi(V) = 0$, and a right ideal I_V, consisting of all φ such that $\varphi(L) \subset V$. In the ring of bounded linear operators in a Banach space, all compact (or completely continuous) operators form a two-sided ideal.

All elements of the form xa with $x \in R$ form a left ideal, and those of the form ay for $y \in R$ a right ideal. For two-sided ideals the corresponding construction is a little more complicated. We treat this at once in a more general form. Let $\{a_\alpha\}$ be a system of elements of a ring R; all sums of the form $x_1 a_{\alpha_1} y_1 + \cdots + x_r a_{\alpha_r} y_r$ with $x_i, y_i \in R$ form a two-sided ideal. It is called the *ideal generated by the system* $\{a_\alpha\}$.

In complete analogy with the commutative case we can define the *cosets* of a two-sided ideal and the ring of these cosets. We preserve the previous notation R/I and the name *quotient ring* for this. For example, if R is the ring of bounded linear operators in a Banach space and I is the two-sided ideal of compact operators, then many properties of an operator φ depend only on its image in R/I. Thus to say that φ satisfies the Fredholm alternative is equivalent to saying that its image in R/I has an inverse.

The homomorphisms theorem is stated and proved in complete analogy with the commutative case (§ 4, Theorem II).

Let $\{\varphi_\alpha\}$ be a system of elements of the noncommuting polynomial algebra $K\langle \xi_1, \ldots, \xi_n \rangle$ and I the two-sided ideal it generates. In the algebra $R = K\langle \xi_1, \ldots, \xi_n \rangle / I$, we write $a_1, \ldots, a_n$ for the images of the elements $\xi_1, \ldots, \xi_n$. These are obviously generators of R; we say that R is defined by generators $a_1, \ldots, a_n$ and relations $\varphi_\alpha = 0$. By the homomorphisms theorem, any algebra with a finite number of generators can be defined by some system of generators and relations. But although the system of generators is by definition finite, it sometimes happens that the system of relations cannot be chosen to be finite.

The commutative polynomial ring $K[x_1, \ldots, x_n]$ has the defining relations $x_i x_j = x_j x_i$. Let R be the ring of differential operators with polynomial coefficients in n variables $x_1, \ldots, x_n$. Generators in this algebra are, for example, the operators q_i of multiplication by x_i (with $q_i(f) = x_i f$) and $p_j = \dfrac{\partial}{\partial x_j}$. It is easy to see that it has the defining relations

$$p_i p_j = p_j p_i, \quad q_i q_j = q_j q_i, \qquad (9)$$

$$p_i q_j = q_j p_i \quad \text{if } i \neq j, \quad \text{and} \quad p_i q_i - q_i p_i = 1.$$

We apply this construction to some other important classes of algebras. Suppose given an n-dimensional vector space L and a symmetric bilinear form, which we denote by (x, y). We consider the algebra having generators in 1-to-1 correspondence with the elements of some basic of L (and denoted by the same letters), and relations of the form

$$xy + yx = (x, y) \quad \text{for } x, y \in L. \tag{10}$$

Thus our algebra is the quotient algebra of the tensor algebra $T(L)$ by the ideal I generated by the elements $(x, y) - xy - yx$.

We consider two extreme cases.

Example 8. Suppose that the bilinear form (x, y) is identically zero. Then it follows from the relation (10) that $x^2 = 0$ (if the characteristic of K is $\neq 2$; if $\operatorname{char} K = 2$ then we need to take $x^2 = 0$ as part of the definition). Any element of the algebra constructed in this way is a linear combination of products $e_{i_1} \cdot \ldots \cdot e_{i_r}$ of basis elements of L. It is easy to see that all such products generate the space $\bigwedge^r(L)$ (see § 5 for the definition of the module $\bigwedge^r(M)$). The whole of our algebra is represented as a direct sum $\bigwedge^0(L) \oplus \bigwedge^1(L) \oplus \cdots \oplus \bigwedge^n(L)$. This algebra is graded and of finite rank 2^n; it is called the *exterior algebra* of L and denoted $\bigwedge(L)$; multiplication in $\bigwedge(L)$ is denoted by $x \wedge y$.

It is easy to see that if $x \in \bigwedge^r(L)$ and $y \in \bigwedge^s(L)$ then

$$x \wedge y = y \wedge x \quad \text{if either } r \text{ or } s \text{ is even} \tag{11}$$

and

$$x \wedge y = -y \wedge x \quad \text{if both } r \text{ and } s \text{ are odd.}$$

This can be expressed in another way. Write $\bigwedge(L) = R$, and set

$$\bigwedge^0(L) \oplus \bigwedge^2(L) \oplus \bigwedge^4(L) \oplus \cdots = R^0, \quad \bigwedge^1(L) \oplus \bigwedge^3(L) \oplus \cdots = R^1.$$

Then $R = R^0 \oplus R^1$, and

$$R^0 \cdot R^0 \subset R^0, \quad R^0 \cdot R^1 \subset R^1, \quad R^1 \cdot R^0 \subset R^1, \quad R^1 \cdot R^1 \subset R^0. \tag{12}$$

A decomposition with properties (12) is called a *$\mathbb{Z}/2$-grading* of the algebra R. For $R = \bigwedge(L)$ we can state (11) by saying that $x \wedge y = y \wedge x$ if either x or $y \in R^0$, and $x \wedge y = -y \wedge x$ if both x and $y \in R^1$. An algebra with a $\mathbb{Z}/2$-grading having these properties is called a *superalgebra*. The exterior algebra $\bigwedge(L)$ is the most important example of these. Interest is superalgebras has been stimulated by quantum field theory. On the other hand, it turns out that purely mathematically, they form a very natural generalisation of commutative rings, and can serve as the basis for the construction of geometric objects, analogues of projective space (superprojective spaces) or of differential and analytic manifolds (supermanifolds). This theory has applications to supergravitation in physics, and it is studied by supermathematicians.

Example 9. The definition of exterior algebra used a basis of the vector space L (the vectors x, y in (10) belong to it). Of course, the construction does not depend on this choice. We can give a completely intrinsic (although less economical) definition, taking x and y in (10) to be all the vectors of L. It is easy to see that we arrive at the same algebra. In this form the definition is applicable to any module M over a commutative ring A. We obtain the notion of the *exterior algebra* of a module:

$$\bigwedge M = \bigoplus_r \bigwedge^r M.$$

If M has a system of n generators then $\bigwedge^r M = 0$ for $r > n$. In particular, the exterior algebra of the module of differential 1-forms Ω^1 on a n-dimensional differentiable manifold is called the *algebra of differential forms*, $\Omega = \bigoplus_{r \leqslant n} \Omega^r$. We will see later important applications of the exterior product of forms.

Example 10. Now consider the other extreme case, when the bilinear form (x, y) in (10) is nondegenerate and corresponds to a quadratic form $F(x)$, that is, $F(x) = \frac{1}{2}(x, x)$ (we suppose that the characteristic of K is $\neq 2$). We can argue in this case just as in the previous one, exchanging factors in the product $e_{i_1} \cdot \ldots \cdot e_{i_r}$ using (10). The only difference is that for $j < i$ the product $e_i e_j$ gives rise to two terms, one containing $-e_j e_i$ and one containing (e_i, e_j), giving a product of $r - 2$ factors. As a result we prove in exactly the same way that the products $e_{i_1} \cdot \ldots \cdot e_{i_r}$ with $i_1 < \cdots < i_r$ form a basis of our algebra, so that it is again of rank 2^n. This algebra is called the *Clifford algebra* of the vector space L and the quadratic form F, and is denoted by $C(L)$. The significance of this construction is that in $C(L)$ the quadratic form F becomes the square of a linear form:

$$F(x_1 e_1 + \cdots + x_n e_n) = (x_1 e_1 + \cdots + x_n e_n)^2. \tag{13}$$

Thus the quadratic form becomes 'a perfect square', but with coefficients in some noncommutative algebra. Suppose that $F(x_1, \ldots, x_n) = x_1^2 + x_n^2$; then by (13), $x_1^2 + \cdots + x_n^2 = (x_1 e_1 + \cdots + x_n e_n)^2$. Using the isomorphism between the ring of differential operators with constant coefficients $\mathbb{R}\left[\frac{\partial}{\partial y_1}, \ldots, \frac{\partial}{\partial y_n}\right]$ and the polynomial ring $R[x_1, \ldots, x_n]$, we can rewrite this relation in the form

$$\frac{\partial^2}{\partial y_1^2} + \cdots + \frac{\partial^2}{\partial y_n^2} = \left(\frac{\partial}{\partial y_1} \cdot e_1 + \cdots + \frac{\partial}{\partial y_n} \cdot e_n\right)^2. \tag{14}$$

It was precisely the idea of taking the square root of a second order operator which motivated Dirac when he introduced a notion analogous to the Clifford algebra in his derivation of the so-called Dirac equation in relativistic quantum mechanics.

The products $e_{i_1} \cdot \ldots \cdot e_{i_r}$ with an even number r of factors generate a subspace C^0 of the Clifford algebra C, those with odd r a subspace C^1; clearly, $\dim C^0 = \dim C^1 = 2^{n-1}$. It is easy to see that $C = C^0 \oplus C^1$ and that this defines a $\mathbb{Z}/2$-grading. In particular C^0 is a subalgebra of C, called the *even Clifford algebra.*

Consider the map which sends a basis element $e_{i_1} \cdot \ldots \cdot e_{i_r}$ of $C(L)$ into the product $e_{i_r} \cdot \ldots \cdot e_{i_1}$ in the opposite order. It is easy to see that this gives an involution of $C(L)$, which we denote by $a \mapsto a^*$.

Example 11. Suppose that $F(x_1, x_2) = x_1^2 + x_2^2$ and that $K = \mathbb{R}$. Then $C(L)$ has rank 4 and basis $1, e_1, e_2, e_1e_2$ with $e_1^2 = e_2^2 = 1$ and $e_2e_1 = -e_1e_2$. It is easy to see that $C(L) \cong M_2(\mathbb{R})$. For this, we set

$$E_{11} = \frac{1 + e_1}{2}, \quad E_{12} = \frac{-e_2 - e_1e_2}{2}, \quad E_{21} = \frac{-e_2 + e_1e_2}{2}, \quad E_{22} = \frac{1 - e_1}{2},$$

and we then need to see that the elements E_{ij} multiply together according to rule (2). The isomorphism of $C(L)$ with $M_2(\mathbb{R})$ sends e_1 to the matrix $\begin{bmatrix} 1 & 0 \\ 0 & -1 \end{bmatrix}$ and e_2 to $\begin{bmatrix} 0 & -1 \\ -1 & 0 \end{bmatrix}$. Then by (14) the Laplace operator $\frac{\partial^2}{\partial x^2} + \frac{\partial^2}{\partial y^2}$ can be written as $\left(\begin{bmatrix} 1 & 0 \\ 0 & -1 \end{bmatrix}\frac{\partial}{\partial x} + \begin{bmatrix} 0 & -1 \\ -1 & 0 \end{bmatrix}\frac{\partial}{\partial y}\right)^2$. If the operator $\mathscr{D} = \begin{bmatrix} 1 & 0 \\ 0 & -1 \end{bmatrix}\frac{\partial}{\partial x} + \begin{bmatrix} 0 & -1 \\ -1 & 0 \end{bmatrix}\frac{\partial}{\partial y}$ acts on the column $\begin{bmatrix} u \\ v \end{bmatrix}$ then the equation $\mathscr{D}\begin{bmatrix} u \\ v \end{bmatrix} = 0$ gives:

$$\frac{\partial u}{\partial x} = \frac{\partial v}{\partial y}, \quad \frac{\partial u}{\partial y} = -\frac{\partial v}{\partial x},$$

that is, the Cauchy-Riemann equations.

Now suppose that $F(x_1, \ldots, x_{2n}) = x_1^2 + \cdots + x_{2n}^2$. We divide the indexes $1, \ldots, 2n$ into n pairs: $(1, 2), (3, 4), \ldots, (2n - 1, 2n)$, and write α, β, etc. to denote any collection $(i_1, \ldots, i_n)$ of indexes such that i_p belongs to the pth pair. If $\alpha = (i_1, \ldots, i_n)$ and $\beta = (j_1, \ldots, j_n)$ then we set

$$E_{\alpha\beta} = E_{i_1j_1}E_{i_2j_2} \cdots E_{i_nj_n},$$

where the E_{ij} are expressed in terms of e_i, e_j as in the case $n = 1$. It is easy to see that the $E_{\alpha\beta}$ again multiply according to rule (2), that is $C(L) \cong M_{2^n}(\mathbb{R})$.

Example 12. If $F(x_1, x_2, x_3) = x_1^2 + x_2^2 + x_3^2$ and $K = \mathbb{R}$, then the even Clifford algebra $C^0(L)$ is isomorphic to the quaternion algebra: e_1e_2, e_2e_3, e_1e_3 multiply according to the rule of Example 5.

In the commutative case fields can be characterised as rings without ideals (other than 0). In the noncommutative case, as usual, the relation is more complicated. One proves just as in the commutative case that the absence of left ideals (other than 0) is equivalent to the fact that every element other than 0 has a left inverse (satisfying $a^{-1}a = 1$), and right ideals relate to right inverses in the same way. Thus division algebras are the rings without left ideals (or without right ideals), other than 0.

What does the absence of two-sided ideals correspond to? A ring not having any two-sided ideal other than 0 is said to be *simple*. We will see later the exceptionally important role played by simple rings in the theory of rings, so that they can, together with division algebras, be considered as a natural extension of fields to the noncommutative domain.

Example 13. We determine the structure of left ideals of the ring R of linear transformations of an n-dimensional vector space L over a division algebra D. Let us show that the construction given earlier (the ideals ${}_VI$ and I_V) describes all of these, restricting ourselves to left ideals. Let $I \subset R$ be such an ideal, and $V \subset L$ the subspace consisting of all elements $x \in L$ such that $\varphi(x) = 0$ for all $\varphi \in I$. If $\varphi_1, \ldots, \varphi_k$ is a basis of I as a vector space over D then $\bigcap \operatorname{Ker} \varphi_i = V$. It is easy to see that if $\varphi \in R$ satisfies $\operatorname{Ker} \varphi = V$ then any linear transformation φ' for which $\operatorname{Ker} \varphi' \supset V$ can be expressed in the form $\psi\varphi$ with $\psi \in R$. It follows easily from this that if $\varphi_1, \varphi_2 \in I$ then I contains a transformation $\bar{\varphi}$ for which $\operatorname{Ker} \bar{\varphi} = \operatorname{Ker} \varphi_1 \cap \operatorname{Ker} \varphi_2$. Applying this remark to the transformations $\varphi_1, \ldots, \varphi_k$, we find an element $\bar{\varphi} \in I$ such that $\operatorname{Ker} \bar{\varphi} = V$, and by what we have said above, this implies that all transformations φ with $\varphi(V) = 0$ are contained in I, that is, $I = \{\varphi \mid \varphi(V) = 0\} = {}_VI$. Right ideals are considered in a similar way.

Suppose finally that I is a two-sided ideal. As a left ideal, I corresponds to some subspace V, such that $I = \{\varphi \mid \varphi(V) = 0\}$. Take $x \in V$ with $x \neq 0$. For $\varphi \in I$ we have $\varphi(x) = 0$. Since I is a right ideal, for any $\psi \in R$ we have $\varphi\psi \in I$ and hence $\varphi(\psi(x)) = 0$. But we could take $\psi(x)$ to be any vector of L, and hence $I = 0$. Thus a ring R isomorphic to $M_n(D)$ is simple.

Another example of a simple ring is the ring R of differential operators with polynomial coefficients. To keep the basic idea clear, set $n = 1$. Interpreting p as the operator $\dfrac{d}{dx}$, it is easy to check the relation $pf(q) - f(q)p = f'(q)$. If $\mathscr{D} = \sum f_i(q)p^i$ is contained in a two-sided ideal I and $\mathscr{D} \neq 0$ then on passing to the expression $p\mathscr{D} - \mathscr{D}p$ a number of times, we find an element $\Delta \in I$ which is a nonzero polynomial in p with constant coefficients, $\Delta = g(p)$. Since the relations (9) do not distinguish between p and q, we have the relation $g(p)q - qg(p) = g'(p)$. Composing expressions of this form several times, we find a nonzero constant in I. Hence $I = R$. (The validity of these arguments requires that the coefficient field has characteristic zero.)

Example 14. Algebras which are close to being simple are the Clifford algebras $C(L)$ and $C^0(L)$ for any vector space L with a quadratic form F (we assume that the ground field has characteristic $\neq 2$). The following results can be verified without difficulty. The algebra $C(L)$ is simple if $n \equiv 0 \bmod 2$ (where $n = \dim L$), and then $Z(C(L)) = K$. The algebra $C^0(L)$ is simple if $n \equiv 1 \bmod 2$, and then $Z(C^0(L)) = K$. The remaining cases are related to properties of the element $z = e_1 \ldots e_n \in C(L)$, where $e_1, \ldots, e_n$ is an orthogonal basis of L. It is easy to see that z is contained in the centre of $C(L)$ if $n \equiv 1 \bmod 2$ and in the centre of $C^0(L)$ if $n \equiv 0 \bmod 2$, and in both cases the centre of these algebras is of the form $K + Kz$. Then $z^2 = a \in K$ and $a = (-1)^{n/2}\mathscr{D}$ or $a = 2 \cdot (-1)^{(n-1)/2}\mathscr{D}$ depending on the parity of n; $\mathscr{D}$ denotes the discriminant of the form F with respect to the basis $e_1, \ldots, e_n$. If a is not a square in K then $K + Kz = K(\sqrt{a})$ and the corresponding algebra is simple with centre $K(\sqrt{a})$. If a is a square then the algebra $K + Kz$ is isomorphic to $K \oplus K$ and the corresponding Clifford algebra is isomorphic to the direct sum of two simple algebras of the same rank with centre K.

§9. Modules over Noncommutative Rings

A *module* over an arbitrary ring R is defined in the same way as in the case of a commutative ring: it is a set M such that for any two elements $x, y \in M$, the sum $x + y$ is defined, and for $x \in M$ and $a \in R$ the product $ax \in M$ is defined, satisfying the following conditions (for all $x, y, z \in M$, $a, b \in R$):

$$\left.\begin{array}{c} x + y = y + x; \\ (x + y) + z = x + (y + z); \\ \text{there exists } 0 \in M \text{ such that } 0 + x = x + 0 = x; \\ \text{there exists } -x \text{ such that } x + (-x) = 0; \\ 1 \cdot x = x; \\ (ab)x = a(bx); \\ (a + b)x = ax + bx; \\ a(x + y) = ax + ay. \end{array}\right\} \quad (1)$$

In exactly the same way, the notions of isomorphism, homomorphism, kernel, image, quotient module and direct sum do not depend on the commutativity assumption. A ring R is a module over itself if we define the product of a (as an element of the ring) and x (as an element of the module) to be equal to ax. The submodules of this module are the left ideals of R. If I is a left ideal then the residue classes mod I form a module R/I over R. The multiplication of x (as an element of the module) on the right by a (as an element of the ring) does not define an R-module structure. In fact, if we denote temporarily this product by $\{ax\} = xa$ (on the left as in the module, on the right as in the ring) then $\{(ab)x\} = \{b\{ax\}\}$, which contradicts the axioms (1). However, we can say that in this way R becomes a module over the opposite ring R'. In this module over R', the submodules correspond to right ideals of R.

The most essential examples of modules over noncommutative rings are first of all the ring itself and its ideals as modules over the ring. We will see shortly just how useful it is to consider these modules for the study of the rings themselves; and secondly, the study of the many modules over group rings is the subject of group representation theory, which will be discussed in detail later.

If R is an algebra over a field K then every R-module M is automatically a vector space over K (possibly infinite-dimensional). The module axioms show that for any $a \in R$ the map $\varphi_a(x) = ax$ for $x \in M$ is a linear transformation of this vector space. Moreover, sending a to the linear transformation φ_a is a homomorphism of the algebra R into the algebra $\mathrm{End}_K M$ of all linear transformations of M as a K-vector space. Conversely, a homomorphism $R \to \mathrm{End}_K L$ (denoted by $a \mapsto \varphi_a$) into the algebra of linear transformations of a vector space L obviously defines an R-module structure on L:

$$ax = \varphi_a(x) \quad \text{for} \quad a \in R \text{ and } x \in L. \tag{2}$$

In this situation, we sometimes use somewhat different terminology from that usual in the general case. In view of the especial importance of this special case we repeat, in the new terminology, the basic definitions given above in the general case.

Restatement of the Definition of a Module. A *representation* of an algebra R over a field K on a K-vector space L is a homomorphism of R to the algebra $\mathrm{End}_K L$ of linear transformations of L.

In other words, a representation of R sends each element $a \in R$ to a linear transformation φ_a so that the following conditions are satisfied:

$$\varphi_1 = E \quad \text{(the identity transformation);} \tag{3}$$

$$\varphi_{\alpha a} = \alpha\varphi_a \quad \text{for } \alpha \in K \text{ and } a \in R; \tag{4}$$

$$\varphi_{a+b} = \varphi_a + \varphi_b \quad \text{for } a, b \in R; \tag{5}$$

$$\varphi_{ab} = \varphi_a\varphi_b \quad \text{for } a, b \in R. \tag{6}$$

Restatement of the Definition of a Submodule. A *subrepresentation* is a subspace $V \subset L$ invariant under all transformations φ_a for $a \in R$, with the representation of R induced by these transformations.

Restatement of the Definition of a Quotient Module. The *quotient representation* by a subrepresentation $V \subset L$ is the space L/V with the representation induced by the transformation φ_a.

If R is an algebra of finite rank over a field K with basis $1 = e_1, \ldots, e_n$ and multiplication table $e_ie_j = \sum c_{ijk}e_k$ then conditions (3)–(6) in the definition of a representation reduce to specifying transformations $\varphi_{e_1}, \ldots, \varphi_{e_n}$ satisfying the relations

$$\varphi_1 = E, \tag{7}$$

$$\varphi_{e_i}\varphi_{e_j} = \sum c_{ijk}\varphi_{e_k}. \tag{8}$$

If $R = K[G]$ is the group algebra of a finite group G then conditions (7), (8) take the form

$$\varphi_1 = E, \tag{9}$$

$$\varphi_{g_1}\varphi_{g_2} = \varphi_{g_1}\varphi_{g_2}. \tag{10}$$

Conditions (9)–(10) guarantee that all transformations are invertible.

If G is an infinite group and the group algebra is defined as the set of linear combinations of elements of the group (see §8, Example 4), then a representation is given by the same conditions (10). If the group algebra is defined as the algebra of functions on the group with the convolution operation, the elements of the group G are only contained in it as delta-functions, and hence the operators φ_g may not exist. On the other hand, if operators φ_g satisfying (9) and (10) exist, then the operator φ_f corresponding to a function f can be defined as the integral of operator functions $f(g)\varphi_g$ taken over the whole group. Hence for group representations, conditions (9) and (10) give more than conditions (3)–(6) for the group algebra, and we take conditions (9) and (10) as the definition of a *group representation.*

If the module M in which the representation of an algebra R is realised is finite-dimensional over a field K then we say that the representation is *finite-dimensional.* In this case the linear transformations φ_a are given by matrixes (once a basis of M has been chosen). Let us reformulate once more the basic notions of representation theory in this language.

A *finite-dimensional representation of an algebra* R is a homomorphism $R \to M_n(K)$, which assigns to each element $a \in R$ a matrix $C_a \in M_n(K)$ satisfying the conditions:

$$C_1 = E,$$

$$C_{\alpha a} = \alpha C_a,$$

$$C_{a+b} = C_a + C_b,$$

$$C_{ab} = C_a C_b.$$

In the case of a representation of a group G these conditions are replaced by

$$C_e = E,$$

$$C_{g_1 g_2} = C_{g_1} C_{g_2}.$$

Restatement of the Notion of Isomorphism of Modules. Two representations $a \mapsto C_a$ and $a \mapsto C'_a$ are *equivalent* if there exists a nondegenerate matrix P such that $C'_a = PC_aP^{-1}$ for every $a \in R$.

Restatement of the Notion of Submodule. A representation $a \mapsto C_a$ has a *subrepresentation* $a \mapsto D_a$ if there exists a nondegenerate matrix P such that the matrixes $C'_a = PC_aP^{-1}$ are of the form

$$C'_a = \begin{bmatrix} D_a & S_a \\ 0 & F_a \end{bmatrix}. \tag{11}$$

Restatement of the Notion of Quotient Module. The matrixes F_a in (11) form the *quotient representation* by the given subrepresentation.

Restatement of the Notion of Direct Sum of Modules. If $S_a = 0$ in (11) we say that the representation C_a is the *direct sum* of the representations D_a and F_a.

Considering in particular an algebra R as a module over itself, we get an important representation of R. It is called the *regular representation*. If R has a finite basis $e_1, \ldots, e_n$ with structure constants c_{ijk} then the element $\alpha_1 e_1 + \cdots + \alpha_n e_n$ corresponds in the regular representation to the matrix (p_{jk}), where $p_{jk} = \sum_i c_{ikj}\alpha_i$.

We now return to modules over an arbitrary ring R (not necessarily an algebra) and consider an important condition having the character of finite dimensionality. This relates to the definition of the dimension of a vector space as the greatest length of a chain of subspaces.

We define the *length* of an R-module M to be the upper bound of the length r of chains of submodules:

$$M = M_0 \supsetneqq M_1 \supsetneqq \cdots \supsetneqq M_r = 0.$$

Of course, the length of a module may be either finite or infinite. Consider a module of finite length r and a chain $M = M_0 \supsetneqq M_1 \supsetneqq \cdots \supsetneqq M_r = 0$ of maximal length. If the module M_i/M_{i+1} contained a submodule N distinct from M_i/M_{i+1} itself and 0, then its inverse image M' under the canonical homomorphism $M_i \to M_i/M_{i+1}$ would be a submodule $M_i \supsetneqq M' \supsetneqq M_{i+1}$; substituting this into our chain, we would get a longer chain. Hence there can be no such submodule in M_i/M_{i+1}. We thus arrive at a very important notion.

A module M is *simple* if it does not have any submodules other than 0 and M. A representation of an algebra (or of a group) is *irreducible* if the corresponding module is simple.

Simpleness is a very strong condition.

Example 1. In the case of vector spaces over a field, only 1-dimensional spaces are simple.

Example 2. Let L be a finite-dimensional complex vector space with a given linear transformation φ, considered as a module over the ring $\mathbb{C}[t]$ (§5, Example 3). Since φ always has an eigenvector, and therefore a 1-dimensional invariant subspace, L is again simple only if it is 1-dimensional.

Example 3. Consider a ring R as a module over itself. To say that R is simple means that R does not have left ideals, that is, R is a division algebra.

Let M and N be two simple modules and $\varphi\colon M \to N$ a homomorphism. By assumption, $\operatorname{Ker}\varphi = 0$ or M, and $\operatorname{Im}\varphi = 0$ or N. If $\operatorname{Ker}\varphi = M$ or $\operatorname{Im}\varphi = 0$ then φ is the zero homomorphism. In the remaining case $\operatorname{Ker}\varphi = 0$ and $\operatorname{Im}\varphi = N$, so that φ is an isomorphism. Thus we have the following result.

I. Schur's Lemma. *A homomorphism of one simple module into another is either zero or an isomorphism.*

We return to the notion of length. We have seen that if M is a module of length r then in a chain $M = M_0 \supsetneqq M_1 \supsetneqq \cdots \supsetneqq M_r = 0$ each of the modules M_i/M_{i+1} must be simple.

A chain $M = M_0 \supsetneqq M_1 \supsetneqq \cdots \supsetneqq M_r = 0$ in which M_i/M_{i+1} is simple is called a *composition series.* The following result holds.

II. The Jordan-Hölder Theorem. *All composite series of the same module have the same length (and in particular they are either all finite or all infinite). If they are finite then the successive quotients M_i/M_{i+1} appearing in them are isomorphic (but possibly occuring in a different order).*

Thus in a module of finite length the longest chains of submodules are exactly the composition series.

We extend the notion of length to rings. The *length* of a ring R is its length as an R-module. Thus a ring has length r if it has a chain $R \supsetneqq I_1 \supsetneqq \cdots \supsetneqq I_r = 0$ of left ideals and no longer chains.

We have already seen that a division algebra is a ring of length 1. Left ideals of a matrix ring $M_n(D)$ over a division algebra D correspond to linear subspaces of the n-dimensional space L over the opposite division algebra D' (§8, Example 13). Hence the length of $M_n(D)$ is n.

Of course, if a ring R is an algebra over a field K then the length of R does not exceed its rank over K.

A module of finite length is finitely generated (or is of finite type), by analogy with the property of Noetherian modules in the case of commutative rings.

If a ring R has finite length then the length of any finitely generated R-module is also finite. This follows from the fact that if elements $x_1, \ldots, x_n$ generate a module M then M is a homomorphic image of the module R^n under the homomorphism $(a_1, \ldots, a_n) \mapsto a_1x_1 + \cdots + a_nx_n$, and is of finite length as a quotient of a module of finite length.

In conclusion, let us consider in more detail a notion which we have often met with in the theory of modules over a commutative ring.

A homomorphism of an R-module M to itself is called an *endomorphism.* All the endomorphisms of a module obviously form a ring; it is denoted by $\mathrm{End}_R M$. An important difference from the commutative case is that it is not possible to define the multiplication of an endomorphism $\varphi \in \mathrm{End}_R M$ by an element $a \in R$. The map $x \mapsto a\varphi(x)$ is not in general an endomorphism over R; that is, multiplication of endomorphisms by elements of R is not defined in $\mathrm{End}_R M$.

Example 4. Consider the ring R as an R-module. What is the ring of endomorphisms $\mathrm{End}_R R$ of this module? By definition, an endomorphism φ is a map of R to itself which satisfies the conditions

$$\varphi(x + y) = \varphi(x) + \varphi(y), \quad \varphi(ax) = a\varphi(x), \quad \text{for } a, x, y, \in R. \tag{12}$$

Setting $\varphi(1) = f$ we get from (12) for $x = 1$ that $\varphi(a) = af$ for any $a \in R$. Thus any endomorphism is given as right multiplication by an element $f \in R$. It follows from this that the ring $\mathrm{End}_R R$ is the opposite ring of R.

Example 5. Suppose that a module M is isomorphic to a direct sum of n isomorphic modules P, that is, $M \cong P^n$. Then M consists of n-tuples $(x_1, \ldots, x_n)$

with $x_i \in P$. The situation is exactly the same as in describing linear transformations of a vector space, and the answer is entirely analogous. For $\varphi \in \mathrm{End}_R M$ and $x \in P$, set

$$\varphi((0,\ldots,x,\ldots,0)) = (\psi_{i1}(x),\ldots,\psi_{in}(x))$$

(where x is in the ith place on the left-hand side). Here ψ_{ij} are homomorphisms of P to P, that is $\psi_{ij} \in \mathrm{End}_R P$. Replacing φ by the matrix (ψ_{ij}) with entries in $\mathrm{End}_R P$, we get an isomorphism

$$\mathrm{End}_R P^n \cong M_n(\mathrm{End}_R P).$$

In the particular case that P is the division algebra D as module over itself, we arrive (using the result of Example 4) at the expression found earlier (§ 8, Theorem 1) for the ring of linear transformations over a division algebra.

Example 6. The ring $\mathrm{End}_R M$ of endomorphism of a simple module M is a division algebra. This is an immediate consequence of Theorem I.

§ 10. Semisimple Modules and Rings

The theory of modules over noncommutative rings, and the study of the structure of the rings themselves, can be taken well beyond the framework of general definitions and almost obvious properties treated in the preceding section, provided we restrict ourselves to objects satisfying the strong, but frequently occuring property of semisimpleness.

A module M is *semisimple* if every submodule of M is a direct summand. This means that for any submodule $N \subset M$, there exists another submodule $N' \subset M$ such that $M = N \oplus N'$.

Obviously, a submodule, holomorphic image and a direct sum of semisimple modules are semisimple. A simple module is semisimple. Any module of finite length contains a simple submodule; hence a semisimple module of finite length is a direct sum of simple modules. It follows from the Jordan-Hölder theorem (or can be deduced even more simply from § 9, Theorem I) that the decomposition of a semisimple module as a sum of simple modules is unique (that is, the simple summands are uniquely determined up to isomorphism). The number of these summands is the length of the module.

If $P \subset M$ is simple and $N \subset M$ is any submodule then either $P \subset N$ or $P \cap N = 0$. From this we deduce the following:

Theorem I. *If a module is generated by a finite number of simple submodules then it is semisimple and of finite length.*

In fact, if $P_1, \ldots, P_n$ are simple submodules generating M and $N \subset M$ but $N \neq M$, then there exists $P_i \not\subset N$. Then $P_i \cap N = 0$ and the submodule generated by P_i and N is isomorphic to the direct sum $P_i \oplus N$. Applying the same argument to this, we arrive after a number of steps at a decomposition $M = N \oplus N'$.

Example 1. Let M be a finite-dimensional vector space L with a linear transformation φ, considered as a module over $K[t]$ (see § 5, Example 3). If M is simple, then L is a 1-dimensional vector space (§ 9, Example 1). Hence M is semisimple if and only if L can be written as a direct sum of 1-dimensional invariant subspaces, that is, φ can be diagonalised. Semisimpleness in the general case is also close in meaning to the 'absence of nondiagonal Jordan blocks'.

Example 2. Suppose that M corresponds to a finite-dimensional representation φ of an algebra R over the field $\mathbb{C}$; assume that M as a vector space over $\mathbb{C}$ has a Hermitian scalar product (x, y), and that the representation φ has the property that for all $a \in R$ there exists $a' \in R$ such that $\varphi_a^* = \varphi_{a'}$ (where φ^* denotes the complex conjugate transformation). Then M is a semisimple module.

Indeed, if $N \subset M$ is a subspace invariant under the transformations φ_a for $a \in R$, then its orthogonal complement N' will be invariant under the transformations φ_a^*, so by the assumption, under all transformations φ_a. Hence $M = N \oplus N'$ as an R-module.

Example 3. Suppose that M corresponds to a finite-dimensional representation φ of a group G over $\mathbb{C}$, again with a Hermitian scalar product defined such that all the operators φ_g are unitary for $g \in G$, that is

$$(\varphi_g(x), \varphi_g(y)) = (x, y). \tag{1}$$

Then M is semisimple. The proof is the same as in Example 2.

The notion of semisimpleness extends to infinite-dimensional representations of groups, with the modification that the module M as a vector space over $\mathbb{C}$ is given a topology or a norm, and the submodule N is assumed to be closed. In particular, if in the situation of Example 3 M is a Hilbert space under the Hermitian product (x, y) then the argument still works.

Example 4. A finite-dimensional representation of a finite group G over $\mathbb{C}$ defines a semisimple module.

The situation can be reduced to the previous example. Introduce on M (considered as a vector space over $\mathbb{C}$) an arbitrary Hermitian scalar product $\{x, y\}$, and then set

$$(x, y) = \frac{1}{|G|} \sum_{g \in G} \{\varphi_g(x), \varphi_g(y)\}, \tag{2}$$

where the sum runs over all elements $g \in G$, and $|G|$ denotes the number of elements of G. It is easy to see that the product (x, y) satisfies the conditions of Example 3.

The same argument can be adapted to representations over an arbitrary field.

Example 5. If G is a finite group of order n not divisible by the characteristic of a field K, then any finite-dimensional representation of G over K defines a semisimple module.

Let M be the space in which the representation φ acts, and $N \subset M$ a submodule. We choose an arbitrary subspace N' such that $M = N \oplus N'$ (as a vector space). Write π' for the projection onto N parallel to N'; that is, if $x = y + y'$ with $x \in M$, $y \in N$ and $y' \in N'$ then $\pi'(x) = y$. We consider the linear transformation

$$\pi = \frac{1}{n} \sum_{g \in G} \varphi_g^{-1} \pi' \varphi_g. \tag{3}$$

It is easy to check that $\pi M \subset N$, $\pi x = x$ for $x \in N$ and $\varphi_g \pi = \pi \varphi_g$ for $g \in G$. It follows from this that π is the projection onto N parallel to the subspace $N_1 = \operatorname{Ker} \pi$ and that N_1 is invariant under all φ_g for $g \in G$; that is, it defines a submodule $N_1 \subset M$ such that $M = N \oplus N_1$.

We carry over the notion of semisimpleness from modules to rings. A ring R is *semisimple* if it is semisimple as a module over itself. From Examples 4–5 it follows that the group algebra of a finite group G over a field K is semisimple if the order of the group is not divisible by the characteristic of the field.

Theorem II. *A simple ring R (see § 8) of finite length is semisimple.*

In fact, consider the submodule I of R generated by all simple submodules. From the fact that R is of finite length it follows that I is generated by a finite number of submodules $P_1, \ldots, P_n$. Obviously I is a left ideal of R. But I is also a right ideal, since for all $a \in R$, $P_i a$ is a left ideal and a simple submodule, that is $P_i a \subset I$, and hence $Ia \subset I$. Since R is simple, $I = R$, that is R is generated by simple submodules P_i, and is semisimple because of Theorem I.

The theory of modules over semisimple rings has a very explicit character.

Theorem III. *If R is a semisimple ring of finite length and*

$$R = P_1 \oplus \cdots \oplus P_n$$

is a decomposition of R (as a module over itself) as a direct sum of simple submodules then the P_i for $i = 1, \ldots, n$ are the only simple modules over R. Any module of finite length is semisimple and is a direct sum of copies of certain of the modules P_i.

In fact, if M is any module and $x_1, \ldots, x_k$ are elements of M then we can define a homomorphism $f \colon R^k \to M$ by

$$f((a_1, \ldots, a_k)) = a_1 x_1 + \cdots + a_k x_k.$$

From the fact that M is of finite length it follows that for some choice of the elements $x_1, \ldots, x_k$ the image of f is the whole of M. Thus M is a homomorphic image of a semisimple module, and is therefore semisimple. If M is simple, rewriting R^k in the form $P_1^k \oplus \cdots \oplus P_n^k$ we see that $f(P_j) = 0$ if P_j is not isomorphic to M, and hence $P_i \cong M$ for some i.

Corollary. *In particular, we see that over a semisimple ring of finite length there are only a finite number of simple modules (up to isomorphism).*

We now turn to the description of the structure of semisimple rings of finite length. As a module over itself, such a ring decomposes as a direct sum of simple submodules:

$$R = P_1 \oplus \cdots \oplus P_k \tag{4}$$

In this decomposition we group together all the terms which are isomorphic to one another as R-modules:

$$R = (P_1 \oplus \cdots \oplus P_{k_1}) \oplus (P_{k_1+1} \oplus \cdots \oplus P_{k_2}) \oplus \cdots \oplus (P_{k_{p-1}+1} \oplus \cdots \oplus P_{k_p}),$$

that is,

$$R = R_1 \oplus R_2 \oplus \cdots \oplus R_p, \tag{5}$$

where

$$R_i = \sum_{k_i \leqslant j < k_{i+1}} P_j; \tag{6}$$

here in (6) all the simple submodules P_j for $k_i \leqslant j \leqslant k_{i+1}$ occuring in the same summand R_i are isomorphic, and all those occuring in different R_i are nonisomorphic.

Any simple submodule $P \subset R$ is isomorphic to one of the P_α, and it is not hard to deduce from this that it is contained in the same summand R_i as P_α. In particular, for $a \in R$ the module $P_\alpha a$ is isomorphic to P_α, and hence $P_\alpha a \subset R_i$ if $P_\alpha \subset R_i$. In other words, the R_i are not only left ideals (as they must be as R-submodules), but also right ideals, that is, they are two-sided ideals. It follows from this that $R_i R_j \subset R_i$ and $R_i R_j \subset R_j$, and hence $R_i R_j = 0$ for $i \neq j$. It is easy to see that the components of 1 in R_i in the decomposition (5) is the unit element of R_i, so that we have a decomposition as a direct sum of rings

$$R = R_1 \oplus R_2 \oplus \cdots \oplus R_p.$$

Here the R_i are defined entirely uniquely: each of them is generated by all simple submodules of R isomorphic to a given one.

Consider now one of the rings R_i. For this it is convenient to rewrite the decomposition (6) in the form

$$R_i = P_{i,1} \oplus P_{i,2} \oplus \cdots \oplus P_{i,q_i},$$

where the prime submodules $P_{i,j}$ are all isomorphic, that is, $R_i \cong N_i^{q_i}$ for some module N_i isomorphic to all of the $P_{i,j}$ for $j = 1, \ldots, q_i$.

Let us find the ring of automorphisms of this module over R. Since $R_s R_i = 0$ for $s \neq i$, we have $\mathrm{End}_R R_i = \mathrm{End}_{R_i} R_i$, and thus $\mathrm{End}_R R_i$ is isomorphic to the opposite ring R_i' skew-isomorphic to R_i (see § 9, Example 4). On the other hand, by § 9, Example 3, $\mathrm{End}_R R_i \cong M_{q_i}(\mathrm{End}_R N_i)$, and $\mathrm{End}_R N_i$ is a division algebra D_i (§ 9, Example 6). Hence $R_i' \cong M_{q_i}(D_i)$, and

$$R_l \cong M_{q_l}(D'_l). \tag{7}$$

We have seen (§ 8, Example 13) that $M_q(D)$ is a simple ring, and hence the rings R_i in (7) are simple. Putting the decomposition (5) together with the isomorphism (7), we obtain the following fundamental theorem.

IV. Wedderburn's Theorem. *A semisimple ring of finite length is isomorphic to direct sum of a finite number of simple rings. A simple ring of finite length is isomorphic to a matrix ring over some division algebra.*

Conversely, we have seen (§ 8, Example 13) that for D a division algebra, the ring $M_q(D)$ is simple, and it is easy to see that a direct sum of semisimple rings is semisimple. Hence Wedderburn's theorem describes completely the range of the class of semisimple rings: they are direct sums of matrix rings over division algebras. As a particular case we get:

Theorem V. *A commutative ring of finite length is semisimple if and only if it is a direct sum of fields.*

For an arbitrary semisimple ring R, its centre $Z(R)$ is commutative. It is easy to see that $Z(R_1 \oplus R_2) \cong Z(R_1) \oplus Z(R_2)$ and that $Z(M_n(D)) \cong Z(D)$.

Corollary. *The centre of a semisimple ring R is isomorphic to a direct sum of fields, and the number of direct summands of the centre is equal to the number of direct summands in the decomposition* (5) *of R as a direct sum of simple rings.*

In particular, we have the result:

Theorem VI. *A semisimple ring of finite length is simple if and only if its centre is a field.*

We now illustrate the general theory by means of the fundamental example of the ring $M_n(D)$. In § 8, Example 13 we saw what the left ideals of $R = M_n(D)$ look like for a division ring D. Any left ideal of this ring is of the form ${}_VI = \{a \in R \mid aV = 0\}$, where a is considered as a D'-linear transformation of a n-dimensional vector space L (over the opposite division algebra D'), and $V \subset L$ is some vector subspace. Simple submodules correspond to minimal ideals. Obviously, if $V \subset V'$ then ${}_VI \supset {}_{V'}I$. Hence we obtain a minimal ideal ${}_VI$ is we take V to be an $(n-1)$-dimensional subspace of L. Choose some basis $e_1, \ldots, e_n$ of L and let V_i be the space of vectors whose ith coordinate in this basis equals zero; the ideal ${}_{V_i}I$ consists of matrixes having only the ith column nonzero. The decomposition $R = N_1 \oplus \cdots \oplus N_n$ corresponds to the decomposition

$$\begin{bmatrix} d_{11} & \cdot & \cdot & d_{1n} \\ \cdot & \cdot & \cdot & \cdot \\ \cdot & \cdot & \cdot & \cdot \\ d_{n1} & \cdot & \cdot & d_{nn} \end{bmatrix} = \begin{bmatrix} d_{11} & 0 & \cdots & 0 \\ \vdots & & & \\ d_{n1} & 0 & \cdots & 0 \end{bmatrix} + \cdots + \begin{bmatrix} 0 & \cdots & 0 & d_{1n} \\ & & & \vdots \\ 0 & \cdots & 0 & d_{nn} \end{bmatrix} \tag{7'}$$

of a matrix. On multiplying this on the left by an arbitrary matrix, the ith column transforms exactly as a vector of L. Thus all the ideals N_i are isomorphic as R-modules, and are isomorphic to L. This is the unique simple module over R.

In the case of an arbitrary semisimple algebra R of finite length, the position is only a little more complicated. If R decomposes as a direct sum

$$R = M_{n_1}(D_1) \oplus \cdots \oplus M_{n_p}(D_p)$$

of p matrix rings then it has p simple modules $N_1, \ldots, N_p$, where N_i is a n_i-dimensional vector space over the opposite division algebra D_i', and R acts on N_i as follows: $M_{n_j}(D_j)$ annihilates N_i for $j \neq i$, and matrixes of $M_{n_i}(D_i)$ act on N_i as befits a matrix acting on a vector.

The remaining part of this section is devoted to examples and applications of the theory we have treated.

We return to an arbitrary simple ring of finite length, which by Wedderburn's theorem is isomorphic to a matrix ring $M_n(D)$ over some division algebra D. We have given a description of the left ideals of this ring: they are in 1-to-1 inclusion-reversing correspondence with subspaces of an n-dimensional subspace over the opposite division algebra D' (that is, $V_1 \subset V_2$ if and only if ${}_{V_1}I \supset {}_{V_2}I$). To what extent is the whole ring (that is, the division algebra D and the dimension n) reflected by this partially ordered set? The set of linear subspaces of a n-dimensional vector space over D coincides with the set of linear subspaces of $(n-1)$-dimensional projective space $\mathbb{P}^{n-1}(D)$ over D. Thus our question is essentially a question about the axiomatic structure of projective geometry. We recall the solution of this problem. (The axioms we give are not independent; we have chosen them as the most intuitively convincing.)

Let $\mathfrak{P}$ be a *partially ordered set*, that is, for some pairs (x, y) of elements of $\mathfrak{P}$ an order relation $x \leqslant y$ is defined, such that the following conditions are satisfied: (a) if $x \leqslant y$ and $y \leqslant z$ then $x \leqslant z$; and (b) $x \leqslant y$ and $y \leqslant x$ if and only if $x = y$.

We assume that the following axioms are satisfied:

1. For any set of elements $x_\alpha \in \mathfrak{P}$ there exists an element y such that $y \geqslant x_\alpha$ for all α, and if $z \geqslant x_\alpha$ for all α then $z \geqslant y$. The element y is called the *sum* of the x_α and is denoted by $\bigcup x_\alpha$. In particular, the sum of all $x \in \mathfrak{P}$ exists (the 'whole projective space'). It is denoted by I or $I(\mathfrak{P})$.

2. For any set of elements $x_\alpha \in \mathfrak{P}$ there exists an element y' such that $y' \leqslant x_\alpha$ for all α, and if $z' \leqslant x_\alpha$ for all α then $z' \leqslant y'$. The element y' is called the *intersection* of the x_α and is denoted by $\bigcap x_\alpha$. In particular, the intersection of all $x \in \mathfrak{P}$ exists (the 'empty set'). It is denoted by $\varnothing$ or $\varnothing(\mathfrak{P})$.

From now on, for $x, y \in \mathfrak{P}$ with $y \leqslant x$ we write x/y for the partially ordered set of all $z \in \mathfrak{P}$ such that $y \leqslant z \leqslant x$. Obviously conditions 1 and 2 are satisfied in x/y for all x and y.

3. For any x and $y \in \mathfrak{P}$ and $a \in x/y$ there exists an element $b \in x/y$ such that $a \cup b = I(x/y)$ and $a \cap b = \varnothing(x/y)$. If $b' \in x/y$ is another element with the same properties and if $b \leqslant b'$ then $b = b'$.

4. Finite length: the length of all chains $a_1 \leqslant a_2 \leqslant \cdots \leqslant a_r$ with $a_1 \neq a_2$, $a_2 \neq a_3, \ldots, a_{r-1} \neq a_r$ is bounded.

An element $a \in R$ is a *point* if $b \leqslant a$ and $b \neq a$ implies that $b = \varnothing$.

5. For any two points a and b there exists a point c such that $c \neq a, c \neq b$ and $c \leqslant a \cup b$.

A partially ordered set satisfying conditions 1–5 is called a *projective space*. It can be proved that the maximum length of a chain starting from $\varnothing$ and ending with $a \in \mathfrak{P}$ defines a dimension function $d(a)$ on points, which satisfies the relation

$$d(a \cap b) + d(a \cup b) = d(a) + d(b).$$

The number $d(I)$ is called the *dimension* of $\mathfrak{P}$. An example of a n-dimensional projective space is the space $\mathbb{P}^n(D)$ of all linear subspaces of a $(n+1)$-dimensional vector space over a division algebra D.

We have the following result.

VII. Fundamental Theorem of Projective Geometry. (a) *For $n \geqslant 2$ the projective space $\mathbb{P}^n(D)$ (as a partially ordered set) determines the number n and the division algebra D; and* (b) *if $\mathfrak{P}$ is a projective space of dimension at least* 3 *then it isomorphic (as a partially ordered set) to the projective space $\mathbb{P}^n(D)$ over some division algebra D.*

The proof is based on artificially introducing a system of coordinates in the projective space (that is, 'coordinatising' it); the basic idea is already present in the calculus of plane intervals (see § 2, Figures 5 and 6). As in this calculus, the set of elements which appear as coordinates can be constructed fairly easily. On this set one defines operations of addition and multiplication; but the hard part is verifying the axioms of a division algebra. The key to it is the following assertion, known as 'Desargues' theorem':

VIII. Desargues' Theorem. *If the* 3 *lines AA', BB', CC' joining corresponding vertexes of two triangles ABC and $A'B'C'$ intersect in a point O then the points of intersection of the corresponding sides are collinear* (*see Figure* 11).

However, this assertion can only be deduced from the axioms of a projective space if the space has dimension $\geqslant 3$. In 2 dimensions, that is in the plane, it does not follow from the axioms, and not every projective plane is isomorphic to $\mathbb{P}^2(D)$. A necessary and sufficient condition for this is that the previous proposition should hold, and one must add this as an extra axiom, *Desargues' axiom*.

The results we have given characterise the role of completely arbitrary division algebras in projective geometry: they allow us to list explicitly all the non-isomorphic realisations of the system of axioms of n-dimensional projective geometry (together with Desargues' axiom if $n = 2$). As one would expect, algebraic properties of division algebras occur as geometric properties of the corresponding geometries. For example, the commutativity of a division algebra D is equivalent to the following assertion in the projective space $\mathbb{P}^n(D)$ for $n \geqslant 2$.

IX. Pappus' 'Theorem'. *If the vertexes of a hexagon $P_1P_2P_3P_4P_5P_6$ lie* 3 *by* 3 *on two lines, then the points of intersection of the opposite sides P_1P_2 and P_4P_5, P_2P_3 and P_5P_6, P_3P_4 and P_6P_1 are collinear* (*Figure* 12).

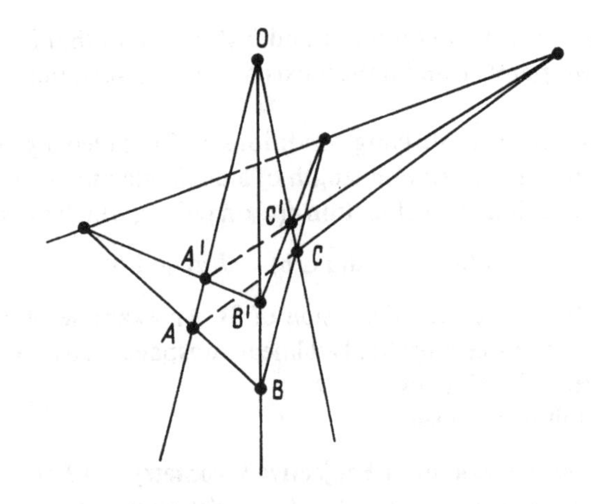

Fig. 11

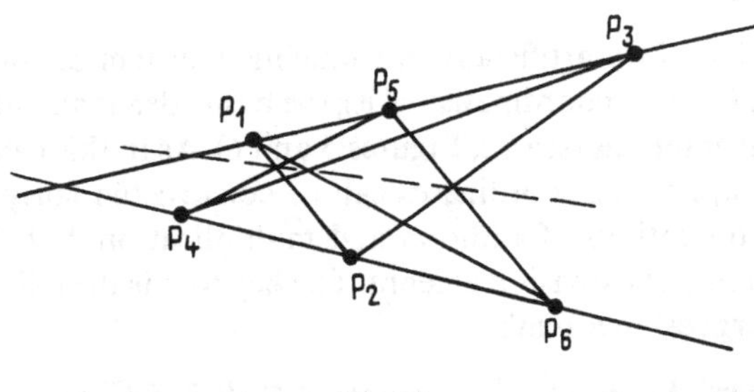

Fig. 12

The condition that the division algebra D is of characteristic 2 is equivalent to the following axiom:

X. Fano's Axiom. *The* 3 *points of intersection of opposite sides AB and DC, AD and BC, AC and BD of a plane quadrilateral ABCD are collinear.*

In Figure 13 this property does not hold, since the real number field has characteristic $\neq 2$.

The finite models of systems of certain geometric axioms which we considered in § 1 (see Figures 1–2) relate to the same circle of ideas; they are finite affine planes over the fields $\mathbb{F}_2$ and $\mathbb{F}_3$, that is, they are obtained from the projective planes $\mathbb{P}^2(\mathbb{F}_2)$ and $\mathbb{P}^2(\mathbb{F}_3)$ by throwing away one line and the points on it (which will then be 'at infinity' from the point of view of the geometry of the remaining points and lines).

There are various infinite-dimensional generalisations of simple rings of finite length and of the theory treated above. One of these starts off from the semisimple

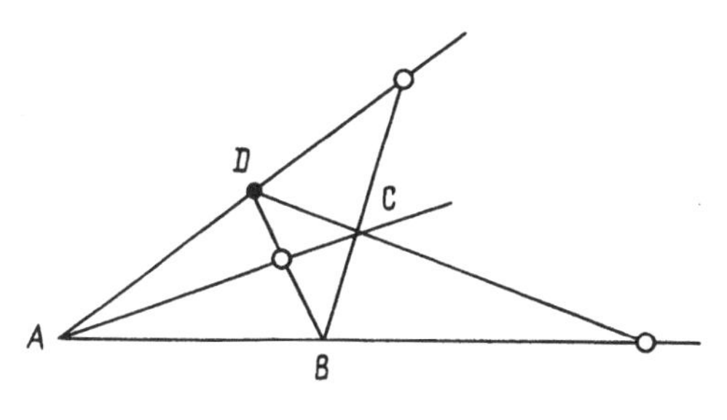

Fig. 13

condition (Example 2) and the criterion VI for a semisimple ring to be simple. This leads to the following definition.

A subring R of the ring of bounded operators on a complex Hilbert space is called a *factor* if together with an operator φ it contains the conjugate operator φ^*, the centre of R consists only of the scalar operators, and R is closed in the natural topology (the so-called weak topology).

Similarly to the way in which a simple ring of finite length defines a projective space satisfying the axioms 1–5, any factor defines a partially ordered set satisfying similar axioms. On this set a dimension function is again defined, but now various cases can occur:

Case I_n. The dimension takes the values 0, 1, 2, ..., n; then the factor is isomorphic to the ring $M_n(\mathbb{C})$.

Case I_∞. The dimension takes the values 0, 1, 2, ..., ∞; in this case the factor is isomorphic to the ring of all bounded operators on an infinite-dimensional Hilbert space.

Case II_1. The dimension takes values in the interval $[0, 1]$.

Case II_∞. The dimension takes values in the interval $[0, \infty]$.

Case III. The dimension takes only the values 0 and ∞.

The partially ordered sets corresponding to II_1, II_∞ and III are highly nontrivial infinite-dimensional analogues of projective planes (we emphasise that the dimensions of subspaces can be any real values), called *continuous geometries*.

From now on we restrict ourselves to considering algebras of finite rank over a field K.

Example 6. We saw in § 8, Example 11 that if L is a real $2n$-dimensional vector space with metric $x_1^2 + \cdots + x_{2n}^2$ then the Clifford algebra $C(L)$ is isomorphic to the matrix algebra $M_{2^{2n}}(\mathbb{R})$. Therefore this algebra has a unique irreducible representation (up to equivalence), of degree 2^{2n}. Thus we can write

$$x_1^2 + \cdots + x_{2n}^2 = (x_1\Gamma_1 + \cdots + x_{2n}\Gamma_{2n})^2,$$

where $\Gamma_1, \ldots, \Gamma_{2n}$ are $2^{2n} \times 2^{2n}$ matrixes, and these matrixes are uniquely determined up to transformations $\Gamma_i \mapsto C\Gamma_i C^{-1}$. No matrixes smaller than $2^{2n} \times 2^{2n}$ can give such a representation.

Now suppose that the field K is algebraically closed. Then the theory of semisimple algebras over K and their representations takes on an especially concrete character. The basis of this is the following simple result.

Theorem XI. *A division algebra of finite rank over an algebraically closed field K is equal to K itself.*

In fact if D has rank n and $a \in D$ then the elements $1, a, a^2, \ldots, a^n$ must be linearly dependent over K. Hence there exists a polynomial $F \in K[t]$ of degree $\leqslant n$, not identically 0, for which $F(a) = 0$. Since K is algebraically closed, $F(t) = \gamma \prod (t - \alpha_i)$, so that $\prod (a - \alpha_i) = 0$. Since D is a division algebra, we must have $a - \alpha_i = 0$ for some i, so that $a \in K$. We thus get the following result.

Theorem XI′. *Over an algebraically closed field K, any simple algebra of finite rank is isomorphic to $M_n(K)$, and any semisimple algebra to a direct sum of such.*

Earlier, in Formula (7), we gave an explicit decomposition of the regular representation of the algebra $M_n(K)$ into irreducible representations. It follows from this that its regular representation is a sum of n equivalent n-dimensional representations (corresponding to n-dimensional space K^n as a module over the matrix ring $M_n(K)$). If $R \cong M_{n_1}(K) \oplus \cdots \oplus M_{n_p}(K)$ then R has p irreducible n_i-dimensional representations N_i (corresponding to the irreducible representations of the matrix algebras $M_{n_i}(K)$), and the regular representation of R is of the form

$$R \cong N_1^{n_1} \oplus N_2^{n_2} \oplus \cdots \oplus N_p^{n_p} \quad \text{where} \quad n_i = \dim N_i.$$

For the decomposition of the centre there is also only one possibility, namely

$$Z(R) \cong K^p.$$

As a result, the theory of representations of semisimple algebras of finite rank over an algebraically closed field reduces to the following:

Theorem XII. *Every representation is a finite sum of irreducible representations.*

Theorem XIII. *All irreducible representations are contained among the irreducible factors of the regular representation. The number of nonequivalent representations among these is equal to the rank of the centre of the algebra.*

Theorem XIV. *Every irreducible representation is contained in the regular representation the same number of times as its dimension.*

XV. Burnside's Theorem. *The sum of squares of the dimensions of irreducible representations equals the rank of the algebra*:

$$n = n_1^2 + \cdots + n_p^2.$$

To specify concretely a representation $\varphi\colon R \to M_n(K)$ we use the traces of the matrixes $\varphi(a)$ for $a \in R$. The function $\operatorname{Tr}(\varphi(a))$ is defined on R and is linear, and is hence determined by its values on the elements of a basis of R. Since $\mathrm{Tr}(CAC^{-1}) = \mathrm{Tr}(A)$, equivalent representations have the same traces. Write $T_\varphi(a)$ for the function $\mathrm{Tr}(\varphi(a))$.

If a representation φ is a direct sum of two others, $\varphi = \varphi_1 \oplus \varphi_2$ then obviously $\mathrm{Tr}_\varphi(a) = \mathrm{Tr}_{\varphi_1}(a) + \mathrm{Tr}_{\varphi_2}(a)$. The traces of irreducible representations are called the *characters* of the algebra R. Let $\chi_1(a), \chi_2(a), \ldots, \chi_p(a)$ be the characters corresponding to the p irreducible representations $\varphi_1, \ldots, \varphi_p$. Any representation φ decomposes as a direct sum of irreducibles, and if χ_i occurs m_i times amongst these, then

$$\mathrm{Tr}_\varphi(a) = m_1\chi_1(a) + \cdots + m_p\chi_p(a). \tag{8}$$

We know that in the decomposition of R as a direct sum of simple algebras $R = R_1 \oplus \cdots \oplus R_p$, the irreducible representation φ_i maps to zero in all the summands R_j for $i \neq j$. Together with (8), this implies the following result.

Theorem XVI. *The representations of a semisimple algebra are uniquely determined by their trace functions* $\mathrm{Tr}_\varphi(a)$.

The results we have given have a very large number of application, most significantly in the special case of a group algebra $R = K[G]$. We will meet these later, but for now we indicate a completely elementary application to the matrix algebra.

XVII. Burnside's Theorem. *An irreducible subalgebra R of a matrix algebra $\mathrm{End}_K L$ over an algebraically closed field K coincides with the whole of* $\mathrm{End}_K L$.

Proof. The hypothesis of the theorem means that if a representation φ of R in an n-dimensional space L is defined by an embedding $R \to \mathrm{End}_K L$ then φ will be irreducible. For any $x \in L$ the map $a \mapsto ax$ defines a homomorphism of R as a module over itself into the simple module L. Take for x the n elements $e_1, \ldots, e_n$ of a basis of L. We get n homomorphisms $f_i\colon R \to L$, or a single homomorphism $f\colon R \to L^n$ given by $f = f_1 + \cdots + f_n$. If $f(a) = 0$ for some $a \in R$ then $f_i(a) = 0$, that is, $ae_i = 0$ and $ax = 0$ for all $x \in L$. But $R \subset \mathrm{End}\, L$, and hence $a = 0$. Hence $R \subset L^n$ as R-module. Since L is a simple module, it follows from this that R is semisimple as a module, and hence the algebra R is semisimple, and as a module $R \cong L^k$ for some k. But according to Theorem XV, $k = n = \dim L$. Hence R has rank n^2, and therefore $R = \mathrm{End}_K L$.

As an illustration, we give a striking application of this result:

XVIII. Burnside's Theorem. *If G is a group of $n \times n$ matrixes over a field K of characteristic* 0, *and if there exists a number $N > 0$ such that $g^N = 1$ for all $g \in G$ then G is finite.*

For the proof we will use some of the very elementary notions of group theory. Obviously, extending the field K if necessary, we can assume that it is algebraically closed. Write R for the set of all combinations of the form

$$\alpha_1 g_1 + \cdots + \alpha_k g_k \quad \text{with} \quad \alpha_i \in K \text{ and } g_i \in G.$$

Obviously, R is an algebra over K and $R \subset M_n(K)$. Consider first the case that R is irreducible. According to the preceding theorem of Burnside we then have $R = M_n(K)$. By assumption, the eigenvalues of any element $g \in G$ are Nth roots of unity. Since the trace $\mathrm{Tr}(g)$ of an $n \times n$ matrix is the sum of n eigenvalues, $\mathrm{Tr}(g)$ can take at most N^n values. It is easy to check that the bilinear form $\mathrm{Tr}(AB)$ on $M_n(K)$ is nondegenerate. Since $R = M_n(K)$, there exists n^2 elements $g_1, \ldots, g_{n^2} \in G$ which form a basis of $M_n(K)$. Let $e_1, \ldots, e_{n^2}$ be the dual basis with respect to the bilinear form $\mathrm{Tr}(AB)$. If $g = \sum_{i=1}^{n^2} \alpha_i e_i$ is the expression of an arbitrary element $g \in G$ in this basis, then $\alpha_i = \mathrm{Tr}(gg_i)$. Thus the coefficients α_i can take only a finite number of values, and hence G is a finite group.

If R is reducible then the matrixes corresponding to elements of G can simultaneously be put in the form

$$\begin{pmatrix} A(g) & C(g) \\ 0 & B(g) \end{pmatrix}.$$

Applying induction, we can assume that $A(g)$ and $B(g)$ have already been proved to form finite groups G' and G''. Consider the homomorphism $f\colon G \to G' \times G''$ given by $f(g) = (A(g), B(g))$; its kernel consists of elements $g \in G$ corresponding to matrixes

$$\begin{pmatrix} E & C(g) \\ 0 & E \end{pmatrix}.$$

It is easy to see that if K is a field of characteristic 0 and $C(g) \neq 0$ then this matrix cannot be of finite order: taking its mth power just multiplies $C(g)$ by m. Hence the kernel of f consists only of the unit element, that is, G is contained in a finite group $G' \times G''$, and hence is itself finite.

§11. Division Algebras of Finite Rank

Wedderburn's theorem entirely reduces the study of semisimple algebras of finite rank over a field K to that of division algebras of finite rank over the same field. We now concentrate on this problem. If D is a division algebra of finite rank over K and L the centre of D then L is a finite extension of K and we can consider D as an algebra over L. Hence the problem divides into two: to study finite field extensions, which is a question of commutative algebra or Galois theory, and that of division algebras of finite rank over a field which is its centre. If an algebra D of finite rank over a field K has K as its centre, then we say that D is a *central algebra* over K.

The question of the existence of central division algebras of finite rank over a given field K and of their structure depends in a very essential way on particular properties of K. We have already met one very simple result in this direction: over an algebraically closed field K there do not exist division algebras of finite rank other than K. In particular this holds for the complex number field.

For the case of the real number field the situation is not much more complicated.

I. Frobenius' Theorem. *The only division algebras of finite rank over the real number field* $\mathbb{R}$ *are* $\mathbb{R}$ *itself, the complex number field* $\mathbb{C}$, *and the quaternion algebra* $\mathbb{H}$.

Here are another two cases when the situation is simple.

II. Wedderburn's Theorem. *Over a finite field* K, *there do not exist any central division algebras of finite rank other than* K *itself.*

In other words, a finite division algebra is commutative. This is of course interesting for projective geometry, since it shows that for finite projective geometries of dimension >2, Pappus' theorem follows from the other axioms (and in dimension 2, from Desargues' theorem).

III. Tsen's Theorem. *If* K *is an algebraically closed field and* C *is an irreducible algebraic curve over* K, *then there do not exist any central division algebras of finite rank over* $K(C)$ *other than* $K(C)$ *itself.*

The three cases in which we have asserted that over some field K there do not exist any central division algebras of finite rank other than K itself can all be unified by one property: a field K is *quasi-algebraically closed* if for every homogeneous polynomial $F(t_1,\dots,t_n) \in K[t_1,\dots,t_n]$ of degree less than the number n of variables, the equation

$$F(x_1,\dots,x_n) = 0$$

has a nonzero solution in K.

It can be shown that if K is any quasi-algebraically closed field, then the only central division algebra of finite rank over K is K itself. On the other hand, each of the fields considered above is quasi-algebraically closed: algebraically closed fields, finite fields and function fields $K(C)$ where K is algebraically closed and C is an irreducible algebraic curve. Tsen's theorem is proved starting from this property, and this is one possible method of proof of Wedderburn's theorem. The theorem that a finite field is quasi-algebraically closed is called Chevalley's theorem. For the case of the field $\mathbb{F}_p$, this is an interesting property of congruences. The property of being quasi-algebraically closed is a direct weakening of algebraically closure; this become obvious if we start with the following characterisation of algebraically closed fields.

Theorem. *A field* K *is algebraically closed if and only if for every homogeneous polynomial* $F(t_1,\dots,t_n) \in K[t_1,\dots,t_n]$ *of degree less than or equal to the number* n *of variables, the equation*

$$F(x_1,\ldots,x_n) = 0 \tag{1}$$

has a nonzero solution in K.

Proof. Obviously, for an algebraically closed field K the equation (1) has a nonzero solution. Suppose that K is not algebraically closed. Then over K there is an irreducible polynomial $P(t)$ of degree $n > 1$. The ring $L = K[t]/(P)$ is a field extension of K of degree n, and is hence an algebra of rank n over K. Considering the regular representation of this algebra, we take each element $x \in L$ to the matrix $A_x \in M_n(K)$. The determinant of the matrix A_x is called the *norm* of x, and is denoted by $N(x)$. From properties of representations (and of determinants) it follows that $N(1) = 1$, and $N(xy) = N(x)N(y)$. From this if $x \neq 0$ then $N(x)N(x^{-1}) = 1$, and hence $N(x) \neq 0$. Consider any basis $e_1, \ldots, e_n$ of L/K (for example the images of the elements $1, t, \ldots, t^{n-1}$ of $K[t]$); we write any element $x \in L$ as $x_1 e_1 + \cdots + x_n e_n$ with $x_i \in K$. It is easy to see that $N(x)$ is a polynomial of degree n in $x_1, \ldots, x_n$, and setting

$$F(x_1,\ldots,x_n) = N(x_1 e_1 + \cdots + x_n e_n)$$

we get an example of an equation of type (1) with no solution.

We proceed now to fields over which there do exist central division algebras of finite rank. Up to now we know one such field, the real number field $\mathbb{R}$, over which there exists a central division algebra of rank 1 ($\mathbb{R}$ itself) and of rank 4 (the quaternions $\mathbb{H}$). These numbers are not entirely fortuitous, as the following result shows.

Lemma. *The rank of a simple central algebra is a perfect square.*

The proof is based on the important notion of extending the ground field. If R is an algebra of finite rank over a field K and L is an arbitrary extension of K, we consider the module $R \otimes_K L$ (see § 5); we define a multiplication on its generators $a \otimes \xi$ by

$$(a \otimes \xi)(b \otimes \eta) = ab \otimes \xi\eta \quad \text{for} \quad a, b \in R, \xi, \eta \in L.$$

It is easy to verify that this turns $R \otimes_K L$ into a ring, which contains L (as $1 \otimes L$), so that this is an algebra over L. If $e_1, \ldots, e_n$ is a basis of R over K then $e_1 \otimes 1, \ldots, e_n \otimes 1$ is a basis of $R \otimes_K L$ over L. Hence the rank of $R \otimes_K L$ over L equals that of R over K. Passing from R to $R \otimes_K L$ is called *extending the ground field.* To put things simply, $R \otimes_K L$ is the algebra over L having the same multiplication table as R.

It is not hard to prove the following assertion:

Theorem IV. *The property that an algebra should be central and simple is preserved under extension of the ground field.*

Now we just have to take L to be the algebraic closure of K; by the general theory $R \otimes_K L \cong M_n(L)$, and hence the rank of R over K, equal to the rank of $R \otimes_K L$ over L is n^2.

Thus the quaternion algebra realises the minimal possible value of the rank of a nontrivial central division algebra. The next case in order of difficulty, of interest mainly in number theory, is the case of the p-adic number field $\mathbb{Q}_p$, introduced in § 7.

V. Hasse's Theorem. *For any n there exist* $\varphi(n)$ *central division algebras of rank* n^2 *over the field* $\mathbb{Q}_p$ (where φ is the Euler function).

The proof of the theorem indicates a method of assigning to each such algebra D a certain primitive nth root of unity, which determines D. This root of unity is called the *invariant* of the division algebra D, and is denoted by $\mu_p(D)$.

We consider the simplest example. For any field K of characteristic $\neq 2$, suppose that $a, b \in K$ are two nonzero elements. Construct the algebra of rank 4 over K with basis $1, i, j, k$ and multiplication table

$$i^2 = a, j^2 = b, ji = -ij = k$$

(the remaining products can be computed from these rules using associativity). The algebra so obtained is called a *generalised quaternion algebra*, and is denoted by (a, b). For example $\mathbb{H} = (-1, -1)$. It is easy to prove that the algebra (a, b) is simple and central, and any simple central algebra of rank 4 can be expressed in this form. Thus by the general theory of algebras, (a, b) is either a division algebra, or isomorphic to $M_2(K)$. Let us determine how to distinguish between these two cases. For this, by analogy with the quaternions, define the *conjugate* of the element $x = \alpha + \beta i + \gamma j + \delta k$ to be $\bar{x} = \alpha - \beta i - \gamma j - \delta k$. It is easy to see that

$$\overline{xy} = \bar{y}\bar{x} \quad \text{and} \quad x\bar{x} = \alpha^2 - a\beta^2 - b\gamma^2 + ab\delta^2 \in K. \tag{2}$$

Set $N(x) = x\bar{x}$. It follows from (2) that $N(xy) = N(x)N(y)$. Hence if $N(x) = 0$ for some nonzero x then (a, b) is not a division algebra: $x\bar{x} = 0$, although $x \neq 0$ and $\bar{x} \neq 0$. If on the other hand $N(x) \neq 0$ for every nonzero x, then $x^{-1} = N(x)^{-1}\bar{x}$ and (a, b) is a division algebra. Thus (a, b) is a division algebra if and only if the equation $\alpha^2 - a\beta^2 - b\gamma^2 + ab\delta^2 = 0$ has only the zero solution for $(\alpha, \beta, \gamma, \delta)$ in K. This equation can be further simplified by writing it in the form

$$\alpha^2 - a\beta^2 = b(\gamma^2 - a\delta^2),$$

or

$$b = \frac{\alpha^2 - a\beta^2}{\gamma^2 - a\delta^2} = \frac{N(\alpha + \beta i)}{N(\gamma + \delta i)} = N\left(\frac{\alpha + \beta i}{\gamma + \delta i}\right) = N(\xi + \eta i) = \xi^2 - a\eta^2,$$

where $\xi + \eta i = \dfrac{\alpha + \beta i}{\gamma + \delta i}$. Thus the condition that (a, b) is a division algebra takes the form: the equation $\xi^2 - a\eta^2 = b$ has no solutions in K. The homogeneous form of the same equation is

$$\xi^2 - a\eta^2 - b\zeta^2 = 0. \tag{3}$$

This should not have any solution in K, otherwise $(a, b) \cong M_2(K)$. It is easy to see that if (3) has a nonzero solution, then it has a solution for which $\xi \neq 0$. It then reduces to the equation

$$ax^2 + by^2 = 1. \tag{4}$$

Now suppose that K is the rational number field $\mathbb{Q}$. Equation (3) is exactly the equation to which Legendre's theorem stated at the end of § 5 relates. This asserts that (3) has a solution in $\mathbb{Q}$ if and only if it has a solution in $\mathbb{R}$ and in all the fields $\mathbb{Q}_p$. In this form the theorem gives us information about the generalised quaternion algebra (a, b) for $a, b \in \mathbb{Q}$. By what we have said above, the algebra $C = (a, b)$ is isomorphic to $M_2(\mathbb{Q})$ if and only if $C \otimes \mathbb{R} \cong M_2(\mathbb{R})$ and $C \otimes \mathbb{Q}_p \cong M_2(\mathbb{Q}_p)$ for all p. But the same line of argument can be carried further, to describe all generalised quaternion algebras over $\mathbb{Q}$. That is, one can show that two such algebras $C = (a, b)$ and $C' = (a', b')$ are isomorphic if and only if $C \otimes \mathbb{R} \cong C' \otimes \mathbb{R}$ and $C \otimes \mathbb{Q}_p \cong C' \otimes \mathbb{Q}_p$ for all p. In other words, consider the invariants μ_p of division algebras of rank 4 over $\mathbb{Q}_p$ (which by definition are equal to -1) and for a simple central algebra C over $\mathbb{Q}$ set

$$\mu_p(C) = \mu_p(C \otimes \mathbb{Q}_p) = -1 \text{ if } C \otimes \mathbb{Q}_p \text{ is a division algebra}$$

$$\mu_p(C) = 1 \text{ if } C \otimes \mathbb{Q}_p \cong M_2(\mathbb{Q}_p);$$

moreover, set

$$\mu_{\mathbb{R}}(C) = -1 \text{ if } C \otimes \mathbb{R} \text{ is a division algebra (that is, } \cong \mathbb{H});$$

$$\mu_{\mathbb{R}}(C) = 1 \text{ if } C \otimes \mathbb{R} \cong M_2(\mathbb{R}).$$

Then the above result can be restated as follows:

Theorem VI. *A division algebra C of rank 4 over $\mathbb{Q}$ is determined by $\mu_{\mathbb{R}}(C)$ and $\mu_p(C)$ for all p, which we call the set of invariants of C.*

What can this set of invariants be? We have seen that not all $\mu_{\mathbb{R}}(C)$ and $\mu_p(C)$ can be equal to 1 (because then, by Legendre's theorem, C would not be a division algebra). Moreover, it is easy to prove that $\mu_p(C) = -1$ holds for only a finite number of primes p. It turns out that there is only one more condition apart from these.

Theorem VII. *An arbitrary set of choices $\mu_{\mathbb{R}}, \mu_p = \pm 1$ (for each prime p) is the set of invariants of some central division algebra of rank 4 over $\mathbb{Q}$ if and only if:* (a) *not all $\mu_{\mathbb{R}}$* and *μ_p are* 1; (b) *only a finite number of them are* -1;

and (c)
$$\mu_{\mathbb{R}} \prod_p \mu_p = 1. \tag{5}$$

Amazingly, the relation (5) turns out just to be a restatement of Gauss' law of quadratic reciprocity, which thus becomes one of the central results of the theory of division algebras over $\mathbb{Q}$.

The results we have treated generalise immediately to arbitrary central division algebras of finite rank over $\mathbb{Q}$. Let C be a central division algebra over $\mathbb{Q}$ of finite rank n^2. The algebra $C \otimes \mathbb{R} = C_{\mathbb{R}}$ is isomorphic either to $M_n(\mathbb{R})$, and then we set $\mu_{\mathbb{R}}(C) = 1$, or to $M_{n/2}(\mathbb{H})$, and then by definition we set $\mu_{\mathbb{R}}(C) = -1$. Similarly, for any prime number p the algebra $C \otimes \mathbb{Q}_p$ is of the form $M_k(C_p)$, where C_p is a central division algebra over $\mathbb{Q}_p$. We set $\mu_p(C) = \mu_p(C_p)$ (see Hasse's theorem, Theorem V above). We have the following results:

VIII. The Hasse-Brauer-Noether Theorem. *$C \cong M_n(\mathbb{Q})$ if and only if $C_{\mathbb{R}} \cong M_n(\mathbb{R})$ and $C \otimes \mathbb{Q}_p \cong M_n(\mathbb{Q}_p)$ for all p, that is $\mu_{\mathbb{R}}(C) = \mu_p(C) = 1$ for all p.*

Hasse's Theorem. *Two central division algebras C and C' over $\mathbb{Q}$ are isomorphic if and only if $C \otimes \mathbb{R} \cong C' \otimes \mathbb{R}$ and $C \otimes \mathbb{Q}_p \cong C' \otimes \mathbb{Q}_p$ for all p, that is $\mu_{\mathbb{R}}(C) = \mu_{\mathbb{R}}(C')$ and $\mu_p(C) = \mu_p(C')$ for all p. A set of numbers $\mu_{\mathbb{R}}$ and μ_p (for all p) can be realised as $\mu_{\mathbb{R}} = \mu_{\mathbb{R}}(C)$ and $\mu_p = \mu_p(C)$ for a central division algebra C over $\mathbb{Q}$ if and only if* (a) *$\mu_p \neq 1$ for only a finite number of primes p;*

and (b)
$$\mu_{\mathbb{R}} \prod_p \mu_p = 1. \tag{6}$$

Entirely similar results hold for finite extensions K of $\mathbb{Q}$, that is, algebraic number fields. They form part of class field theory. The analogue of relation (6) for any algebraic number field is a far-reaching generalisation of Gauss' law of quadratic reciprocity.

The sketch given here of the structure of division algebras over the rational number field can serve as an example of just how closely the structure of division algebras over a field K relate to delicate properties of K.

We give one more example: the structure of central division algebras of finite rank over the field $\mathbb{R}(C)$ where C is a real algebraic curve. In this case, any central division algebra is a generalised quaternion algebra, and is even of the form $(-1, a)$ for $a \in \mathbb{R}(C)$, $a \neq 0$. The algebra $(-1, a)$ is isomorphic to $M_2(\mathbb{R}(C))$ if and only if $a(x) \geqslant 0$ for every point $x \in C$ (including points at infinity in the projective plane). The function $a(x)$ on C changes sign at a finite number of points of C, $x_1, \ldots, x_n$; (Figure 14 illustrates the case of the curve

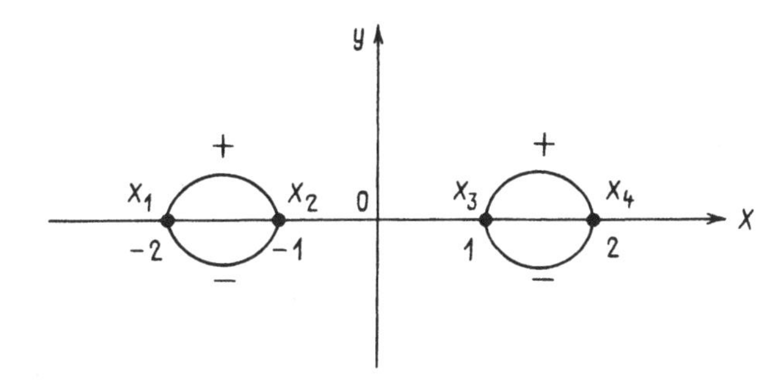

Fig. 14

$$y^2 + (x^2 - 1)(x^2 - 4) = 0$$

and the function $a = y$). The division algebra $(-1, a)$ is determined by these points $x_1, \ldots, x_n$ at which the sign changes. More complicated, but also more interesting is the example of the field $\mathbb{C}(C)$ where C is an algebraic surface. The structure of division algebras in this case reflects very delicate geometric properties of the surface. We will return to these questions in § 12 and § 22.

§ 12. The Notion of a Group

We start with the notion of a transformation group: the notion of a group first arose in this form, and it is in this form that it occurs most often in mathematics and mathematical physics.

A *transformation* of a set X is a 1-to-1 map $f: X \to X$ from X to itself, that is, a map for which there exists an inverse map $f^{-1}: X \to X$, satisfying $f^{-1} \circ f = f \circ f^{-1} = e$. Here $f \circ g$ denotes the product of maps (that is, the composite, the map obtained by performing g and f successively):

$$(f \circ g)(x) = f(g(x)) \quad \text{for } x \in X, \tag{1}$$

and e is the identity map

$$e(x) = x \quad \text{for } x \in X.$$

We say that a set G of transformations of X is a *transformation group* if it contains the identity transformation e and contains together with any $g \in G$ the inverse g^{-1}, and together with any $g_1, g_2 \in G$ the product $g_1 g_2$.

Usually these conditions are fulfilled in an obvious way, because G is defined as the set of all transformations preserving some property. For example, the transformations of a vector space preserving scalar multiplication and the addition of vectors (that is, $g(\alpha x) = \alpha g(x)$ and $g(x + y) = g(x) + g(y)$); these form the group of nondegenerate linear transformations of the vector space. The transformations preserving the distance $\rho(x, y)$ between points of a Euclidean space (that is, such that $\rho(g(x), g(y)) = \rho(x, y)$) form the *group of motions*. If the transformations are assumed to preserve a given point, then we have the *group of orthogonal transformations*.

The group of transformations of one kind or another which preserve some object can often be interpreted as its set of symmetries. For example, the fact that an isosceles triangle is more symmetric that a nonisosceles triangle, and an equilateral triangle more symmetric that a nonequilateral isosceles triangle can be quantified as the fact that the number of motions of the plane taking the triangle to itself is different for these three types of triangles. It consists (a) of just the identity map for a nonisosceles triangle, (b) of the identity map and the reflection in the axis of symmetry for an isosceles triangle, and (c) of 6 transformations for the equilateral triangle, the identity, the rotations through 120° and 240° through the centre 0 and the reflections in the three axes of symmetry (Figure 15).

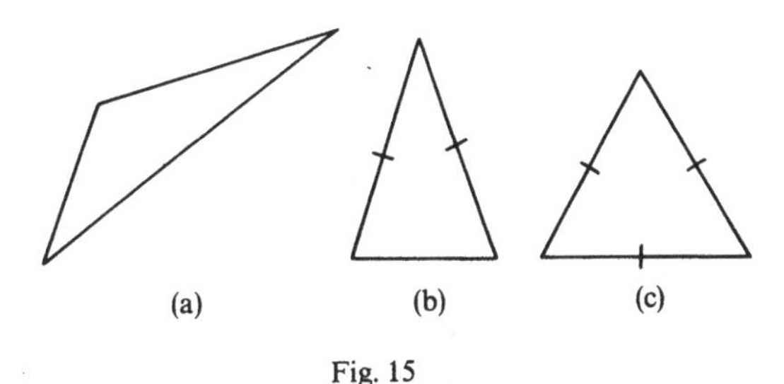

Fig. 15

We give a number of typical examples of different types of symmetry.

The symmetry group of a pattern in the plane consists of all motions of the plane that take it to itself. For example, the symmetry group of the pattern depicted in Figure 16 consists of the following motions: translations in the vector OA, translation in OB followed by reflection in the axis OB, and all composites of these.

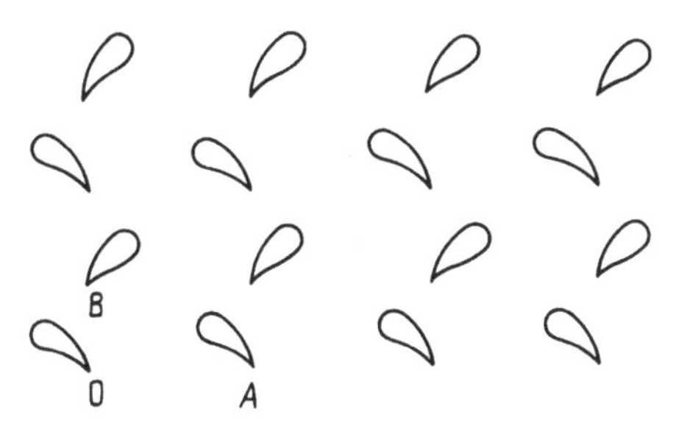

Fig. 16

By a *symmetry of a molecule* we understand a motion of space which takes every atom of the molecule to an atom of the same element, and preserves all valency bonds between atoms. For example, the phosphorus molecule consists of 4 atoms situated at the vertexes of a regular tetrahedron, and its symmetry group coincides with the symmetry group of the tetrahedron, which we will discuss in detail in the following section § 13.

Crystals have a large degree of symmetries, so that the symmetry group of a crystal is an important characteristic of the crystal. Here by a symmetry we understand a motion of space which preserves the position of the atoms of the crystal and all relations between them, taking each atom into an atom of the same element. We describe as an example the symmetry group of the crystal of cooking salt NaCl, depicted in Figure 17. This consists of adjacent cubes, with alternate vertexes occupied by atoms of sodium Na(•) and chlorine Cl(o):

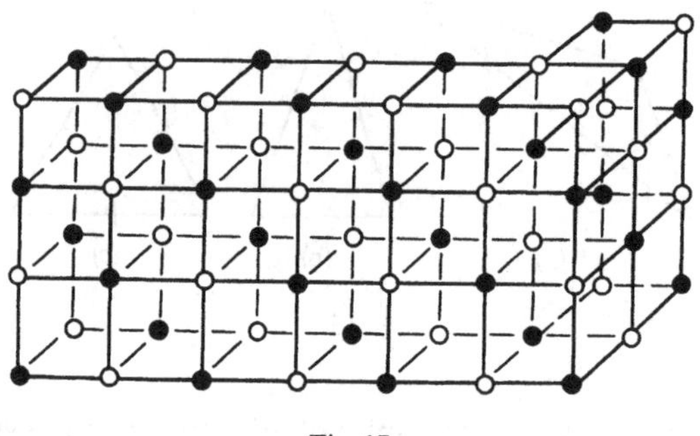

Fig. 17

The set of all symmetries is given (in a coordinate system chosen with the origin in one of the atoms, and axes along the sides of the cubes, which are considered to be of length 1) by permutations of the coordinate axes, reflections in the coordinate planes, and translations in vectors with integer coordinates. It can be expressed by the formulas

$$x_1' = \varepsilon x_{i_1} + k,$$

$$x_2' = \eta x_{i_2} + l,$$

$$x_3' = \zeta x_{i_3} + m,$$

where (i_1, i_2, i_3) is a permutation of $(1, 2, 3)$, each of $\varepsilon, \eta, \zeta = \pm 1$, and $(k, l, m) \in \mathbb{Z}^3$.

Algebraic or analytic expressions may also have symmetries: a *symmetry* of a polynomial $F(x_1, \ldots, x_n)$ is a permutation of the unknowns $x_1, \ldots, x_n$ which preserves F. From this, we get for example the term *symmetric function*, one preserved by all permutations. On the other hand, the function $\prod_{i<k} (x_i - x_k)$ is preserved only by even permutations. In general if F is a function defined on a set X, then a transformation of X which preserves F can be considered as a symmetry of F. In the preceding example, X was the finite set $\{x_1, \ldots, x_n\}$.

A function on 3-space of the form $f(x^2 + y^2 + z^2)$ has all orthogonal transformations as symmetries. Physical phenomena often reflect symmetries of this type. For example, by E. Noether's theorem, if a dynamical system on a manifold X is described by a Lagrangian $\mathscr{L}$ which has a 1-parameter groups $\{g_t\}$ of transformations of X as symmetries, then the system has a first integral that can easily be written down. Thus in the case of the motion of a system of point bodies, invariance of $\mathscr{L}$ under translations leads to the conservation of momentum and invariance under rotations to the conservation of angular momentum.

For any of the types of algebraic objects considered up to now, fields, rings, algebras and modules, the symmetries are transformations of the corresponding sets preserving the basic operations. In this case they are called *automorphisms.*

Thus the automorphisms of an n-dimensional vector space F over a field K are nondegenerate linear transformations; the group of these is denoted by $\mathrm{GL}(F)$ or $\mathrm{GL}(n, K)$. When a basis is chosen, they are given by nondegenerate $n \times n$ matrixes. Similarly, the automorphisms of the free module A^n over a commutative ring A form a group $\mathrm{GL}(n, A)$ and are given by matrixes with entries in A whose determinant is invertible in A. The group consisting of matrixes of determinant 1 is denoted by $\mathrm{SL}(n, A)$.

An automorphism σ of a ring K (in particular, of a field) is a transformation σ such that

$$\sigma(a + b) = \sigma(a) + \sigma(b) \quad \text{and} \quad \sigma(ab) = \sigma(a)\sigma(b). \tag{2}$$

For example, if $R = M_n(K)$ then a nondegenerate matrix $c \in \mathrm{GL}(n, K)$ defines an automorphism $\sigma(a) = cac^{-1}$ of R.

The transformation $\sigma(z) = \bar{z}$ is an automorphism of the complex number field $\mathbb{C}$ as an $\mathbb{R}$-algebra, that is, an automorphism of the field extension $\mathbb{C}/\mathbb{R}$. In a similar way, any field extension L/K of degree 2 (with the characteristic of $K \neq 2$) has exactly 2 automorphisms. For it is easy to see that $L = K(\gamma)$ with $\gamma^2 = c \in K$. Each automorphism is uniquely determined by its action on γ, since $\sigma(a + b\gamma) = a + b\sigma(\gamma)$. But σ preserves the field operations in L and fixes the elements of K, and therefore $(\sigma(\gamma))^2 = \sigma(\gamma^2) = \sigma(c) = c$. Hence $\sigma(\gamma) = \pm\gamma$. Thus the only automorphism are the identity $\sigma(a + b\gamma) = a + b\gamma$ and the automorphism with $\sigma(\gamma) = -\gamma$, that is $\sigma(a + b\gamma) = a - b\gamma$. Extensions having 'less symmetry' also exist. For example, the extension $K = \mathbb{Q}(\gamma)$ where $\gamma^3 = 2$, is of degree 3 over $\mathbb{Q}$, and has only the identity automorphism. For as above, any automorphism σ is determined by its action on γ, and $(\sigma(\gamma))^3 = 2$. If $\sigma(\gamma) = \gamma_1 \neq \gamma$ then $(\gamma_1/\gamma)^3 = 1$, so that $\varepsilon = \gamma_1/\gamma$ satisfies $\varepsilon^3 - 1 = 0$; since $\varepsilon \neq 1$, it satisfies $\varepsilon^2 + \varepsilon + 1 = 0$, and hence $(2\varepsilon + 1)^3 = -3$. Hence K must contain the field $\mathbb{Q}(\sqrt{-3})$. But $K/\mathbb{Q}$ and $\mathbb{Q}(\sqrt{-3})/\mathbb{Q}$ have degree 3 and 2 respectively, and this contradicts the fact that the degree of an extension is divisible by that of any extension contained in it (§ 6, Formula (3)).

Finally the symmetries of physical laws (by this, we understand coordinate transformations which preserve the law) are very important characteristics of these laws. Thus the laws of mechanics should be preserved on passing from one inertial coordinate system to another. The corresponding coordinate transformations (for motion on a line) are of the form

$$x' = x - vt, \qquad t' = t \tag{3}$$

in the mechanics of Galileo and Newton, and

$$x' = \frac{x - vt}{\sqrt{1 - (v/c)^2}}, \qquad t' = \frac{t - (v/c^2)x}{\sqrt{1 - (v/c)^2}}, \tag{4}$$

in the mechanics of special relativity, where c is the speed of light in a vacuum.

The symmetry group given by the transformation formulas (3) is called the *Galileo-Newton group*, and that given by formulas (4) the *Lorentz group*.

An example of another type of symmetry of physical laws is the so-called parity law, according to which all physical laws should preserve their form if we simultaneously invert the sign of time, the sign of electrical charges, and the orientation of space. Here we have a group consisting of just two symmetries.

We mention some very simple notions related to transformation groups. The *orbit* of an element $x \in X$ under a transformation group G of X is the set Gx of elements of the form $g(x)$, as g runs through G. The *stabiliser subgroup* of x is the set G_x of elements of G which fix x, that is, $G_x = \{g | g(x) = x\}$. The stabiliser subgroup of an element is itself a transformation group contained in G.

Consider the relation between two elements $x, y \in X$ that there should exist a transformation $g \in G$ such that $g(x) = y$; this is an equivalence relation, that is, it is reflexive, symmetric and transitive. To check this is just a rephrasing of the three properties in the definition of a transformation group. All elements equivalent to one another form an orbit of G. Hence X breaks up as the disjoint union of orbits; this set of orbits or *orbit space* is denoted by $G\backslash X$. If there is just one orbit, that is, if for any $x, y \in X$ there exists $g \in G$ with $y = g(x)$ then we say that G is *transitive*.

We now proceed to the notion of a group. This formalises only certain aspects of transformation groups: the fact that we can multiply transformations (Formula (1)), and the associativity law $(fg)h = f(gh)$ for this multiplication (which can be verified immediately from the definition).

A *group* is a set G with an operation defined on it (called multiplication), which sends any two elements $g_1, g_2 \in G$ into an element $g_1g_2 \in G$, and which satisfies:

associativity: $(g_1g_2)g_3 = g_1(g_2g_3)$;

existence of an identity: there exists an element $e \in G$ such that $eg = ge = g$ for all $g \in G$ (it is easy to prove that e is unique);

existence of an inverse: for any $g \in G$ there exists an element $g^{-1} \in G$ such that $gg^{-1} = g^{-1}g = e$ (it is easy to prove that g^{-1} is unique).

From the associativity law, one proves easily that for any number of elements $g_1, g_2, \ldots, g_n$, their product (in the given order) does not depend on the position of brackets, and can therefore be written $g_1g_2 \cdots g_n$. The product $g \cdots g$ of g with itself n times is written as g^n, and $(g^{-1})^n$ as g^{-n}.

If multiplication is commutative, that is, $g_2g_1 = g_1g_2$ for any $g_1, g_2 \in G$, we say that G is an *Abelian* (or *commutative*) group. In essence, we have already seen this notion, since it is equivalent to that of module over the ring of integers $\mathbb{Z}$. To emphasise this relation, the group operation in an Abelian group is usually called the sum, and written $g_1 + g_2$. We then write 0 instead of e, $-g$ instead of g^{-1}, and ng instead of g^n. For $n \in \mathbb{Z}$ and $g \in G$, ng is a product, defining a $\mathbb{Z}$-module structure on G.

If a group G has a finite number of elements, we say that G is *finite*. The number of elements is called the *order* of G, and is denoted by $|G|$.

An *isomorphism* of two groups G_1 and G_2 is a 1-to-1 correspondence $f: G \to G_2$ such that $f(g_1 g_2) = f(g_1) f(g_2)$. We say that two groups are *isomorphic* if there exists an isomorphism between them, and we then write $G_1 \cong G_2$.

A finite group $G = \{g_1, \ldots, g_n\}$ can be specified by its 'multiplication table', the so-called *Cayley table*. This is a square matrix with the product $g_i g_j$ in the ith row and jth column. For example, if G is a group of order 2 then it consists of the identity element e and $g \neq e$, and it is easy to see that g^2 can only be e. Hence its Cayley table is of the form

	e	g
e	e	g
g	g	e.

If G is of order 3 with identity element e and $g \neq e$ then it is easy to see that $g^2 \neq e$ and $g^2 \neq g$, so that $G = \{e, g, h\}$ with $h = g^2$. Just as easily, we see that $gh = e$. Hence the Cayley table of G is of the form

	e	g	h
e	e	g	h
g	g	h	e
h	h	e	g.

An isomorphism of two groups means that (up to notation for the elements), their Cayley tables are the same. Of course, Cayley tables can only conveniently be written out for groups of fairly small order. There are other ways of specifying the operation in a group. For example, let G be the group of symmetries of an equilateral triangle (Figure 15, (c)). Let s denote the reflection in one of the heights, and t a rotation through 120° about the centre. Then t^2 is a rotation through 240°. It is easy to see that s, st and st^2 are reflections in the three heights. Thus all the elements of G can be written in the form

$$e,\ t,\ t^2,\ s,\ st,\ st^2. \tag{5}$$

Obviously,

$$s^2 = e, \quad t^3 = e. \tag{6}$$

Furthermore, it is easy to check that

$$ts = st^2. \tag{7}$$

These rules already determine the multiplication of the group. For example

$$(st)^2 = stst = sst^2t = s^2t^3 = e,$$

and

$$t^2s = tts = tst^2 = st^2t^2 = st.$$

This method of describing a group is called specifying it by *generators and relations*; it will be described more precisely in § 14. In this notation, an isomorphism of G with a group G' means that G' also has elements s' and t', in terms of which all the other elements of G' can be expressed as in (5), and which satisfy conditions (6) and (7). Consider for example the group $GL(2, \mathbb{F}_2)$ of nondegenerate 2×2 matrixes with entries in the field $\mathbb{F}_2$. It is easy to write them all out, since their columns are necessarily of the form

$$\begin{bmatrix}1\\0\end{bmatrix}, \begin{bmatrix}0\\1\end{bmatrix} \text{ or } \begin{bmatrix}1\\1\end{bmatrix}, \tag{8}$$

and nonproportional columns (to give a nondegenerate matrix) means in the present case that they are distinct. We get 6 matrixes

$$\begin{bmatrix}1 & 0\\0 & 1\end{bmatrix}, \begin{bmatrix}0 & 1\\1 & 1\end{bmatrix}, \begin{bmatrix}1 & 1\\1 & 0\end{bmatrix}, \begin{bmatrix}0 & 1\\1 & 0\end{bmatrix}, \begin{bmatrix}1 & 1\\0 & 1\end{bmatrix}, \begin{bmatrix}1 & 0\\1 & 1\end{bmatrix}.$$

One sees easily that this list is precisely of the form (5) with $t' = \begin{bmatrix}0 & 1\\1 & 1\end{bmatrix}$ and $s' = \begin{bmatrix}0 & 1\\1 & 0\end{bmatrix}$, and that the relations (6) and (7) hold. Therefore this group is isomorphic to the symmetry group of an equilateral triangle. The isomorphism we have found seems quite mysterious. However, it can be described in a more meaningful way: for this we need to observe that symmetries of a triangle permute the 3 vertexes, and that in our case all of the 6 possible permutations are realised. The nondegenerate matrixes over $\mathbb{F}_2$ act on the 3 column vectors (8), and also realise all possible permutations of these. Thus each of the two groups is isomorphic to the group of all permutations of 3 elements.

From this and from many other examples, we see that isomorphisms may occur between groups whose elements are completely different in nature, and which arise in connection with different problems. The notion of isomorphism focuses attention just on the multiplication law in the groups, independently of the concrete character of their elements. We can imagine that there exists a certain 'abstract' group, the elements of which are just denoted by some symbols, for which a composition law is given (like a Cayley table), and that concrete groups are 'realisations' of it. Quite surprisingly, it often turns out that properties of the abstract group, that is, properties purely of the group operation, often have a lot to offer in understanding the concrete realisation, which the abstract group 'coordinatises'. A classic example is Galois theory, which will be discussed later. However, in the majority of applications of group theory, properties of the abstract group are mixed up together with properties of the concrete realisation.

In the above, we have only given examples of transformation groups, and so as not to give the reader the impression that this is the only way in which groups arise naturally, we give here a number of examples of a different nature. The homotopy groups $\pi_n(X)$ and the homology and cohomology groups $H_n(X)$ and $H^n(X)$ of a topological space X are examples of this kind, but we postpone a discussion of these until §§ 20–21.

Example 1. The Ideal Class Group. The product of ideals in an arbitrary commutative ring A was defined in § 4. This operation is associative and has an identity (the ring A) but the inverse of an ideal almost never exists (even in $\mathbb{Z}$). We identify together in one class all nonzero ideals which are isomorphic as A-modules (see §5). The product operation can be transferred to classes, and for many rings A, these already form a group; this means simply that for each ideal $I \neq (0)$ there exists an ideal J such that IJ is a principal ideal. In this case we speak of the *ideal class group* of A; it is denoted by $\mathrm{Cl}\,A$. This happens for example in the case of the ring of algebraic integers of a finite extension $K/\mathbb{Q}$ (see §7). Here the group $\mathrm{Cl}\,A$ is finite. For the ring A of numbers of the form $a + b\sqrt{-5}$ with $a, b \in \mathbb{Z}$ (§3, Example 12), $\mathrm{Cl}\,A$ is of order 2. Thus all non-principal ideals are equivalent (that is, isomorphic as A-modules), and the product of any two of them is a principal ideal.

Example 2. The Group of Extensions $\mathrm{Ext}_R(A, B)$. Suppose that A and B are modules over a ring R. We say that a module C is an *extension* of A by B if C contains a submodule B_1 isomorphic to B, and C/B_1 is isomorphic to A; the isomorphisms $u: B \cong B_1$ and $v: C/B_1 \cong A$ are included in the notion of extension. The *trivial extension* is $C = A \oplus B$. The group $\mathbb{Z}/p^2\mathbb{Z}$ is a nontrivial extension of $\mathbb{Z}/p\mathbb{Z}$ by $\mathbb{Z}/p\mathbb{Z}$. In exactly the same way, a 2-dimensional vector space together with a linear transformation given in some basis by the matrix $\begin{bmatrix} \lambda & 1 \\ 0 & \lambda \end{bmatrix}$ (a 2×2 Jordan block with eigenvalue λ) defines a module C over the ring $K[x]$ which is a nontrivial extension of the module A by itself, where A corresponds to the 1×1 matrix λ.

We say that two extensions C and C' of A by B are *equivalent* if there exists an isomorphism $\varphi: C \cong C'$ taking $B_1 \subset C$ into $B_1' \subset C'$ and compatible with the isomorphisms $B_1 \cong B \cong B_1'$ and $C/B_1 \cong A \cong C'/B_1'$. The set of all extensions of A by B, considered up to equivalence, is denoted by $\mathrm{Ext}_R(A, B)$. For semisimple rings R it consists of the trivial extension only; in the general case it measures the failure of this typically semisimple situation.

We can make $\mathrm{Ext}_R(A, B)$ into a group. Let $C, C' \in \mathrm{Ext}_R(A, B)$, with $f: B \cong B_1 \subset C$ and $g: C/B_1 \cong A$ the isomorphisms in the definition of C, and f', g' the same for C'. Consider the submodules $D \subset C \oplus C'$ consisting of pairs (c, c') for which $g(c) = g'(c')$, and $E \subset C \oplus C'$ of pairs $(f(b), -f'(b))$ for $b \in B$; then set $C'' = D/E$. The homomorphism f'' given by $f''(b) = (f(b), 0) + E = (0, f'(b)) + E$ defines an isomorphism of B with a submodule $B_1'' \subset C''$; and g'' given by $g''(c, c') = g(c) = g'(c')$ for $(c, c') \in D$ defines a homomorphism of D onto A which is equal to 0 on E, that is, a homomorphism $g'': E'' \to A$, which one easily checks defines an isomorphism of C''/B_1'' with A. Thus C'' is an extension of A by B. The corresponding element $C'' \in \mathrm{Ext}_R(A, B)$ is called the *sum* of the extensions C and C'. It is not hard to check all the group axioms; the zero element is the trivial extension.

Example 3. The Brauer Group. We define a group law on the set of central division algebras of finite rank over a given field K (see §11). Let R_1 and R_2 be two simple central algebras of finite rank over K. We define a multiplication in the module $R_1 \otimes_K R_2$ by

$$(a_1 \otimes a_2)(b_1 \otimes b_2) = a_1 b_1 \otimes a_2 b_2$$

on generators; then $R_1 \otimes_K R_2$ becomes an algebra over K, and it is not hard to prove that it is also simple and central. For example $M_{n_1}(K) \otimes M_{n_2}(K) \cong M_{n_1 n_2}(K)$. If D_1 and D_2 are two central division algebras of finite rank over K, then from what we have said, by Wedderburn's theorem (§ 10, Theorem IV), $D_1 \otimes_K D_2 \cong M_n(D)$, where D is some central algebra; we say that D is the *product* of D_1 and D_2. It can be shown that this defines a group; the inverse element of a division algebra D is the opposite algebra D'. This group is called the *Brauer group* of K, and is denoted by $\operatorname{Br} K$. It can be shown that the description of division algebras over $\mathbb{Q}_p$ and $\mathbb{Q}$ given in § 11, Theorems V and VIII also give the group law of the Brauer group of these field: we need only view the roots of unity appearing there as elements of the group of roots of unity. For example, $\operatorname{Br} \mathbb{Q}_p$ is isomorphic to the group of all roots of unity, and $\operatorname{Br} \mathbb{Q}$ to the group of collections $(\mu_{\mathbb{R}}, \mu_p, \ldots)$ of roots of unity satisfying the conditions of § 11, Theorem VIII. The group $\operatorname{Br} \mathbb{R}$ has order 2.

The following sections §§ 13–15 of this book are devoted to a review of examples of the most commonly occuring types of groups and the most useful concrete groups. This gives a (necessarily very rough) survey of the different ways in which the general notion of group is related to the rest of mathematics (and not only to mathematics). But first of all, we must treat some of the simplest notions and properties which belong to the definition of group itself, and without which it would be awkward to consider the examples.

A *subgroup* of a group G is a subset which contains together with any element its inverse, and together with any two elements their product. An example of a subgroup is the stabiliser subgroup of any point in a transformation group. Let $\{g_\alpha\}$ be an arbitrary set of elements of G; the set of all products of the g_α and their inverses (taken in any order) is easily seen to form a subgroup of G. This is called the subgroup *generated* by $\{g_\alpha\}$. If this group is the whole of G then we say that the g_α are *generators* of G. For example, 1 is a generator of the group $\mathbb{Z}$ (written additively). In the symmetry group of the equilateral triangle, s and t are generators (see (5)).

A *homomorphism* of a group G to G' is a map $f\colon G \to G'$ such that

$$f(g_1 g_2) = f(g_1) f(g_2).$$

A homomorphism f of G to the transformation group of a set X is called an *action* of G on X. To specify an action, we need to define for every element $g \in G$ the corresponding transformation $f(g)$ of X, that is, $f(g)(x)$ for any $x \in X$. Writing $f(g)(x)$ in the short form $g \cdot x$, we see that giving an action of G on X is the same thing as assigning to any $g \in G$ and $x \in X$ an element $g \cdot x \in X$ (that is, a map $G \times X \to X$ given by $(g, x) \mapsto g \cdot x$)) satisfying the condition:

$$(g_1 g_2) \cdot x = g_1 \cdot (g_2 \cdot x)$$

(which is equivalent to f being a homomorphism from G to the group of all transformations of X).

We say that two actions of the same group G on two different sets X and X' are *isomorphic* if we can establish a 1-to-1 correspondence $x \leftrightarrow x'$ between elements of X and X' such that $x \leftrightarrow x'$ implies $g \cdot x \leftrightarrow g \cdot x'$ for any $g \in G$.

Example 4. Consider the group G of real 2×2 matrixes with determinant 1. Such a matrix $g = \begin{bmatrix} \alpha & \beta \\ \gamma & \delta \end{bmatrix}$ acts on the upper half-plane $\mathbb{C}^+ = \{z | \operatorname{Im} z > 0\}$ of one complex variable by the rule $z \mapsto \dfrac{\alpha z + \gamma}{\beta z + \delta}$. We obviously get an action of G on $\mathbb{C}^+$. On the other hand, G acts on the set S of 2×2 positive definite symmetric matrixes, with g taking s into gsg^*, where g^* is the transpose matrix. If we write s as $\begin{bmatrix} a & b \\ b & c \end{bmatrix}$, we can characterise S by the conditions $a > 0$, $ac - b^2 > 0$, which in the projective plane with homogeneous coordinates $(a : b : c)$ define the interior $\mathfrak{K}$ of the conic with equation $ac = b^2$. Obviously G also acts on $\mathfrak{K}$ by projective transformations taking this conic to itself. A positive definite quadratic form $ax^2 + 2bxy + cy^2$ can be written in the form $a(x - zy)(x - \bar{z}y)$ with $z \in \mathbb{C}^+$. It is easy to see that taking the matrix $\begin{bmatrix} a & b \\ b & c \end{bmatrix}$ to z, we get a 1-to-1 correspondence between the sets $\mathfrak{K}$ and $\mathbb{C}^+$, defining an isomorphism of the two actions of G. As is well known, $\mathbb{C}^+$ and $\mathfrak{K}$ define two interpretations of Lobachevsky plane geometry, the *Poincaré* and *Cayley-Klein models*. In either of these interpretations, G defines the group of proper (orientation-preserving) motions of the Lobachevsky plane.

There are three very important examples of an action of a group on itself:

$$g \cdot x = gx,$$

$$g \cdot x = xg^{-1},$$

$$g \cdot x = gxg^{-1}$$

(the left-hand sides are the actions, the right-hand sides are in terms of multiplication in G). These are called the *left regular*, the *right regular* and the *adjoint action.* The left regular and right regular actions are isomorphic: an isomorphism is given by the map $x \mapsto x^{-1}$ of G to itself.

An action of a group G on a set X defines of course an action of any subgroup $H \subset G$. In particular, the left regular action defines an action of any subgroup $H \subset G$ on the whole of G. The orbits of the transformation group obtained in this way are called the *left cosets* of G under H. Thus a left coset consists of all elements of the form hg, where $g \in G$ is some fixed element, and h runs through all possible elements of H; it is denoted by Hg. By what we have said above concerning orbits, any element $g_1 \in Hg$ defines the same coset, and the whole group decomposes as a disjoint union of cosets. Similarly, the right regular action of a subgroup $H \subset G$ defines the *right cosets*, of the form gH. The orbits of the adjoint representation are called the *conjugacy classes* of elements, and elements belonging to one orbit are *conjugate*; g_1 and g_2 are conjugate if $g_2 = gg_1g^{-1}$ for some $g \in G$. If $H \subset G$ is a subgroup then for fixed $g \in G$, all the elements ghg^{-1} with $h \in H$ are easily seen to form a subgroup; this is called a *conjugate* subgroup of H, and is denoted by gHg^{-1}. For example, if G is a transformation group of a set X and $g \in G$, $x \in X$ and $y = gx$, then the stabilisers of x and y are conjugate: $G_y = gG_xg^{-1}$. The number of left cosets of a subgroup $H \subset G$ (finite or otherwise) is called the *index* of H, and denoted by $(G : H)$. If G is a finite group, then the number of elements in each coset of H is equal to the order $|H|$, and hence

$$|G| = |H| \cdot (G : H). \tag{9}$$

In particular, $|H|$ divides $|G|$.

Suppose that the action of a group G on a set X is transitive. Then for any x, $y \in X$ there exists $g \in G$ such that $g(x) = y$, and all such elements form a right coset gG_x under the stabiliser of x. We get in this way a 1-to-1 correspondence between elements of X and cosets of G_x in G. If X is finite and we write $|X|$ for the number of elements of X, then we see that

$$|X| = (G : G_x).$$

If the action of G is not transitive, let X_α be its orbits; then $X = \bigcup X_\alpha$. Choosing a representative x_α in each orbit X_α we get a 1-to-1 correspondence between the elements of the orbit and cosets of G_{x_α} in G. In particular, if X is finite, then

$$|X| = \sum (G : G_\alpha), \tag{10}$$

where G_α are the stabilisers of the representatives chosen in each orbit.

The *image* $\operatorname{Im} f$ of a homomorphism $f\colon G \to G'$ is the set of all elements of the form $f(g)$ with $g \in G$; it is a subgroup of G'. The *kernel* $\operatorname{Ker} f$ of f is the set of elements $g \in G$ such that $f(g) = e$. The kernel is of course a subgroup, but satisfies an additional condition:

$$g^{-1}hg \in \operatorname{Ker} f \quad \text{if} \quad h \in \operatorname{Ker} f \text{ and } g \in G. \tag{11}$$

The verification of this is trivial. A subgroup $N \subset G$ satisfying condition (11) is called a *normal subgroup*.[3] In other words, a normal subgroup must be invariant under the adjoint action, $g^{-1}Ng = N$. We write $N \lhd G$ to denote the fact that N is a normal subgroup of G. The definition of a normal subgroup is equivalent to the fact that any left coset of N is also a right coset, $gN = Ng$. Thus, although the left and right regular actions of a normal subgroup on the group will not in general be the same, they have the same orbits.

The decomposition of G into cosets of a normal subgroup N is compatible with multiplication: if we replace g_1 or g_2 by an element in the same coset then

[3] A literal translation of the Russian term, derived from the German *Normalteiler*, would be *normal divisor*, containing the idea of a factor in a product (compare § 16).

g_1g_2 also remains in the same coset. This is a tautological reformulation of the definition of a normal subgroup. It follows from this that multiplication can be carried over to cosets, which then form a group, called the *quotient group* by N. This is denoted by G/N, and taking an element $g \in G$ into the coset gN defines a homomorphism $G \to G/N$, called the *canonical homomorphism*.

We have relations which are already familiar from the theory of rings and modules.

Homomorphisms Theorem. *The image of a homomorphism is isomorphic to the quotient group by its kernel.*

Any homomorphism f reduces to a canonical one: there exists an isomorphism of $G/\mathrm{Ker}\, f$ and $\mathrm{Im}\, f$ such that composing with the canonical homomorphism $G \to G/\mathrm{Ker}\, f$ gives f.

Example 5. Consider the group G of all motions of the Euclidean plane E, that is, of all transformations preserving the distance between points. This action can be extended to the vector space L of all free vectors of E: if x, $y \in E$ and $\overrightarrow{xy}$ denotes the vector with origin x and end point y then $\tilde{g}(\overrightarrow{xy})$ is by definition the vector $\overrightarrow{g(x)g(y)}$. It is easy to check that if $\overrightarrow{xy} = \overrightarrow{x_1y_1}$ (that is, xy and x_1y_1 are equal and parallel) then also $\overrightarrow{g(x)g(y)} = \overrightarrow{g(x_1)g(y_1)}$, so that the transformation $\tilde{g}$ is defined without ambiguity. In L the transformation $\tilde{g}$ is orthogonal. The map $g \mapsto \tilde{g}$ is a homomorphism from the group of motions to the group of orthogonal transformations. The image of this homomorphism is the whole of the group of orthogonal transformations. Its kernel consists of translations of E in vectors $u \in L$. Thus the translation group is a normal subgroup. This can also be verified directly: if T_u is a translation in a vector u, then one sees easily that $gT_ug^{-1} = T_{g(u)}$. We see that the group of orthogonal transformations is isomorphic to the quotient group of the group of motions by the normal subgroup of translations. In the present case this is easy to see directly, by choosing some point $O \in E$. Then any motion can be written in the form $g = T_u \circ g'$, where $g' \in G_O$ (the stabiliser of the point O). Obviously, G_O is isomorphic to the group of orthogonal transformations, and taking the coset T_ug' into g' gives our homomorphism.

Let g be an element of G. The map $\varphi_g\colon \mathbb{Z} \to G$ given by $\varphi_g(n) = g^n$ is obviously a homomorphism; its image consists of all powers of g, and is called the *cyclic subgroup* generated by g, and is denoted by $\langle g \rangle$. If there exists an element g such that $G = \langle g \rangle$ (that is, all elements of G are powers of g), then we say that G is *cyclic*, and g is a *generator*. An example of a cyclic group is the group $\mathbb{Z}$ of integers under addition; it has the generator 1. Every subgroup of $\mathbb{Z}$ consists either of 0 only, or is the subgroup $k\mathbb{Z}$ of all multiples in $\mathbb{Z}$ of the smallest positive number k contained in it, that is, it is also cyclic. Returning to the homomorphism $\varphi_g\colon \mathbb{Z} \to G$, we can say that either $\mathrm{Ker}\, \varphi_g = 0$ (which means that all the powers g^n of g are distinct), or $\mathrm{Ker}\, \varphi_g = k\mathbb{Z}$; this means that $g^k = e$, and $\langle g \rangle$ is isomorphic to the cyclic group $\mathbb{Z}/k\mathbb{Z}$ of order k. In the first case we say that g is an element of *infinite order*, and in the second, of *order* k. If G is a finite group then by (9), k divides $|G|$.

Of all the methods of constructing groups, we mention here the simplest. Suppose that G_1 and G_2 are two groups. Consider the set of all pairs (g_1, g_2) with $g_1 \in G_1$ and $g_2 \in G_2$. The operation on pairs is defined element-by-element: $(g_1, g_2)(g_1', g_2') = (g_1 g_1', g_2 g_2')$. It is easy to see that we get in this way a group. It is called the *direct product* of G_1 and G_2 and denoted by $G_1 \times G_2$. If e_1 and e_2 are the identity elements of G_1 and G_2 then the maps $g_1 \mapsto (g_1, e_2)$ and $g_2 \mapsto (e_1, g_2)$ are isomorphisms of G_1 and G_2 with subgroups of $G_1 \times G_2$. Usually we identify elements of G_1 and G_2 with their images under these isomorphisms, that is, we write g_1 for (g_1, e_2) and g_2 for (e_1, g_2). Then G_1 and G_2 are subgroups of $G_1 \times G_2$; in $G_1 \times G_2$ we have $g_2 g_1 = g_1 g_2$ if $g_1 \in G_1$ and $g_2 \in G_2$. If G_1 and G_2 are Abelian (and written additively) then we obtain the operation of direct sum of $\mathbb{Z}$-modules which we know from § 5, Example 5.

§ 13. Examples of Groups: Finite Groups

In the same way that the general notion of a group relates to transformation groups of an arbitrary set, finite groups relate to groups of transformations of a finite set; in this case, transformations are also called *permutations.*

Example 1. The group of all transformations of a finite set X consisting of n elements $x_1, \ldots, x_n$ is called the *symmetric group* on $x_1, \ldots, x_n$; it is denoted by $\mathfrak{S}_n$. The stabiliser $(\mathfrak{S}_n)_{x_1}$ of x_1 is obviously isomorphic to $\mathfrak{S}_{n-1}$. Each coset $g(\mathfrak{S}_n)_{x_1}$ consists of all permutations that take x_1 into a given element x_i. Therefore the number of cosets is equal to n; hence $|\mathfrak{S}_n| = n \cdot |\mathfrak{S}_{n-1}|$, and by induction $|\mathfrak{S}_n| = n!$.

For $i = 1, \ldots, n-1$, write σ_i for the transformation which transposes x_i and x_{i+1} and leaves fixed the remaining elements. Obviously

$$\sigma_i^2 = e. \tag{1}$$

Since any permutation of $x_1, \ldots, x_n$ can be realised by successively transposing neighbouring elements, any element $g \in \mathfrak{S}_n$ is a product of $\sigma_1, \ldots, \sigma_{n-1}$, that is, $\sigma_1, \ldots, \sigma_{n-1}$ are generators of $\mathfrak{S}_n$. Obviously

$$\sigma_i \sigma_j = \sigma_j \sigma_i \quad \text{if} \quad |i - j| > 1, \tag{2}$$

since then σ_i and σ_j transpose disjoint pairs of elements. The product $\sigma_i \sigma_{i+1}$ cyclically permutes the 3 elements x_i, x_{i+1} and x_{i+2}, and therefore

$$(\sigma_i \sigma_{i+1})^3 = e \quad \text{for} \quad 1 \leqslant i \leqslant n-2. \tag{3}$$

Fact. *It can be shown that the multiplication in $\mathfrak{S}_n$ is entirely determined by the relations* (1), (2) *and* (3) *between the generators $\sigma_1, \ldots, \sigma_{n-1}$.*

The precise meaning of this assertion will be made clear in § 14.

Let $\sigma \in \mathfrak{S}_n$ be an arbitrary permutation, and $H = \langle \sigma \rangle$ the cyclic subgroup generated by σ. Under the action of H, the set X breaks up into k orbits $X_1, \ldots,$

X_k; write $n_1, \ldots, n_k$ for the number of elements in these. Obviously

$$n = n_1 + \cdots + n_k. \tag{4}$$

The group $H = \langle\sigma\rangle$ cyclically permutes the elements within each orbit X_i. Specifying the partition $X = \bigcup X_i$ and the cyclic permutation of the elements within each X_i (for example, writing $X_i = \{x_{\alpha_1}, x_{\alpha_2}, \ldots, x_{\alpha_{n_i}}\}$ where $\sigma x_{\alpha_j} = x_{\alpha_{j+1}}$ for $j \leqslant n_i - 1$ and $\sigma x_{\alpha_{n_i}} = x_{\alpha_1}$) uniquely determines the permutation σ. This data is called the decomposition of σ into cycles. The numbers $n_1, \ldots, n_k$ are called the *cycle type* of the permutation.

If $\sigma' = g\sigma g^{-1}$ is a conjugate element then its decomposition into cycles is of the form $X = \bigcup gX_i$ where $gX_i = \{gx_{\alpha_1}, \ldots, gx_{\alpha_{n_i}}\}$, and so the numbers $n_1, \ldots, n_k$ remain unchanged. Conversely, if σ' is any permutation of the same cycle type $n_1, \ldots, n_k$, then it is easy to construct a permutation g for which $\sigma' = g^{-1}\sigma g$. Thus two permutations are conjugate if and only if they have the same cycle type. In other words, the conjugacy classes of elements of $\mathfrak{S}_n$ are in 1-to-1 correspondence with collections of natural numbers $n_1, \ldots, n_k$ satisfying (4); in particular, the number of conjugacy classes of elements of $\mathfrak{S}_n$ is equal to the number of partitions of n as a sum of positive integers.

Example 2. Suppose now that the elements $x_1, \ldots, x_n$ are independent variables in the ring $\mathbb{Z}[x_1, \ldots, x_n]$, and consider the polynomial

$$\Delta = \prod_{i<j} (x_i - x_j).$$

It is obvious that under a permutation σ, the polynomial Δ is either unchanged, or changes sign:

$$\Delta(\sigma(x_1), \ldots, \sigma(x_n)) = \varepsilon(\sigma)\Delta(x_1, \ldots, x_n), \quad \text{with} \quad \varepsilon(\sigma) = \pm 1;$$

and $\sigma \mapsto \varepsilon(\sigma)$ is a homomorphism of $\mathfrak{S}_n$ into the group $\{\pm 1\}$ of order 2. The kernel of this group is called the *alternating group*, and is denoted by $\mathfrak{A}_n$; a permutation is *even* if $\sigma \in \mathfrak{A}_n$ and *odd* if $\sigma \notin \mathfrak{A}_n$. Obviously the index of $\mathfrak{A}_n$ in $\mathfrak{S}_n$ equals $(\mathfrak{S}_n : \mathfrak{A}_n) = 2$, and therefore $|\mathfrak{A}_n| = n!/2$.

In many questions it is important to have a list of all normal subgroups of the groups $\mathfrak{S}_n$ and $\mathfrak{A}_n$. The answer is as follows.

Theorem I. *For $n \neq 4$ the group $\mathfrak{S}_n$ has no normal subgroups other than $\{e\}$, $\mathfrak{A}_n$ and $\mathfrak{S}_n$; and $\mathfrak{A}_n$ has none other than $\{e\}$ and $\mathfrak{A}_n$. When $n = 4$ there exists in addition a normal subgroup (in both $\mathfrak{S}_n$ and $\mathfrak{A}_n$) of order 4, consisting of e and of all permutations of cycle type $(2, 2)$.*

Another way in which important examples of finite groups arise is in the study of finite subgroups of certain well-known groups. Let us treat a classical example, the finite subgroups of the group of orthogonal transformations of Euclidean space.

The groups of most interest from the geometric and physical points of view are the groups acting on 3-dimensional space. But to have a simple model, we treat first the analogous results for the plane. Orthogonal transformations that preserve the orientation will be called rotations; these also form a group.

Example 3. Finite Groups of Rotations of the Plane.

Theorem II. *Finite groups of rotations of the plane are cyclic; any such group of order n consists of all rotations about a fixed point through angles of* $\frac{2k\pi}{n}$ *for* $k = 0, 1, \ldots, n-1$.

We denote this group by C_n. It can be characterised as the group of all symmetries of an oriented regular n-gon (see Figure 18 for the case $n = 7$).

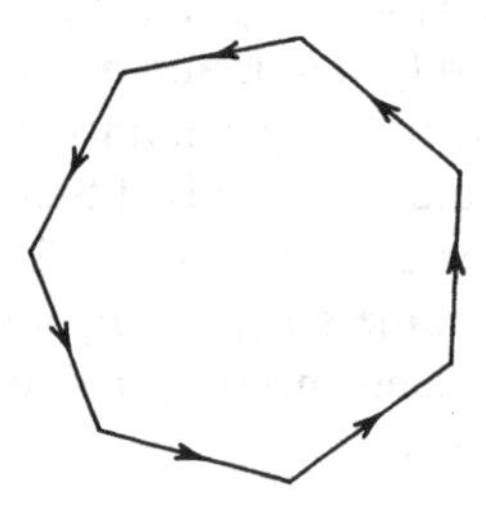

Fig. 18

Theorem III. *A finite subgroup of the group of orthogonal transformations of the plane containing reflections is the group of all symmetries of a regular n-gon; this group is denoted by* D_n. *It has order 2n, and consists of the transformations in* C_n *and n reflections in the n axes of symmetry of the regular n-gon.*

The cases $n = 1$ *and 2 are somewhat exceptional:* D_1 *is the group of order 2 generated by a single reflection, and* D_2 *the group of order 4 generated by reflections in the x- and y-axes.*

Example 4. The classical examples of finite groups of rotations of 3-space relate to regular polyhedrons: each regular polyhedron M has an associated group G_M, consisting of all rotations preserving M. The regularity of a polyhedron is reflected in the presence of a large number of symmetries. Suppose that M is a convex bounded n-dimensional polyhedron in n-space. Define a *flag* of M to be a set $F = \{M_0, M_1, \ldots, M_{n-1}\}$, where M_i is an i-dimensional face of M and $M_i \subset M_{i+1}$. We say that a polyhedron M is *regular* if its symmetry group G_M acts transitively on the set of all flags of M. In particular, for $n = 3$ the group G_M should act transitively on the set of pairs consisting of a vertex and an edge out of it. It is easy to see that the stabiliser of such a pair consists only of the identity transformation, so that the order of G_M equals the product of the number of vertexes of the regular polyhedron by the number of edges out of any vertex. All the regular polyhedrons were determined in antiquity (and they are sometimes called the *Platonic solids*). They are the *tetrahedron, cube, octahedron, dodecahedron* and *icosahedron*. To each regular polyhedron there is a related dual polyhedron,

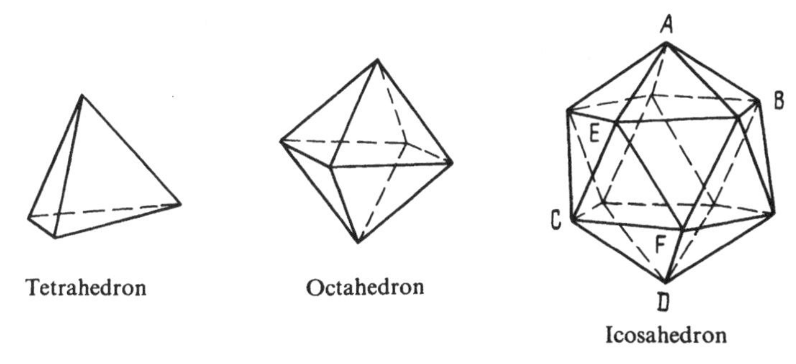

Fig. 19

whose vertexes are the midpoints of the faces of the first; obviously both of these have the same group G_M. The tetrahedron is self-dual, the cube is dual to the octahedron, and the dodecahedron is dual to the icosahedron (see Figure 19). The corresponding groups are denoted by T, O, Y respectively; from what was said above, their orders are given by

$$|T| = 12, \quad |O| = 24, \quad |Y| = 60.$$

In addition to these groups, there exist other obvious example of finite subgroups of the rotation group: the cyclic group C_n of order n, consisting of rotations around an axis l through angles of $\frac{2k\pi}{n}$ for $k = 0, 1, \ldots, n-1$, and the dihedral group D_n of order $2n$, which contains C_n and in addition n rotations through an angle of π around axes lying in a plane orthogonal to l, meeting l and making angles with one another multiples of $\frac{2\pi}{n}$. The group D_n can be viewed as the group of rotations of a degenerated regular polyhedron, a plane n-gon (Figure 20); in this form we have already met it in Example 3.

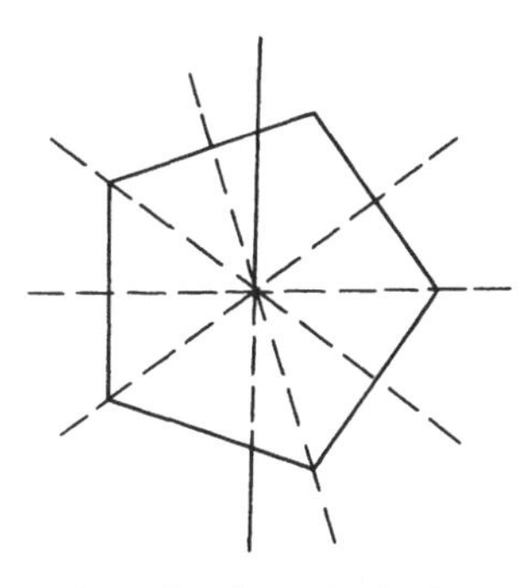

The axes of rotation through π in the group D_5

Fig. 20

Theorem IV. *The cyclic and dihedral groups and the groups of the tetrahedron, octahedron and icosahedron are all the finite subgroups of the group of rotations of 3-space.*

The precise meaning of this assertion is that every such group G is either cyclic, or there exists a regular polyhedron M such that $G = G_M$. Since all regular polyhedrons of one type can be obtained from one another by a rotation and a uniform dilation, it follows from this that the corresponding groups are conjugate subgroups of the group of rotations.

The group of rotations of the regular tetrahedron acts on the set of vertexes. It only realises even permutations of them; this becomes obvious if we write the volume of the tetrahedron in the form

$$V = \frac{1}{6} \det \begin{vmatrix} 1 & 1 & 1 & 1 \\ x_1 & x_2 & x_3 & x_4 \\ y_1 & y_2 & y_3 & y_4 \\ z_1 & z_2 & z_3 & z_4 \end{vmatrix},$$

where (x_i, y_i, z_i) are the coordinates of its vertexes. From this it is clear that the tetrahedral group T is isomorphic to the alternating group $\mathfrak{A}_4$. Algebraically, the action of T on the set of vertexes corresponds to its action by conjugacy on the set of subgroups of order 3 (each such subgroup consists of the rotations about an axis joining a vertex to the centre of the opposite side).

Entirely similarly, subgroups of order 3 of the octahedral group O correspond to axes joining the centres of opposite faces of the octahedron. There are 4 such axes, hence 4 such subgroups, and the action of O on them defines an isomorphism $O \cong \mathfrak{S}_4$.

In the icosahedral group Y, consider first the elements of order 2; these are given by rotations through 180° about axes joining the midpoints of opposite edges. Since the number of edges is 30, the number of such axes is 15, and hence there are 15 elements of order 2. It can be shown that for each axis of order 2 there exist another two orthogonal to it and to one another (for example the axis joining the midpoints of AB and CD in Figure 19, and the two others obtained by rotating this about an axis of order 3 passing through the centre of the triangle CFE). The 3 elements of order 2 defined by these axes, together with the identity element, form an Abelian group of order 4 isomorphic to $\mathbb{Z}/2 \times \mathbb{Z}/2$; therefore the icosahedral group Y contains 5 such subgroups. The action of the groups Y by conjugacy on these 5 subgroups (or on the 5 triples of mutually orthogonal axes of order 2) defines an isomorphism $Y \cong \mathfrak{A}_5$.

One important relation between these groups often plays an important role. From the isomorphisms $O \cong \mathfrak{S}_4$ and $T \cong \mathfrak{A}_4$ and from Theorem I it follows that the group O contains a unique normal subgroup of index 2 isomorphic to T. We can see this inclusion if we truncate the angles of a regular tetrahedron in such a way that the plane sections divide its sides in half (Figure 21). What is left is a regular octahedron; all the symmetries of the tetrahedron form the group T and preserve the inscribed octahedron, and hence are contained in O.

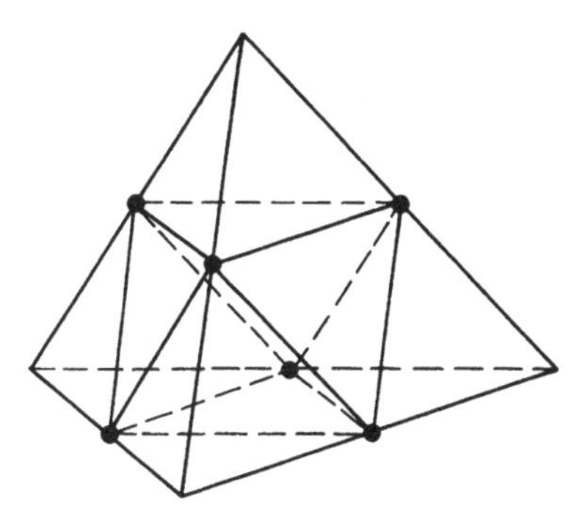

Fig. 21

The groups of regular polyhedrons occur in nature as the symmetry groups of molecules. For example (Figure 22), the molecule $H_3C\text{–}CCl_3$ has symmetry group C_3, the ethane molecule C_2H_6 symmetry group D_3, the methane molecule CH_4 the tetrahedral group T (the atom C is at the centre of the tetrahedron, and the atoms H at the vertexes), and uranium hexafluoride UF_6 the octahedral group O (the atom U is at the centre of the octahedron, and the atoms F at the vertexes).

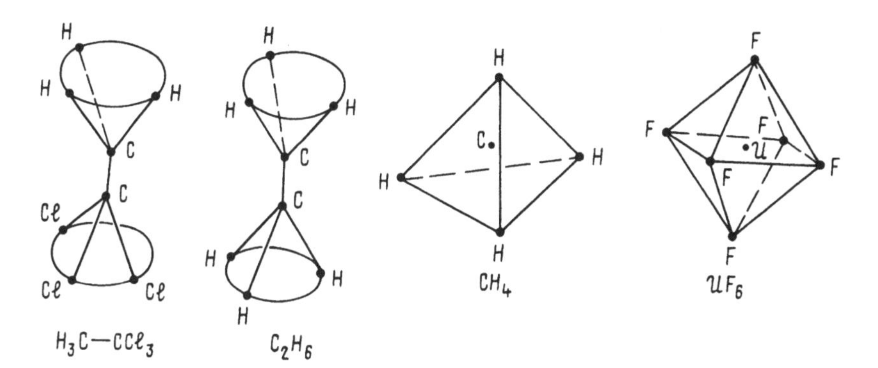

Fig. 22

A classification of finite subgroups of the group of all orthogonal transformations follows easily from Theorem IV. The group Γ' of all orthogonal transformations is the direct product $\Gamma \times \{e, e'\}$ where Γ is the group of rotations of 3-space, and $Z = \{e, e'\}$ is the group of order 2 consisting of the identity transformation and the central reflection e' (with $e'(x) = -x$). It is a not too difficult algebraic problem to investigate the subgroups of a direct product $\Gamma \times H$ if we know all the subgroups of Γ and of H. In the simplest case, when as in our case $H = Z$ is a group of order 2, the answer is as follows. A subgroup $G \subset \Gamma \times Z$ is either entirely contained in Γ, or is of the form $G_0 \times Z$ where $G_0 \subset \Gamma$, or is obtained from a subgroup $\bar{G} \subset \Gamma$ by the following trick: choose a subgroup $G_0 \subset \bar{G}$ of index 2 in $\bar{G}$, and let $\bar{G} = G_0 \cup V$ be its decomposition into cosets under G_0; it is easy to check that the set of elements $g_0 \in G_0$ and $e'v$ with $v \in V$ then forms a group, which we must take for G. For example, in the 2-dimensional case, the group D_n is obtained in this way from $\bar{G} = C_{2n}$ and $G_0 = C_n$. The construction of groups of this last type requires a listing of all rotation groups $\bar{G}$ and of their subgroups G_0 of index 2. The corresponding group of orthogonal transformations G is denoted by $\bar{G}G_0$. We thus obtain the following result.

Theorem V. *The finite groups of orthogonal transformations of 3-dimensional space not consisting only of rotations are the following:*

$$C_n \times Z, D_n \times Z, T \times Z, O \times Z, Y \times Z, C_{2n}C_n, D_{2n}D_n, D_nC_n, OT.$$

The final group arises because of the inclusion $T \subset O$ illustrated in Figure 21.

Recall that for any finite group $G \subset \mathrm{GL}(n, \mathbb{R})$ there exists a positive definite quadratic form f on $\mathbb{R}^n$ invariant under G(§ 10, Example 4). From the fact that the form f can be reduced to a sum of squares by a nondegenerate linear transformation φ, it follows, as we can see easily, that the group $\varphi^{-1}G\varphi$ consists of orthogonal transformations. Hence Theorems III and V give us also a classification of finite subgroups of $\mathrm{GL}(2, \mathbb{R})$ and $\mathrm{GL}(3, \mathbb{R})$.

Example 5. Every finite group of rotations of 3-space preserves the sphere S centred at the origin, and hence can also be interpreted as a group of motions of spherical geometry. If we identity the sphere with the Riemann sphere of a complex variable z, then the fractional-linear transformations

$$z \mapsto \frac{\alpha z + \beta}{\gamma z + \delta} \quad \text{with} \quad \alpha, \beta, \gamma, \delta \in \mathbb{C} \quad \text{and} \quad \alpha\delta - \beta\gamma \neq 0, \tag{5}$$

are realised as conformal transformations of the sphere S. The motions of S form part of the conformal transformations, and hence the finite groups of motion of the sphere we have constructed provide finite subgroups of the group of fractional-linear transformations.

Theorem VI. *All finite subgroups of the group of fractional-linear transformations* (5) *are obtained in this way. It can be shown moreover that the subgroups corresponding to regular polyhedrons of the same type are conjugate in the group of fractional-linear transformations.*

One of the applications of this result is as follows. Let

$$\frac{d^2w}{dz^2} + p(z)\frac{dw}{dz} + q(z)w = 0$$

be a differential equation with rational functions $p(z)$, $q(z)$ as coefficients, and

w_1, w_2 two linearly independent solutions. The function $v = w_2/w_1$ is a many-valued analytic function of the complex variable z. Moving in the z-plane around the poles of $p(z)$ and $q(z)$ replaces the functions w_1 and w_2 by linear combinations, so that v transforms according to formula (5), that is $v \mapsto \dfrac{\alpha v + \beta}{\gamma v + \delta}$. Suppose now that v is an algebraic function. Then v has only a finite number of sheets, and hence we obtain a finite group of transformations of the form (5). Since we know all such groups, we can describe all the second order linear differential equations having algebraic solutions.

Example 6. The Symmetry Group of Plane Lattices. The ring of integers $\mathbb{Z}$ is a discrete analogue of the real number field $\mathbb{R}$, the module $\mathbb{Z}^n$ a discrete analogue of the vector space $\mathbb{R}^n$ and the group $\mathrm{GL}(n, \mathbb{Z})$ a discrete analogue of $\mathrm{GL}(n, \mathbb{R})$. Following this analogy, we now study the finite subgroups of $\mathrm{GL}(2, \mathbb{Z})$, and in the following example, of $\mathrm{GL}(3, \mathbb{Z})$. We will be interested in classifying these groups up to conjugacy in the groups $\mathrm{GL}(2, \mathbb{Z})$ and $\mathrm{GL}(3, \mathbb{Z})$ containing them. In the following section § 14 we will see that this problem has physical applications in crystallography.

The question we are considering can be given the following geometric interpretation. We think of $\mathbb{Z}^n$ as a group C of vectors of the n-dimensional space $\mathbb{R}^n$; a group $C \subset \mathbb{R}^n$ of this form is called a *lattice*. Any group $G \subset \mathrm{GL}(n, \mathbb{Z})$ is realised as a group of linear transformations of $\mathbb{R}^n$ preserving a lattice C. For any finite group G of linear transformations of $\mathbb{R}^n$ there exists an invariant metric, that is, a positive definite quadratic form $f(x)$ on $\mathbb{R}^n$ such that $f(g(x)) = f(x)$ for all $g \in G$ (see § 10, Example 4). A quadratic form defines on $\mathbb{R}^n$ the structure of a Euclidean space, and the group G becomes a finite group of orthogonal transformations taking C into itself. Our problem then is equivalent to classifying the symmetry groups of lattices in Euclidean space $\mathbb{R}^n$. By a symmetry we mean of course an orthogonal transformation taking the lattice into itself. It is easy to see that the group of all symmetries of any lattice is finite. The groups G_1 and G_2 of symmetries of lattices C_1 and C_2 will correspond to conjugate subgroups in the group of integral matrixes with determinant ± 1 if there exists a linear transformation γ which takes C_1 into C_2 and G_1 into G_2. That is, $C_2 = \varphi(C_1)$ and $G_2 = \varphi G_1 \varphi^{-1}$, and φ takes the action of G_1 on C_1 into the action of G_2 on C_2. We will say that such lattices are *equivalent*.

Lattices having nontrivial symmetries (other than the central reflection) are called *Bravais lattices*, and their symmetry groups *Bravais groups*.

We now study this problem in the case of plane lattices. Our investigation breaks up into two stages. First of all, we must determine which finite groups of orthogonal transformations preserve some lattice (that is, consist of symmetries of it). For reasons which will become clear in the next section, these groups are called *crystal classes* (or crystallographic classes). They are of course contained in the list given by Theorem III. The basic tool in sorting out which ones occur is the following assertion, which is elementary to prove: a plane lattice can only go into itself under a rotation about one of its points through an angle of 0, π, $\frac{2\pi}{3}, \frac{\pi}{2}$ or $\frac{\pi}{3}$.

Theorem VII. *There are* 10 2-*dimensional crystal classes,*

$$C_1, C_2, C_3, C_4, C_6, D_1, D_2, D_3, D_4, D_6.$$

Figure 23 illustrates the fundamental parallelograms of lattices having the various symmetry groups; under each, we indicate the symmetry groups which the corresponding lattices admit. We have (from left to right): an arbitrary parallelogram, an arbitrary rectangle, an arbitrary rhombus, a square and a parallelogram composed of two equilateral triangles. The corresponding lattices are called: general, rectangular, rhombic, square and hexagonal, and are denoted by $\Gamma_{\text{gen}}, \Gamma_{\text{rect}}, \Gamma_{\text{rhomb}}, \Gamma_{\text{sq}}$ and Γ_{hex}.

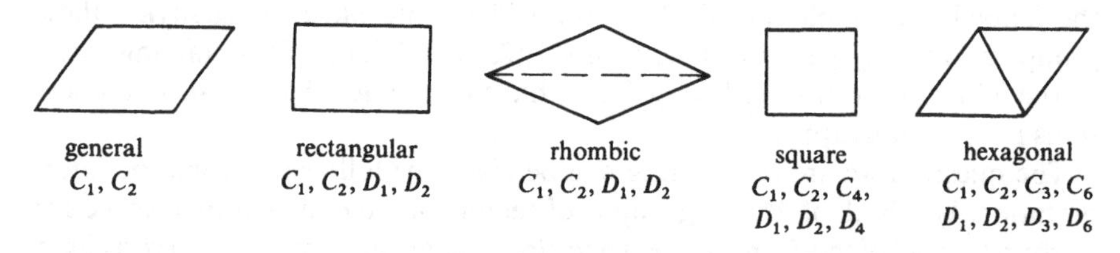

Fig. 23

However, Theorem VII does not quite solve our problem. Inequivalent symmetry groups may belong to the same crystal class. Algebraically, this means that two groups G and $G' \subset \mathrm{GL}(2, \mathbb{Z})$ may be conjugate in the group of orthogonal transformations, but not in $\mathrm{GL}(2, \mathbb{Z})$. As an example we have the groups G and G' of order 2, where G is generated by the matrix $\begin{bmatrix} 1 & 0 \\ 0 & -1 \end{bmatrix}$ and G' by $\begin{bmatrix} 0 & 1 \\ 1 & 0 \end{bmatrix}$. Geometrically these correspond to symmetries of order 2 of lattices whose fundamental parallelogram are the rectangle and the rhombus in Figure 23. The symmetry consists in the first case of a reflection in the horizontal side of the rectangle, and in the second of a reflection in the horizontal diagonal of the rhombus. They are inequivalent since the lattice corresponding to the rectangle has a basis of vectors which the symmetry multiplies by 1 and -1, whereas that corresponding to the rhombus is not generated by vectors invariant under the symmetry and vectors multiplied by -1 by the symmetry. However, this phenomenon does not occur very often.

Theorem VIII. *There are* 13 *inequivalent symmetry groups of plane lattices.*

$$C_1(\Gamma_{\text{gen}}), C_2(\Gamma_{\text{gen}}), C_4(\Gamma_{\text{sq}}), C_3(\Gamma_{\text{hex}}), C_6(\Gamma_{\text{hex}}),$$

$$D_1(\Gamma_{\text{rect}}), D_1(\Gamma_{\text{rhomb}}), D_2(\Gamma_{\text{rect}}), D_2(\Gamma_{\text{rhomb}}),$$

$$D_4(\Gamma_{\text{sq}}), D_3'(\Gamma_{\text{hex}}), D_3''(\Gamma_{\text{hex}}), D_6(\Gamma_{\text{hex}}).$$

We have indicated in brackets the lattices on which the given groups are realised as symmetry groups.

We have already treated the example of the group D_1 realised on two lattices Γ_{rect} and Γ_{rhomb}. The group D_2 is realised on the same lattices, and is obtained by adding to D_1 the central reflection. The most delicate are the realisations D_3' and D_3'' of D_3 as a symmetry group of Γ_{hex}. Both of them are contained in D_6, and are symmetry groups of equilateral triangles (as one expects of D_3), but these triangles are inscribed in different ways in the hexagon preserved by D_6 (Figure 24).

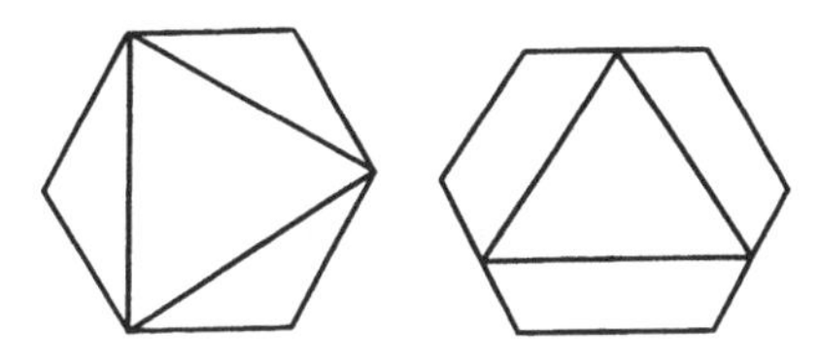

Fig. 24

Example 7. The Symmetry Groups of Space Lattices. We consider finite subgroups of $\mathrm{GL}(3, \mathbb{Z})$, using without further explanation the terminology introduced in Example 6.

Theorem IX. *There are* 32 3-*dimensional crystal classes*:

Crystal system	Crystal classes
Triclinic	$C_1 \times Z, C_1$
Monoclinic	$C_2 \times Z, C_2, C_2C_1$
Orthorhombic	$D_2 \times Z, D_2, D_2C_2$
Trigonal	$D_3 \times Z, D_3, D_3C_3, C_3 \times Z, C_3$
Tetragonal	$D_4 \times Z, D_4, D_4C_4, D_4D_2, C_4Z, C_4C_2$
Hexagonal	$D_6 \times Z, D_6, D_6C_6, D_6C_3, C_6Z, C_6, C_6C_3$
Cubic	$O \times Z, O, T, OT, T \times Z, T$

The notation for the groups is taken from Theorem V. The groups are arranged in the table so that each row consists of the subgroups of the first group in the row.

The series of groups are called *crystal classes* (or crystallographic classes) in crystallography, and have the exotic names given in the table. Each crystal class is characterised as a set of symmetries of some polyhedron. These polyhedrons are listed in Figure 25. (Their analogues in the plane are the parallelogram, rectangle, rhombus, square, and equilateral triangle.)

The crystal classes can be represented in a very intuitive way, as in Figure 26, Table 1, taken from the book [Delone, Padurov and Aleksandrov **32** (1934)]; the notation used in Table 1 is given below in Table 2.

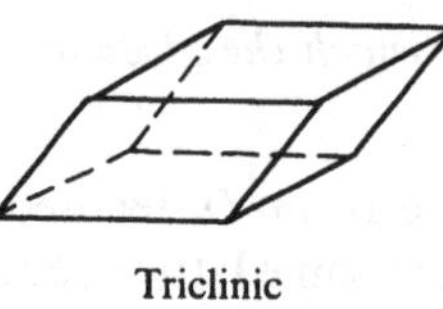

Triclinic
(arbitrary parallelipiped)

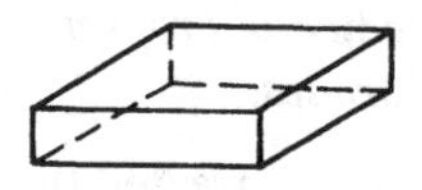

Monoclinic
(rectangular prism)

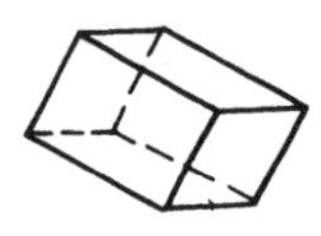

Orthorhombic (arbitrary rectangular parallelipiped)

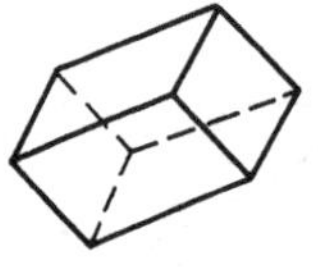

Trigonal (cube compressed along a space diagonal)

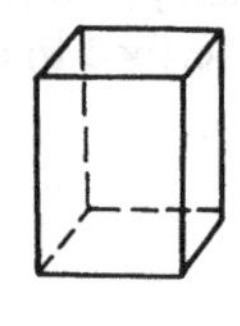

Tetragonal (rectangular prism with square base)

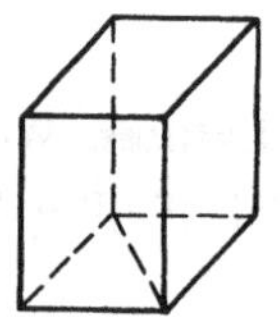

Hexagonal (rectangular prism with base made up of two equilateral triangles)

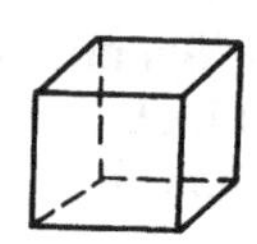

Cubical
(cube)

Fig. 25

We will not classify all types of inequivalent symmetry groups of 3-dimensional lattices. There are 72 different types.

Higher-dimensional generalisations of the 2- and 3-dimensional constructions just treated cannot of course be studied in such detail. Here there are only a few general theorems.

X. Jordan's Theorem. *For any n, a finite group G of motions of n-space has an Abelian normal subgroup A whose index $(G : A)$ in G is bounded by a constant $\psi(n)$ depending only on n.*

In the 3-dimensional case, the theorem is well illustrated by the dihedral group D_n, which contains the cyclic group C_n as a normal subgroup of index 2.

For the analogues of the Bravais groups one proves easily the following result.

Theorem XI. *For given n there are a finite number of nonisomorphic finite subgroups in the group of integral matrixes of determinant ± 1.*

Thus the problem reduces to describing up to conjugacy the subgroups of $GL(n, \mathbb{Z})$ isomorphic to a given group G. We meet here a problem which is analogous to the problem of representations of finite groups which we discussed in § 10 and will discuss further in § 17. The difference is that now, in place of linear transformations of a vector space (and nondegenerate matrixes) we have automorphisms of the module $\mathbb{Z}^n$ (and integral $n \times n$ matrixes of determinant ± 1). The corresponding notion (which we will not define any more precisely) is an n-dimensional integral representation. A basic result is another theorem of Jordan.

Theorem XII. Jordan's Theorem. *Every finite group has only a finite number of inequivalent integral representations of given dimension.*

Example 8. The symmetric groups are a special case of an important class of groups, *finite groups generated by reflections.* Choose an orthonormal basis $e_1, \ldots, e_n$ in n-dimensional Euclidean space $\mathbb{R}^n$; we send a permutation σ of the set $\{1, \ldots, n\}$ into the linear transformation $\hat{\sigma}$ which permutes the vectors of this basis: $\hat{\sigma}(e_i) = e_{\sigma(i)}$. The map $\sigma \mapsto \hat{\sigma}$ is an isomorphism of $\mathfrak{S}_n$ with a certain subgroup S of the group of orthogonal transformations of $\mathbb{R}^n$; obviously, S is generated by the transformations $\hat{\sigma}_i$ corresponding to the transpositions σ_i. The set of vectors fixed by $\hat{\sigma}_i$ is the linear subspace $L \subset \mathbb{R}^n$ with basis $e_1, \ldots, e_{i-1}$, $e_i + e_{i+1}, e_{i+2}, \ldots, e_n$; clearly, $\dim L = n - 1$. If we consider a vector f orthogonal to the hyperplane L, for example $f = e_i - e_{i+1}$ then $\hat{\sigma}_i$ is given by the formulas

$$\hat{\sigma}_i(x) = x \quad \text{for} \quad x \in L;$$
$$\hat{\sigma}_i(f) = -f \quad \text{for} \quad (f, L) = 0.$$

Any transformation s given by these formulas (for some choice of hyperplane L) is called a *reflection.* Obviously, $s^2 = e$. A group of orthogonal transformations which has a system of generators consisting of reflections is called a *group generated by reflections.*

The basic results of the theory of finite groups generated by reflections are as follows.

Theorem XIII. *For any finite group G generated by reflections in a Euclidean space E there exists a uniquely determined decomposition*

$$E = E_0 \oplus E_1 \oplus \cdots \oplus E_p$$

of E as a direct sum of pairwise orthogonal subspaces E_i invariant under G with the following properties:

(1) *E_0 consists of vectors $x \in E$ with $g(x) = x$ for all $g \in G$; for $i = 1, \ldots, p$, each E_i has no subspaces invariant under G, apart from* 0 and E_i.

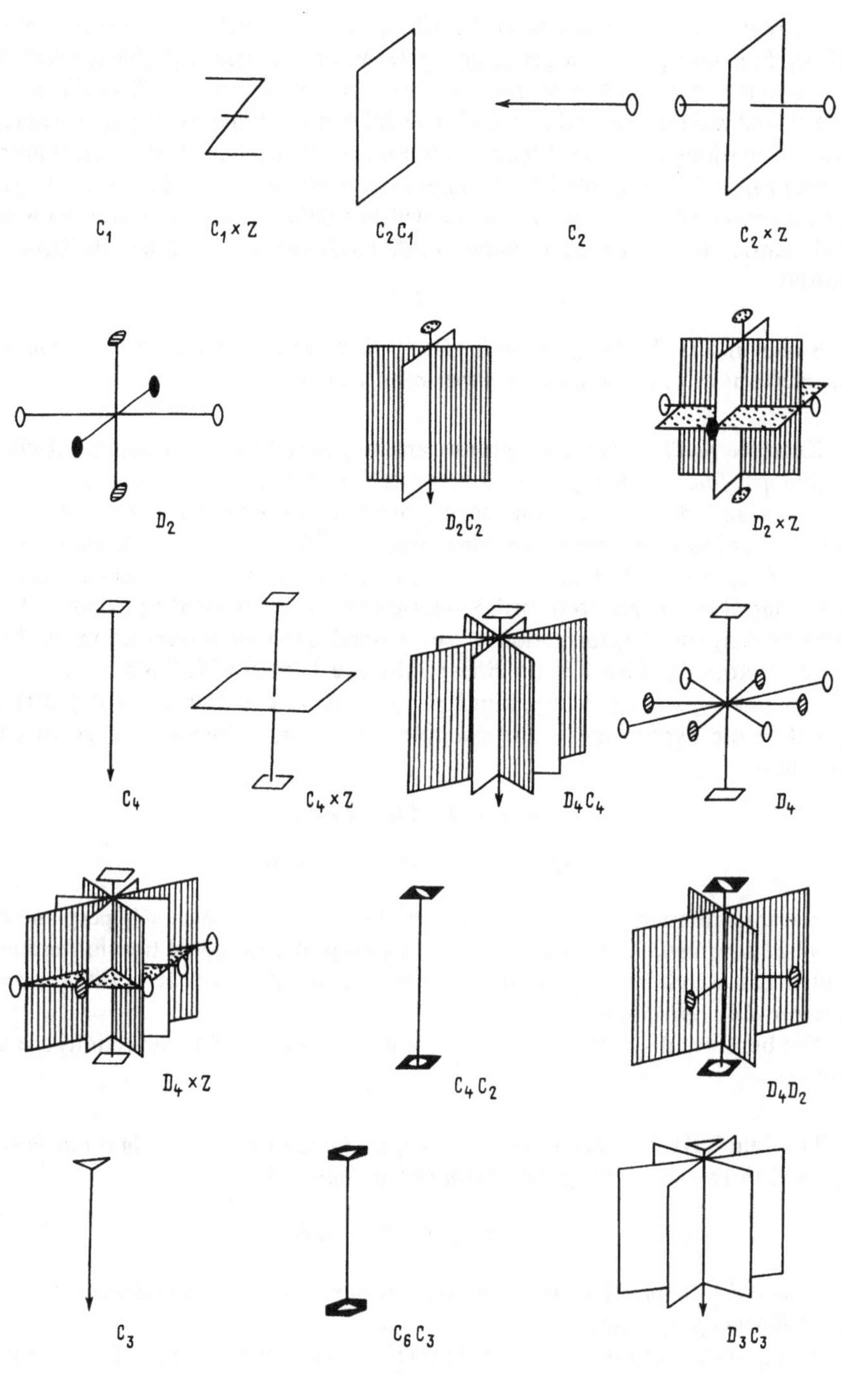

Fig. 26, Table 1

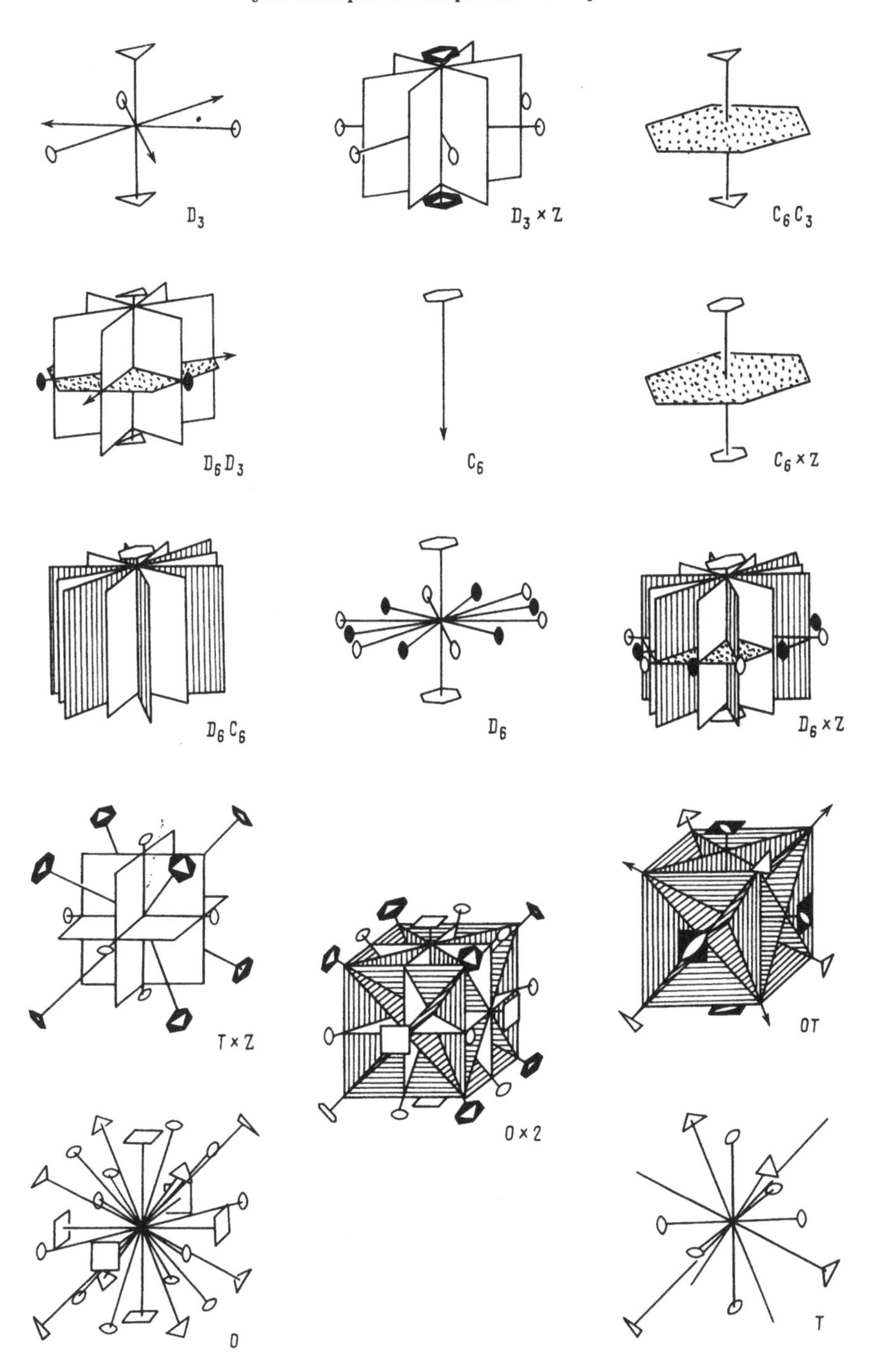

Fig. 26, Table 1 (*continued*)

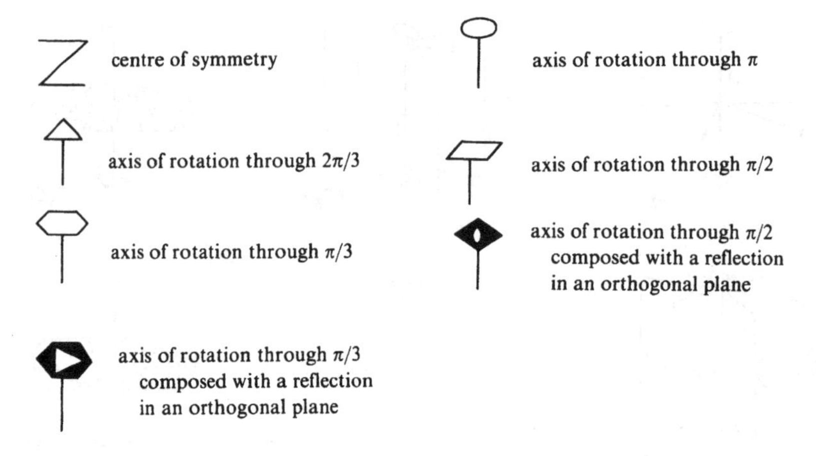

Fig. 26, Table 2

(2) *The group G is a direct sum of groups G_i for $i = 1, \ldots, p$, where G_i consists of all transformations $g \in G$ fixing $x \in E_j$ for $j \neq i$, and G_i is also a group generated by reflections.*

For example, for the action of the symmetric group $\mathfrak{S}_n$ on n-dimensional space $\mathbb{R}^n$ described above, we have the decomposition $\mathbb{R}^n = E_0 \oplus E_1$, where

$$E_0 = \{\alpha(e_1 + \cdots + e_n)\}$$

$$E_1 = \{\alpha_1 e_1 + \cdots + \alpha_n e_n | \alpha_1 + \cdots + \alpha_n = 0\}.$$

A group generated by reflections G is said to be *irreducible* if E does not have subspaces invariant under G and distinct from 0 and E.

Let $\sigma_1, \ldots, \sigma_k$ be a set of reflections. Obviously

$$\sigma_i^2 = e. \tag{7}$$

There is a convenient way of describing certain other special relations between the generators, namely those of the form

$$(\sigma_i \sigma_j)^{m_{ij}} = e. \tag{8}$$

For this we draw a graph with vertexes corresponding to the reflections $\sigma_1, \ldots, \sigma_k$, and with two vertexes joined by an edge if relation (8) holds with $m_{ij} > 2$. If $m_{ij} > 3$ then we write the number m_{ij} over the corresponding edge. It is easy to see that relation (8) with $m_{ij} = 2$ just means that σ_i and σ_j commute.

Theorem XIV. *In each irreducible finite group G generated by reflections, there is a system of generators $\sigma_1, \ldots, \sigma_k$ that are themselves reflections, connected by* *relations described in one of the graphs listed below. These relations define the group G.*

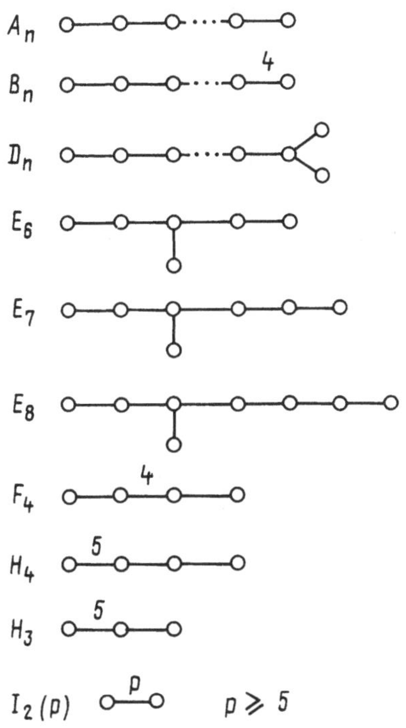

A_n

B_n 4

D_n

E_6

E_7

E_8

F_4 4

H_4 5

H_3 5

$I_2(p)$ p $p \geqslant 5$

The subscript n (in A_n, B_n and so on) indicates the number of vertexes of the graph, and also the dimension of the space in which G acts.

The graph A_n corresponds to the example we already know of the group $\mathfrak{S}_{n+1}$ acting in the space E_1 of the decomposition (6). It can be interpreted as the group of symmetries of the regular n-dimensional simplex, given in coordinates by the conditions $\alpha_1 + \cdots + \alpha_{n+1} = 1$ and $\alpha_i \geqslant 0$.

The graph B_n corresponds to the group consisting of all permutations and sign changes of the vectors of some orthonormal basis of a n-dimensional vector space. This is the group of symmetries of the n-dimensional cube (or octahedron); it has order $2^n n!$. For $n = 3$ it is $O \times Z$. The graph D_n corresponds to a subgroup of index 2 in the group corresponding to B_n. It consists of the permutations and multiplication of basis elements by numbers $\varepsilon_i = \pm 1$ such that $\prod \varepsilon_i = 1$. For $n = 3$ it is $OT \subset O \times Z$ (see Figure 21). The graph H_3 corresponds to the symmetry group of the icosahedron, and $I_2(p)$ to the dihedral group D_p. The graphs H_4 and F_4 correspond to the symmetry groups of certain regular 4-dimensional polyhedrons.

All the groups listed in Theorem XIV are crystal classes, except for H_3, H_4 and $I_2(p)$ for $p = 5$ and $p \geqslant 7$.

Example 9. There exists another method of constructing finite groups which we describe in detail later, and for the moment only hint at with the following example. Consider the group $\mathrm{GL}(n, \mathbb{F}_q)$ consisting of nondegenerate $n \times n$ matrixes with entries in a finite field $\mathbb{F}_q$. It is isomorphic to the group $\mathrm{Aut}_{\mathbb{F}_q} \mathbb{F}_q^n$ of linear transformations of the vector space $\mathbb{F}_q^n$; each such transformation is determined by a choice of basis in $\mathbb{F}_q^n$. Hence $|\mathrm{GL}(n, \mathbb{F}_q)|$ is equal to the number of bases $e_1, \ldots, e_n$ in $\mathbb{F}_q^n$. We can take e_1 to be any of the $q^n - 1$ nonzero vectors of $\mathbb{F}_q^n$; for given e_1, we can take e_2 to be any of the $q^n - q$ vectors not proportional to e_1; for given e_1 and e_2, we can take e_3 to be any of the $q^n - q^2$ vectors which are not linear combinations of e_1 and e_2, and so on. Therefore

$$|\mathrm{GL}(n, \mathbb{F}_q)| = (q^n - 1)(q^n - q)(q^n - q^2) \ldots (q^n - q^{n-1}). \tag{9}$$

One of the applications of the groups $\mathrm{GL}(n, \mathbb{F}_q)$ is the proof of Theorem XI. We fix a prime number $p \neq 2$ and consider the homomorphism $\varphi_p \colon \mathbb{Z} \to \mathbb{Z}/(p) = \mathbb{F}_p$. It defines a homomorphism of matrix groups

$$\varphi_p \colon \mathrm{GL}(n, \mathbb{Z}) \to \mathrm{GL}(n, \mathbb{F}_p),$$

the kernel of which consists of matrixes of the form $A = E + pB$, and $\det A = \pm 1$. Let us prove that any finite group $G \subset \mathrm{GL}(n, \mathbb{Z})$ is mapped isomorphically by φ_p onto some subgroup of $\mathrm{GL}(n, \mathbb{F}_p)$. Since there are only a finite number of subgroups of $\mathrm{GL}(n, \mathbb{F}_p)$, the assertion will follow from this, and (9) will give an estimate for $|G|$. The kernel of $G \to \mathrm{GL}(n, \mathbb{F}_p)$ is $G \cap \operatorname{Ker} \varphi_p$, and we need to prove that this subgroup consists of the identity element only. For this we prove that $\operatorname{Ker} \varphi_p$ does not contain any elements of finite order other than E. Let $A = E + p^r B$ with $B \in M_n(\mathbb{Z})$ and not all elements of B divisible by p, and suppose that $A^n = E$. By the binomial theorem

$$np^r B + \sum_{k=2}^{n} \binom{n}{k} p^{rk} B^k = 0.$$

But an elementary arithmetic argument shows that for $p > 2$ and $k > 1$, all the numbers inside the summation sign are divisible by a bigger power of p than np^r, which gives a contradiction.

§ 14. Examples of Groups: Infinite Discrete Groups

We proceed now to consider infinite groups. Of course, the purely negative characteristic of not being finite does not reflect the situations which really arise. Usually the infinite set of elements of a group is defined by some constructive process or formula. This formula may contain some parameters, which may take integer values, or be real numbers, or even points of a manifold. This is the starting point of an informal classification: groups are called discrete in the first case, and continuous in the second. The simplest example of a discrete group is the infinite cyclic group, whose elements are of the form g^n where n runs through all the integers.

In any case, discrete groups often arise as transformation groups which are discrete, now in a more precise sense of the word. Thus, the group of integers is isomorphic to the group of those translations of the line which preserve the function $\sin 2\pi x$, consisting of translations $x \mapsto x + d$ by an integer d. First of all, let's consider this situation.

Let X be a topological space; in all the examples, X will be assumed to be locally compact and Haussdorff, and most frequently will be a manifold, either differentiable or complex analytic. A group G of automorphisms of X is *discrete* (or *discontinuous*) if for any compact subset $K \subset X$ there exist only a finite set of transformations $g \in G$ for which $K \cap gK$ is nonempty. We can introduce a topology on the set of orbits $G \backslash X$, by taking open sets to be the subsets whose inverse image under the canonical map $f\colon X \to G \backslash X$ are open. If the stabiliser of every point $x \in X$ is just e, then we say that G acts *freely*. In this case any point $\xi \in G \backslash X$ has a neighbourhood whose inverse image under the canonical map $f\colon X \to G \backslash X$ breaks up as a disjoint union of open sets, each of which is mapped homeomorphically by f. In other words, X is an *unramified cover* of the space $G \backslash X$. In particular, if X was a differentiable or analytic manifold, then $G \backslash X$ will again be a manifold of the same type.

If some group $\mathfrak{G}$ is simultaneously a manifold (cases of this will be considered in the following § 15) then a subgroup $G \subset \mathfrak{G}$ is said to be *discrete* if it is discrete under the left regular action on $\mathfrak{G}$.

The construction of quotient spaces $G \backslash X$ is an important method of constructing new topological spaces. An intuitive way of representing them is related to the notion of a *fundamental domain*. By this we mean a set $\mathscr{D} \subset X$ such that the orbit of every point $x \in X$ meets $\mathscr{D}$, and that of an interior point x of $\mathscr{D}$ only meets $\mathscr{D}$ in x itself. Then different points of one orbit belonging to the closure $\bar{\mathscr{D}}$ of $\mathscr{D}$ can only lie on the boundary of $\mathscr{D}$, and we can visualise the space $G \backslash X$ as obtained by glueing $\bar{\mathscr{D}}$ together, identifying points on the boundary that belong to one orbit. For example, the above group of translations of the line, consisting of transformations $x \mapsto x + d$, has the interval $[0, 1]$ as a fundamental domain. Identifying the boundary points, we get a circle. The space $G \backslash X$ is compact if and only if G has a fundamental domain with compact closure.

Example 1. Discrete subgroups of the group of vectors of an n-dimensional real vector space $\mathbb{R}^n$.

Theorem I. *Any discrete subgroup of the group $\mathbb{R}^n$ is isomorphic to $\mathbb{Z}^m$ with $m \leqslant n$, and consists of all linear combinations with integer coefficients of m linearly independent vectors $e_1, \ldots, e_m$.*

A group of this form is called a *lattice*. A fundamental domain of a lattice can be constructed by completing the system of vector $e_1, \dots, e_m$ to a basis $e_1, \dots, e_n$ and then setting

$$\mathscr{D} = \{\alpha_1 e_1 + \cdots + \alpha_n e_n | 0 \le \alpha_1, \dots, \alpha_m \le 1\}.$$

The space $G \backslash \mathbb{R}^n$ is compact if and only if $m = n$. In this case the fundamental domain $\mathscr{D}$ is the parallelipiped constructed on the vectors $e_1, \dots, e_n$.

In this paragraph, we assume that the reader has met the notion of a Riemann surface and its genus. If $n = 2$ then the plane $\mathbb{R}^2$ can be viewed as the plane of one complex variable; as such it is denoted by $\mathbb{C}$. If G is a lattice in $\mathbb{C}$ then the quotient space $G \backslash \mathbb{C}$ inherits from $\mathbb{C}$ the structure of a 1-dimensional complex manifold, that is, it is a compact Riemann surface. Its genus is equal to 1, and it can be shown that all compact Riemann surfaces of genus 1 can be obtained in this way. Meromorphic functions on the Riemann surface $G \backslash \mathbb{C}$ are meromorphic functions $f(z)$ of one complex variable, invariant under translation $z \mapsto z + \alpha$ with $\alpha \in G$, that is, *elliptic functions* with elements of G as periods.

Theorem. *Two Riemann surfaces $G_1 \backslash \mathbb{C}$ and $G_2 \backslash \mathbb{C}$ constructed as above are conformally equivalent if and only if the lattices G_1 and G_2 are similar.*

Example 2. Crystallographic Groups. This is a direct generalisation of Example 1 (or more precisely, of the case $m = n$). The atoms of a crystal are arranged in space in a discrete and extremely symmetrical way. This is seen in the fact that their relative position in space repeats itself indefinitely. More precisely, there exists a bounded domain $\mathscr{D}$ such that any point of space can be taken into a point of $\mathscr{D}$ by a symmetry of the crystal, that is, by a motion of space that preserves the physical properties of the crystal (taking every atom into an atom of the same element, and preserving all relations between atoms). In other words, the symmetry group G of the crystal is a discrete group of motions of 3-space $\mathbb{R}^3$ and the space $G \backslash \mathbb{R}^3$ is compact. In this connection, a *crystallographic group* is a discrete group G of motions of n-dimensional Euclidean space $\mathbb{R}^n$ for which the quotient space $G \backslash \mathbb{R}^n$ is compact.

The main result of the theory of crystallographic groups is the following:

II. Bieberbach's Theorem. *The translations contained in a crystallographic group G form a normal subgroup A such that $A \backslash \mathbb{R}^n$ is compact, and the index $(G:A)$ is finite.*

In the case $n = 3$, this means that for every crystal there exists a parallelipiped Π (a fundamental domain of the subgroup $A \lhd G$, where G is the symmetry group of the crystal) such that all properties of the crystal in Π and in its translates $g\Pi$ (for $g \in A$) are identical: and these translates fill the whole of space. Π is called the *repeating parallelipiped* of the crystal.

In the general case, by Theorem I, A consists of translations in vectors of some lattice $C \cong \mathbb{Z}^n$. The finite group $F = G/A$ is a symmetry group of C. From this, using Jordan's Theorem (§ 13, Theorem XII), we can deduce:

Theorem III. *The number of crystallographic groups in a space of given dimension n is finite.*

In this, two groups G_1 and G_2 are considered to be the same if one can be taken into the other by an affine transformation of $\mathbb{R}^n$. It can be shown that this property is equivalent to G_1 and G_2 being isomorphic as abstract groups.

The crystallographic groups, arising in connection with crystallography, also have a very natural group-theoretical characterisation: they are exactly the groups G which contain a normal subgroup of finite index isomorphic to $\mathbb{Z}^n$, and not contained in any bigger Abelian subgroup.

For crystallography it is extremely important to have a list of all the types of crystallographic groups in 3-dimensional space. Indeed, for any crystal, if we can indicate its group, its fundamental domain and the position of its atoms inside this, then we have determined the whole crystal, however far it grows. This gives a method of representing crystals in finite terms, which is actually used in compiling crystallographic tables. The list of all crystallographic groups is too long for us to give here, but some idea of it can be gained from the 2-dimensional case.

Theorem IV. *There are* 17 *different crystallographic groups in the plane.*

Each of these has a normal subgroup $A \lhd G$ which consists of translations in the vectors of some lattice C. The transformations $g \in G$ of course take this lattice into itself, with elements $g \in A$ acting just by translations. Hence the finite group $F = G/A$ is a symmetry group of this lattice, and belong to one of thirteen types described in § 13, Theorem VIII. It can happen however that two distinct groups G have the same lattice C and determine the same symmetry group of C. To give an example, consider the rectangular lattice Γ_{rect} and the symmetry group $\mathscr{D}$ of Γ_{rect} consisting of the identity and the reflection in the OB axis (Figure 27).

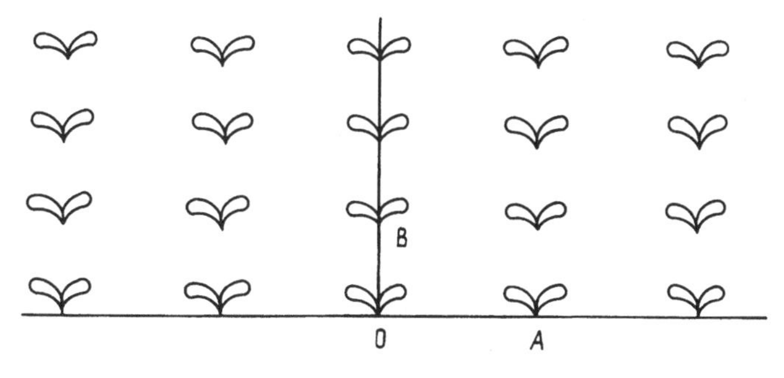

Fig. 27

We can consider the group G generated by the group T of translations in vectors of C and the above group $\mathscr{D}$ of orthogonal transformations. The group G can be characterised as the symmetry group of the pattern illustrated in Figure 27. On the other hand, consider the motion s consisting of translation in the vector $\frac{1}{2}\overrightarrow{DB}$ together with a reflection in the axis OB, and the translation t in the vector $\overrightarrow{OA}$. Write G_1 for the group generated by s and t. Then since s^2 is the translation in $\overrightarrow{OB}$, the group T and the lattice C will be the same in the two cases, and the groups of symmetries of C generated by the motions of G and G_1 will also coincide. However, the groups G and G_1 are not isomorphic: G contains reflections, but G_1, as one checks easily, only contains translations and motions in which a reflection is combined with a translation along one of the vertical lines of the lattice. In particular, the group G contains an element of order 2, and G_1 does not. The group G_1 coincides with the symmetry group of the pattern illustrated in Figure 16.

As in this example, from the 13 groups of symmetries listed in §13, Theorem VIII we can form 13 2-dimensional crystallographic groups generated by translations in vectors of the corresponding lattice and orthogonal transformations which are symmetries of it (acting as in the construction of the group G in the above example). In this case the stabiliser G_x for x any point of the lattice will be isomorphic to the symmetry group of one of the 13 types which we chose. But in some cases a more delicate construction is possible (as the construction of G_1 in the example). Then the fixed subgroup will be smaller than the symmetry group, since some symmetries will only occur in G in combination with translations (like the transformation s in the example). Thus we can construct a new group for the symmetry group $\mathscr{D}_1(\Gamma_{\text{rect}})$ (the group G_1 we have constructed above), two groups for $\mathscr{D}_2(\Gamma_{\text{rect}})$ and one for $\mathscr{D}_4(\Gamma_{\text{sq}})$. This gives 17 groups.

We conclude with the example of the 'new' group corresponding to the symmetry $\mathscr{D}_4(\Gamma_{\text{sq}})$. We include in it the group C_4 of rotations of the plane about the point O through angles of $0, \pi/2, \pi, 3\pi/2$ and the translation along an axis l through O combined with a reflection in this axis. The group G is generated by these transformations. If σ is the rotation through $\pi/2$ then $s' = \sigma s \sigma^{-1}$ is a transformation similar to s, but having axis l' orthogonal to l. The subgroup of translations is generated by the translations s^2 and $(s')^2$ along the axes l and l'. The group G we have constructed is the symmetry group of the pattern of Figure 28.

The groups we have constructed are also called the 'ornament groups' (or 'wallpaper pattern groups' in English textbooks), since they can be interpreted as groups of symmetries of patterns in the plane. A complete list of ornaments corresponding to each of the 17 groups is contained, for example, in the book [Mal'tsev **80** (1956)]. Figures 16, 27, 28 are examples of such patterns, especially thought up to characterise some of the groups. However, the ornaments created from purely aesthetic considerations are of course much more interesting. An example is the ornament of Figure 29, taken from [Speiser **98** (1937)]. It is interesting in that it is taken from a tomb in Thebes and was created by ancient Egyptian artisans. This shows that a deep understanding of the idea of symmetry, axiomatised in the notion of a group, developed very long ago.

The situation in the 3-dimensional case is much more complicated.

Fig. 28

Fig. 29

Theorem V. *The number of different crystallographic groups in* 3-*space is equal to* 219. *As in Theorem* III, *we consider groups to be the same if they are isomorphic, or* (*what is the same*) *if they can be taken into one another by an affine transformation of space.*

All 219 groups are realised as the symmetry groups of genuine crystals.

In crystallography, the number of different crystallographic groups is often given as 230. This comes from the fact that there, groups are only considered to be the same if they are taken into one another by a transformation which preserves the orientation of space; in the plane, the two notions of equivalence lead to the same classification.

The theory of crystallographic groups explains the role of finite symmetry groups of lattices which we considered in § 13, Example 7. Symmetries of a crystal are given by the whole crystallographic group G, but because of the fact that the distances between atoms are very small, the group which is more noticeable from the macroscopic point of view is not the translation group A but the quotient group G/A, a symmetry group of A. It is interesting to note that in the list of groups of § 13, Theorem VII, we only meet groups containing rotations through $\pi/2$, $\pi/3$ or multiples of these. Hence only these rotations can occur as symmetries of crystals. It is all the more astonishing that in real life we often meet other symmetries. For example, everyone knows the flowers of the geranium and the bluebell (*campanula*) whose petals have symmetry of order 5. In Figure 30, taken from the book [*The life of plants* **1** (1981)], we can see the 5-fold symmetry of the flowers of *campanula* (bluebell) (a) and *Stapelia variegata* (variegated carrion-flower) (b), and the 7-fold symmetry of the position of the leaves of the baobab tree (c).

Fig. 30a–c

Example 3. Non-Euclidean Crystallography. Discrete groups of motions are of interest not just for Euclidean spaces, but also for other spaces. Here we discuss the case of the Lobachevsky plane Λ; we will only consider discrete groups of motions G of the plane Λ preserving the orientation, satisfying the two conditions: (1) A motion $g \in G$ with $g \neq e$ does not fix any point of Λ; and (2) the space $G \backslash \Lambda$ is compact. In the case of the Euclidean plane the only groups satisfying these conditions are the groups of translations in vectors of a lattice.

The interest in groups G of this type arose in connection with the fact that under condition (1) the space $G \backslash \Lambda$ is a manifold, in the present case a surface. If we use Poincaré's interpretation of the Lobachevsky plane in the upper half-plane $\mathbb{C}^+$ of the plane of one complex variable, then the surface $G \backslash \Lambda$ inherits the complex structure of the upper half-plane, and (assuming condition (2)) is a compact Riemann surface. Meromorphic functions on the Riemann surface $G \backslash \mathbb{C}^+$ are meromorphic functions on $\mathbb{C}^+$ invariant under G. They are called *automorphic functions*. This can be compared with the situation in Example 1, where we considered the space $G \backslash \mathbb{C}$. In that case we obtained compact Riemann surfaces of genus 1. It is proved that in the case we are now considering we obtain precisely all compact Riemann surfaces of genus >1 (the Poincaré-Koebe uniformisation theorem). Thus both of these cases together give a group-theoretical description of all compact Riemann surfaces (the remaining case of genus 0 is the Riemann sphere).

As fundamental domain of the group G of the type under consideration we can take a $4p$-gon in the Lobachevsky plane with alternate pairs of sides equal: that is $a_1, b_1, a'_1, b'_1, \ldots, a_p, b_p, a'_p, b'_p$ where a_i and a'_i, b_i and b'_i are equal intervals. The transformations taking the side a_i into a'_i or b_i into b'_i (the directions of the sides which are identified under these are indicated in Figure 31) are the generators of G.

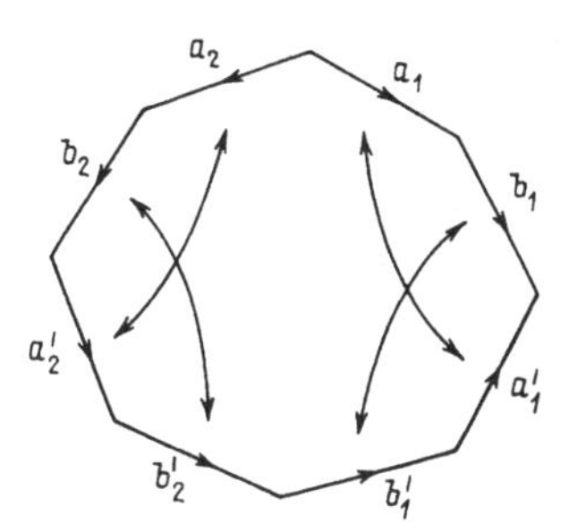

Fig. 31

The unique relation which the polygon must satisfy is of course that the sides a_i, a'_i and b_i and b'_i are equal, and that the sum of its angles is 2π (this relates to the fact that in $G \backslash \Lambda$ all the vertexes are glued together as one point.

Example 3a. An important particular case of Example 3 arises if we consider the Cayley-Klein (rather than the Poincaré) interpretation of Lobachevsky space.

Let $f(x, y, z)$ be an indefinite quadratic form with integer coefficients. Consider the group $G \subset \mathrm{SL}(3, \mathbb{Z})$ consisting of integral transformations preserving f. Interpreting x, y, z as homogeneous coordinates on the projective plane, we realise G as a group of projective transformations of the set $f > 0$, that is, a group of motions of the Lobachevsky plane Λ in the Cayley-Klein interpretation. It can be shown that $G \backslash \Lambda$ is compact if and only if the equation $f(x, y, z) = 0$ does not have any rational solutions other than $(0, 0, 0)$. (A criterion for this equation to have solutions is given by Legendre's theorem, §7, Theorem III.) In this case condition (1) of Example 3 may not be satisfied; this will be the case if G contains an element of finite order other than e. But then there exists a subgroup $G' \subset G$ of finite index satisfying (1): applying the argument given at the end of §13, Example 9, we can show that we can take the subgroup consisting of matrixes $g \in G$ with $g \equiv E \bmod p$ (for any choice of prime $p \neq 2$).

Example 4. The group $\mathrm{SL}(2, \mathbb{Z})$ consisting of 2×2 integral matrixes with determinant 1. The significance of this group is related to the fact that in a 2-dimensional lattice two bases e_1, e_2 and f_1, f_2 are related by

$$f_1 = ae_1 + ce_2, \quad f_2 = be_1 + de_2,$$

with

$$a, b, c, d \in \mathbb{Z} \quad \text{and} \quad ad - bc = \pm 1, \quad \text{that is} \quad \begin{bmatrix} a & b \\ c & d \end{bmatrix} \in \mathrm{GL}(2, \mathbb{Z}).$$

If we also require that the direction of rotation from f_1 to f_2 is the same as that from e_1 to e_2; then $ad - bc = 1$, that is $\begin{bmatrix} a & b \\ c & d \end{bmatrix} \in \mathrm{SL}(2, \mathbb{Z})$. A problem which crops up frequently is the classification of lattices in the Euclidean plane up to similarity. In Example 1 we saw that, for example, the classification of compact Riemann surfaces of genus 1 reduces to this. As there, we realise our plane as the complex plane $\mathbb{C}$: then similarities are given by multiplication by nonzero complex numbers. Let z_1, z_2 be a basis of a lattice $C \subset \mathbb{C}$. We will suppose that the angle between z_1 and z_2 is $\leqslant \pi$, and choose the order of the vectors so that the rotation from z_1 to z_2 is anticlockwise. Applying a similarity, which can be expressed as multiplication by z_1^{-1}, we get a similar lattice C' with basis $1, z$, where $z = z_1^{-1} z_2$, and, in view of the assumptions we have made, z lies in the upper half-plane $\mathbb{C}^+$. Then two bases 1, z and 1, w define similar lattices if the basis $(bz + d, az + c)$ for some $a, b, c, d \in \mathbb{Z}$ with $ad - bc = 1$ can be taken into $(1, w)$ by a similarity. This similarity must be given by $(bz + d)^{-1}$, and hence $w = \dfrac{az + c}{bz + d}$. Thus we define the action of the group $\mathrm{SL}(2, \mathbb{Z})$ on the upper half-plane $\mathbb{C}^+$ as follows: for $g = \begin{bmatrix} a & b \\ c & d \end{bmatrix} \in \mathrm{SL}(2, \mathbb{Z})$, $gz = \dfrac{az + c}{bz + d}$.

Here the matrix $\begin{bmatrix} -1 & 0 \\ 0 & -1 \end{bmatrix}$ acts as the identity, so that we have an action of the group $\mathrm{SL}(2, \mathbb{Z})/N$, where $N = \left\{ \begin{bmatrix} 1 & 0 \\ 0 & 1 \end{bmatrix}, \begin{bmatrix} -1 & 0 \\ 0 & -1 \end{bmatrix} \right\}$. This quotient group is denoted by PSL(2, $\mathbb{Z}$); it is called the *modular group*. We see that the set of lattices up to similarity can be represented as $G\backslash\mathbb{C}^+$, where G is the modular group. The modular group acts discretely on the upper half-plane. A fundamental domain for it is given by the shaded region of Figure 32.

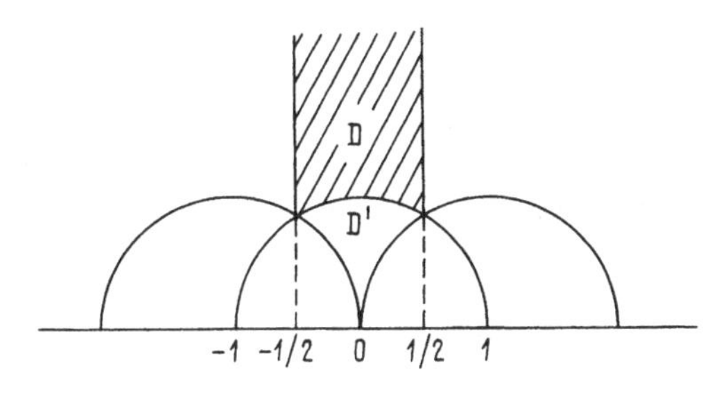

Fig. 32

This region $\mathscr{D}$ is called the *modular figure*. It is not bounded, but it has another important property. As is well known, the upper half-plane is a model of the Lobachevsky plane, and motions in it which preserve the orientation are given as transformations $z \mapsto \dfrac{\alpha z + \gamma}{\beta z + \delta}$, where $\alpha, \beta, \gamma, \delta \in \mathbb{R}$ with $\alpha\delta - \beta\gamma = 1$. Thus the modular group is a discrete group of motions of the Lobachevsky plane. Now in the sense of Lobachevsky geometry, the modular figure is of finite area. In view of this the surface $G\backslash\mathbb{C}^+$ is not compact, but in the natural metric it has finite area.

The modular group is analogous to the groups considered in Example 3, but is not one of them: to start with, some of its transformations have fixed points (for example $z \mapsto -1/z$), and secondly, $G\backslash\Lambda$ is not compact. The analogy with Example 3 will be clearer if we think of the modular figure from the point of view of Lobachevsky geometry. It is a triangle with one vertex at infinity, and the sides converging to this vertex become infinitely close to one another. This is more visible for the equivalent region $\mathscr{D}'$ of Figure 32.

Example 5. Let $G = \mathrm{GL}(n, \mathbb{Z})$. This is a discrete subgroup of $\mathrm{GL}(n, \mathbb{R})$, and acts on the same spaces as $\mathrm{GL}(n, \mathbb{R})$. Of particular importance is its action on the set $\mathscr{H}_n$ of real positive definite matrixes A defined up to a positive multiple: $g(A) = gAg^*$ (see § 12, Example 4 for the case $n = 2$). This action expresses the notion of integral equivalence of quadratic forms. A fundamental domain here is also noncompact, but has bounded volume (in the sense of the measure invariant under the action of $\mathrm{GL}(n, \mathbb{R})$).

The group $\mathrm{GL}(n, \mathbb{Z})$ belongs to the important class of *arithmetic groups*, which we will discuss in the next section.

Example 6. Free Groups. Consider a set of symbols $s_1, \ldots, s_n$ (for simplicity we will think of this as a finite set, although the arguments do not depend on this). To each symbol s_i we assign another symbol s_i^{-1}. A *word* is a sequence of symbols s_i and s_j^{-1} in any order (written down next to one another), for example $s_1 s_2 s_2 s_1^{-1} s_3$. The empty word e is also allowed, in which no symbol appears. A word is *reduced* if it does not at any point contain the symbols s_i and s_i^{-1} or s_i^{-1} and s_i adjacent. The *inverse* of a word is the word in which the symbols are written out in the opposite order, with s_i replaced by s_i^{-1} and s_i^{-1} by s_i. The *product* of two words A and B is the word obtained by writing B after A, and then cancelling out all adjacent pairs of s_i and s_i^{-1} until we get a reduced word (possibly the empty word). The set of reduced words with this operation of multiplication forms a group, as is checked without difficulty. This group is the *free group* on n generators, and is denoted by $\mathscr{S}_n$. Obviously, the words $s_1, \ldots, s_n$, consisting of just one symbol are generators, s_i^{-1} is the inverse of s_i, and any word can be thought of as a product of the s_i and s_i^{-1}.

The free group $\mathscr{S}_2$ on generators x and y can be realised as a group of transformations of a 1-dimensional complex, that is, of a topological space consisting of points and segments joining them. For this, for the points we take all the different elements of $\mathscr{S}_2$, and we join two points corresponding to reduced words A and B if B can be obtained from A by multiplying on the right with x, y, x^{-1} or y^{-1} (see Figure 33).

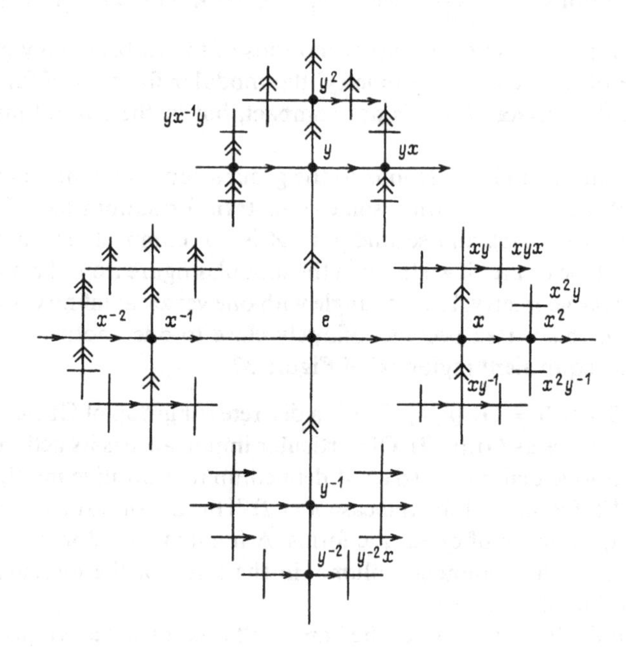

Fig. 33

Obvio usly, if two words A and B are represented by points joined by a segment, then the same is true for CA and CB for any $C \in \mathscr{S}_2$. Hence the left regular action (see § 12) of $\mathscr{S}_2$ defines an action of $\mathscr{S}_2$ on this complex. If we introduce a 'biorientation' on the complex, marking each segment with one of two types of arrow (right and upwards, as in Figure 33), it is easy to see that the group $\mathscr{S}_2$ will be the full group of automorphisms of this bioriented complex.

Consider any group G having n generators $g_1, \ldots, g_n$. It is easy to see that the correspondence which takes a reduced word in $s_1, \ldots, s_n$ into the same expression in $g_1, \ldots, g_n$ is a homomorphism of $\mathscr{S}_n$ onto G. Hence every group is a homomorphic image of a free group, so that free groups play the same role in group theory as free modules in the theory of modules and the noncommutative polynomial ring in the theory of algebras (see § 5 and § 8).

Let

$$G = \mathscr{S}_n/N$$

be a presentation of a group G as a quotient group of a free group by a normal subgroup N. Elements $r_1, \ldots, r_m$ which, together with their conjugates, generate N are called *defining relations* for G. Obviously, the relations

$$r_1 = e, \ldots, r_m = e$$

hold in G (where the r_i are viewed as words in the generators $g_1, \ldots, g_n$ of G). Specifying defining relations uniquely determines the normal subgroup N, and hence G. This gives a precise meaning to the statement that a group is *defined by relations*, which we have already used. A group having a finite number of generators is *finitely generated*, and if it can also be presented by a finite number of relations, then it is *finitely presented*. For example, § 13, (1), (2) and (3) are defining relations of the symmetric group $\mathfrak{S}_n$, and § 13, (7) and (8) are those of a finite group generated by reflections. It can be shown that the group PSL(2, $\mathbb{Z}$) of Example 4 is generated by the matrixes $s = \begin{bmatrix} 0 & 1 \\ -1 & 0 \end{bmatrix}$ and $t = \begin{bmatrix} 0 & -1 \\ 1 & -1 \end{bmatrix}$, and that the defining relations of the group are of the form

$$s^2 = e, \quad t^3 = e.$$

Can we accept a presentation of a group in terms of generators and relations as an adequate description (even if the number of generators and relations is finite)? If $g_1, \ldots, g_n$ are generators of a group G, then to have some idea of the group itself, we must know when different expressions of the form $g_{i_1}^{\alpha_{i_1}} g_{i_2}^{\alpha_{i_2}} \ldots g_{i_m}^{\alpha_{i_m}}$ determine (in terms of the defining relations) the same element of the group. This question is called the *word problem* or the *identity problem*. It is trivial for free groups, and has been solved for certain very special classes of groups, for example for groups given by one relation, but in the general case it turns out to be impossibly hard. The same can be said of another problem of this type, that of knowing whether two groups given by generators and relations are isomorphic (this is called the *isomorphism problem*).

Both of these problems were raised to a new plane when mathematical logicians created a precise definition of an algorithm. Up to this point, one could only solve the identity problem and put forward a procedure, called an algorithm, for establishing the identity of two expressions given in terms of generators. Now however, it turns out that there is a well-posed problem: are the identity and isomorphism problems solvable?

This was quickly settled. It turns out that, among groups given by generators and relations, there exist some in which the identity problem is not solvable, and groups for which the isomorphism problem is not solvable even with the identity group.

Perhaps the most striking example of the need as a matter of principle to apply notions of mathematical logic to the study of purely group-theoretical problems is the following result.

Higman's Theorem. *A group with a finite number of generators and infinite number of defining relations is isomorphic to a subgroup of a group defined by a finite number of relations if and only if the set of its relations is recursively enumerable. (The latter term, also relating to mathematical logic, formalises the intuitive notion of an inductive method of getting hold of all elements of some set by constructing them one by one.)*

Presentations of groups by generators and relations occur most frequently in topology.

Example 7. The Fundamental Group. Let X be a topological space. Its fundamental group consists of closed paths, considered up to continuous deformation. A *path* with starting point $x \in X$ and end point $y \in Y$ is a continuous map $f: I \to X$ of the interval $I = [0 \leqslant t \leqslant 1]$ into X for which $f(0) = x$ and $f(1) = y$. A path is *closed* if $x = y$. The *composite* of two paths $f: I \to X$ with starting point x and end point y and $g: I \to X$ with starting point y and end point z is the map $fg: I \to X$ given by

$$(fg)(t) = f(2t) \quad \text{for} \quad 0 \leqslant t \leqslant 1/2,$$

$$(fg)(t) = g(2t-1) \quad \text{for} \quad 1/2 \leqslant t \leqslant 1.$$

Two paths $f: I \to X$ and $g: I \to X$ with the same starting point x and end point y are *homotopic* if there exists a continuous map $\varphi: J \to X$ of the square $J = \{0 \leqslant t, u \leqslant 1\}$ such that

$$\varphi(t,0) = f(t), \quad \varphi(t,1) = g(t), \quad \varphi(0,u) = x, \quad \varphi(1,u) = y.$$

Closed paths starting and ending at x_0, considered up to homotopy, form a group under the multiplication defined as the composite of paths. This group is called the *fundamental group* of X; it is denoted by $\pi(X)$; (and also by $\pi_1(X)$, in view of the fact that groups $\pi_n(X)$ for $n = 1, 2, 3, \ldots$, also exist, and will be defined in § 20). In the general case, the fundamental group depends on the choice of the point x_0 and is denoted by $\pi(X, x_0)$, but if any two points of X can be joined by a path (we will always assume this in what follows), then the groups $\pi(X, x_0)$ for $x_0 \in X$ are all isomorphic. A space X is *simply connected* if $\pi(X) = \{e\}$.

If X is a cell complex (that is, a union of disjoint 'cells', the images of balls of different dimensions) and has a single 0-dimensonal cell, then the fundamental group $\pi(X)$ has as generators the paths corresponding to 1-dimensional cells, and defining relations corresponding to 2-dimensional cells. For example, a 1-dimensional complex does not have any 2-dimensional cells, and its fundamental group is therefore free. The fundamental group of a 'bouquet' of n circles (see Figure 34 for the case $n = 4$) is a free group with n generators.

Fig. 34

An oriented compact surface homeomorphic to a sphere with p handles (see Figure 35 for the case $p = 2$) can be obtained by glueing a $4p$-gon along alternate pairs of sides, as illustrated (for $p = 2$) in Figure 31. It is therefore a cell complex with a single 0-dimensional, $2p$ 1-dimensional and one 2-dimensional cell. Hence

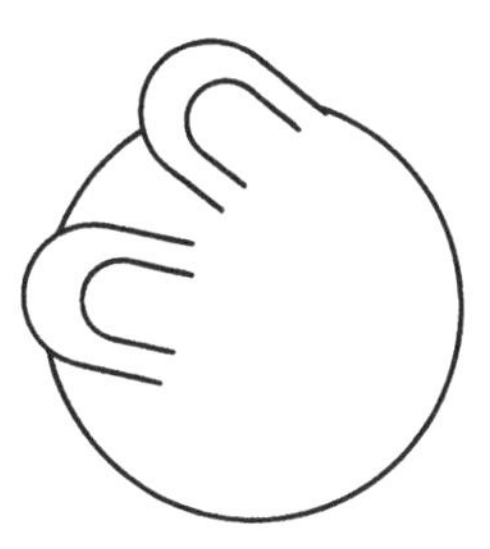

Fig. 35

its fundamental group has $2p$ generators: $s_1, t_1, s_2, t_2, \ldots, s_p, t_p$ where s_i is obtained from the paths a_i and a_i', and t_i from b_i and b_i'. There is one relation between them, which corresponds to going round the perimeter of the $4p$-gon keeping track of the direction of the sides:

$$s_1 t_1 s_1^{-1} t_1^{-1} \dots s_p t_p s_p^{-1} t_p^{-1} = e. \tag{1}$$

The undecidability of the isomorphism problem for groups allows us to prove (by constructing manifolds with these fundamental groups) that the homeomorphism problem for manifolds of dimension $\geqslant 4$ is undecidable.

The fundamental group is closely related to the discrete transformation groups we considered earlier. In fact, if X is a space in which any two points can be joined by a path, then there exists a connected and simply connected space $\hat{X}$ and a group G isomorphic to $\pi(X)$ acting on it, in such a way that $X = G\backslash\hat{X}$. The space $\hat{X}$ is called the *universal cover* of X. Conversely, if $\hat{X}$ is connected and simply connected, and a group G acts discretely and freely on $\hat{X}$ then $\hat{X}$ is the universal cover of $X = G\backslash\hat{X}$, and G is isomorphic to $\pi(X)$. Thus in Example 3, a Riemann surface X is represented as $G\backslash\mathbb{C}^+$; hence $\mathbb{C}^+$ is the universal cover of X and $G \cong \pi(X)$. From this we get that G is defined by the relation (1). The complex $\hat{X}$ illustrated in Figure 33 is obviously simply connected and the free group $\mathscr{S}_2$ acts freely on it. It is easy to see that a fundamental domain for this action is formed by two segments ex and ey. Hence $\mathscr{S}_2\backslash\hat{X}$ is a bouquet of two circles (as in Figure 34, but with $n = 2$), obtained by identifying e with x and with y, and $\hat{X}$ is the universal covering of this bouquet.

Example 8. The Group of a Knot. A *knot* is a smooth closed curve Γ in 3-dimensional space which does not intersect itself. The problem is that of classifying knots up to isotopy—a continuous deformation of space. The main invariant for this is the *knot group* of Γ, that is, the fundamental group of the complement, $\pi(\mathbb{R}^3 \setminus \Gamma)$. To get a pictorial representation of the knot, one projects it into the plane, indicating in the resulting diagram which curve goes over and which goes under at each crossover point (Figure 36).

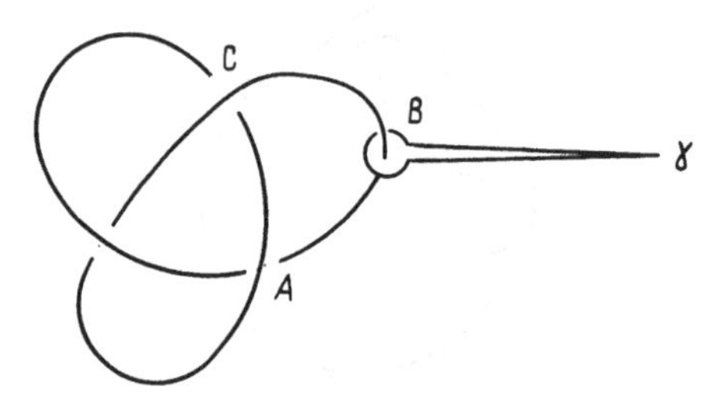

Fig. 36

The generators of the knot group correspond to the segments into which the points of intersection divide the resulting curve (for example, the path γ of Figure 36 corresponds to the interval ABC). It can be shown that the defining relations correspond to the crossover points. The simplest knot, an unknotted circle, can be projected without self-intersection, so that its knot group is the infinite cyclic group. The role of the knot group is illustrated for example by the following result.

Theorem. *A knot is isotopic to the unknotted circle if and only if its knot group is isomorphic to the infinite cyclic group.*

Here we run into an example where a substantial topological question leads to a particular case of the isomorphism problem.

Example 9. Braid Groups. We consider a square $ABCD$ in 3-space, and put on each of the sides AB and CD a collection of n points: $P_1, \ldots, P_n$ and $Q_1, \ldots, Q_n$. A *braid* is a collection of n smooth disjoint curves contained in a cube constructed on $ABCD$ with starting points $P_1, \ldots, P_n$ and end points $Q_1, \ldots, Q_n$ (but possibly in a different order (see Figure 37, (a)).

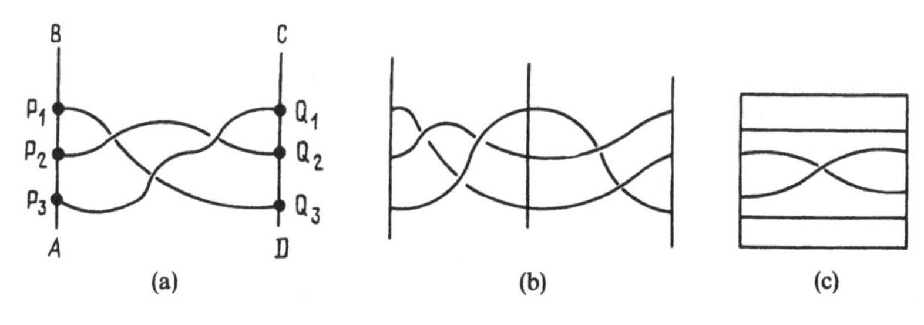

(a) (b) (c)

Fig. 37

A braid is considered up to isotopy. Multiplication of braids is illustrated in Figure 37, (b). Identifying the points P_i and Q_i we obtain *closed braids*. The classes of closed braids up to isotopy form the *braid group* Σ_n. Generators of Σ_n are the braids σ_i for $i = 1, \ldots, n-1$ in which only two threads are interchanged (Figure 37, (c)). The defining relations are of the form

$$\begin{aligned} \sigma_i \sigma_j &= \sigma_j \sigma_i \quad \text{for} \quad j \neq i \pm 1, \\ \sigma_i \sigma_{i+1} \sigma_i &= \sigma_{i+1} \sigma_i \sigma_{i+1}. \end{aligned} \tag{2}$$

The question of when two braids are isotopic can be restated as the identity problem in the braid group, defined by the relations (2). In this particular case the identity problem is solvable and in fact solved—this is one application of group theory to topology.

The significance of the braid group consists in the fact that a braid can be thought of as defining a motion of an unordered set of n points on the plane which are not allowed to come together. The exact result is as follows. Let D denote the set of points $(z_1, \ldots, z_n) \in \mathbb{C}^n$ for which $z_i = z_j$ for some $i \neq j$. The symmetric group $\mathfrak{S}_n$ acts on $\mathbb{C}^n$ by permuting the coordinates, and preserves D.

Write X_n for the manifold $\mathfrak{S}_n \backslash (\mathbb{C}^n \setminus D)$. The braid group Σ_n is the fundamental group of this space: $\Sigma_n = \pi(X_n)$.

A point $\xi \in X_n$ is a unordered set of n distinct complex numbers $z_1, \ldots, z_n$. It can be specified by giving the coefficients $a_1, \ldots, a_n$ of the polynomial

$$t^n + a_1 t^{n-1} + \cdots + a_n = (t - z_1) \ldots (t - z_n).$$

Thus we can also say that $\Sigma_n \cong \pi(\mathbb{C}^n \setminus \Delta)$, where $\mathbb{C}^n$ is the space of the variables $a_1, \ldots, a_n$, and Δ is obtained by setting to zero the discriminant of the polynomial with coefficients $a_1, \ldots, a_n$.

§ 15. Examples of Groups: Lie Groups and Algebraic Groups

We now turn to the consideration of groups whose elements are given by continuously varying parameters; in other words, these are groups, occuring frequently in connection with questions of geometry or physics, whose set of elements itself has a geometry. This geometry may sometimes be very simple, but at other times far from trivial.

For example, the group of translations $x \mapsto x + \alpha$ (for $\alpha \in \mathbb{R}$) of the line, which reflects the coordinate change involved in changing the origin, is obviously isomorphic to the group of real numbers under addition, and is parametrised by points of the line. In the group of rotations of the plane about a fixed point O, each element is determined by the angle of rotation φ, and two values of φ determine the same rotation if they differ by an integer multiple of 2π. Hence our group is isomorphic to $\mathbb{R}/2\pi\mathbb{Z}$, and is parametrised by points of a circle with centre at O: if we fix some starting point P on the circle, a rotation is determined by the point to which it takes P. We can view the same circle as a fundamental domain of the group $2\pi\mathbb{Z}$, the interval $[0, 2\pi]$ with its end points identified. However, from examples as simple as these we do not as yet get a feeling for the specific nature of the situations arising here.

Example 1. The Group of Rotations of 3-Space. This group arises in connection with the description of the motion of a rigid body with one point fixed; we will assume that the body is 3-dimensional, not contained in a plane. We now attach rigidly to the body a coordinate system with centre at the fixed point O. Then a motion of the body defines a motion of the whole of space, namely, that for which the coordinates of each point in the moving coordinate system do not change; that is, 3-space moves together with the body. If we compare the position of all points at times $t = 0$ and $t = t_0$ then obviously they move in such a way that the distances between them do not change. In other words, passing from the initial position to the position at time $t = t_0$ is an *orthogonal transformation* φ_t of

3-space, fixing the origin O. However, since the transformation φ_t depends on t, and does so in a continuous way, it must preserve the orientation of 3-space. An orthogonal motion of 3-space preserving its orientation is called a *rotation*. It is in fact realised by a rotation through a definite angle around some axis; this is *Euler's theorem*, which can be proved by elementary geometric considerations. Alternatively, it follows from the fact that the characteristic polynomial $\det(\lambda E - A)$ of our transformation A, being a polynomial of degree 3 with negative constant term ($\det A > 0$, since A is orientation-preserving), must have a positive root, which must be 1 since A is orthogonal; the corresponding eigenvector defines the axis of rotation.

Thus the elements of the group of rotations describe all the possible positions occupied by a rigid body moving with a fixed point O, and any actual motion of this body is described by a curve in this group (with time t as parameter); the group of motions is the *configuration space* of a moving rigid body with a fixed point. What is this group like geometrically? To see it, let us specify a rotation about an axis l through an angle φ by a vector pointing in the direction of l and of length φ with $-\pi \leqslant \varphi \leqslant \pi$. Vectors of this form fill out a ball of radius π centred at O. However, points of the boundary sphere corresponding to the same axis l but with different values $\varphi = -\pi$ and $\varphi = \pi$ define the same rotation. Thus the group of rotations can be described as the ball in 3-space $\mathbb{R}^3$ with diametrically opposite points of the boundary identified. As is well known, under this identification we get 3-dimensional projective space $\mathbb{P}^3$. This is the geometric description of the group of rotations.

The same description of the group of rotations of 3-space can be obtained in another way. Consider the group G consisting of quaternions q of modulus 1 (see §8, Example 5). Writing $q = a + bi + cj + dk$, this is given by the equation

$$a^2 + b^2 + c^2 + d^2 = 1,$$

that is, it is the 3-dimensional sphere S^3. Let $\mathbb{H}^-$ be the 3-dimensional vector space of purely imaginary quaternions, defined by $\operatorname{Re} x = 0$. The group G acts on $\mathbb{H}^-$ by $x \mapsto qxq^{-1}$ for $x \in \mathbb{H}^-$ and $q \in G$. Since $|qxq^{-1}| = |q| \cdot |x| \cdot |q|^{-1} = |x|$, the action gives rise to an orthogonal transformation of $\mathbb{H}^-$. It is easy to see that q acts as the identity only if $q = \pm 1$, so that we obtain a homomorphism f of G into the group of orthogonal transformations of 3-space, with kernel ± 1. Since G is connected, the image of f must be contained in the group of rotations, and by comparing dimensions it is easy to see that it must coincide with it. In other words, we have the following result.

Theorem I. *The group of rotations of* 3*-space is isomorphic to the quotient* $G/\{\pm 1\}$ *of the group of quaternions of modulus* 1 *by the subgroup* $\{\pm 1\}$.

Since G is a 3-dimensional sphere, it follows that the group of rotations of 3-space is obtained from the sphere S^3 by identifying diametrically opposite points. We have thus once more obtained an identification of this group with projective 3-space.

We thus meet examples of transformation groups (the group of translations of the line, of rotations of the plane, and of rotations of 3-space) each of which has elements naturally parametrised in a 1-to-1 way by points of a certain manifold X (the line, the circle, projective 3-space). The next step is to abstract away from the specification of our group as a transformation group, and to assume that a manifold X adequately describes the set of elements of a group, and that a group law is specified on this set X. We thus arrive at the notion of a Lie group, which has two versions according as to whether we suppose that X is a differentiable or complex analytic manifold; the resulting group is called a differentiable or complex analytic Lie group. The definition is as follows:

Definition. A group G which is at the same time a differentiable (or complex analytic) manifold is a *Lie group* if the maps

$$G \to G \quad \text{given by} \quad g \mapsto g^{-1}$$

and

$$G \times G \to G \quad \text{given by} \quad (g_1, g_2) \mapsto g_1 g_2$$

are differentiable (or complex analytic).

The fact that the set of elements of a Lie group G is a manifold provides it with a geometry. The algebra (that is, the presence of a group law) means that this geometry is homogeneous. The elements of the left regular representation are called *translations* by elements of the group and they define a (differentiable or complex analytic) transitive transformation group of G. These allow us, for example, to start from a tangent vector τ at the identity $e \in G$, and to use left translation by $g \in G$ to get a tangent vector τ_g at any point g, that is a vector field on the whole of G. Vector fields of this form are said to be *left-invariant*. In the same way we can construct *left-invariant* (or right-invariant) *differential forms* on G. Finally, by the same method we can construct a *left-invariant* (or right-invariant) *Riemannian metric* on G.

Theorem I, which describes the group of rotations of 3-space, gives an example of geometric properties that are typical of many Lie groups. Firstly, the homomorphism $G \to G/\{\pm 1\}$, where G is the group of quaternions of modulus 1 is obviously an unramified cover. Since G is diffeomorphic to the 3-sphere, it is connected and simply connected, and hence it is the universal cover of the group of rotations (see § 14, Example 7). It follows from this that the group of rotations has fundamental group π of order 2. We can get not just topological, but also differential geometric information on the group of rotations. Its invariant Riemannian metric can be made compatible with that of the group G. But G is the sphere S^3, and is hence a manifold of positive Riemannian curvature. Hence the same is true of the group of rotations. It can be shown that for any compact Lie group the invariant Riemannian metric has nonnegative curvature, and that the directions in which the curvature is zero correspond to Abelian subgroups.

A closed submanifold $H \subset G$ of a Lie group G which is at the same time a subgroup is a *Lie subgroup* of G. In this case one can show that the set of cosets $H\backslash G$ is again a manifold, and that the quotient map $G \to H\backslash G$ and the action of G on $H\backslash G$ are differentiable (or complex analytic). Since the action of G on $H\backslash G$ is transitive, $H\backslash G$ is a homogeneous manifold with respect to G. We have the relation

$$\dim G = \dim H + \dim H\backslash G, \tag{1}$$

which corresponds to § 12, (9).

In what follows we will treat in more detail two types of Lie groups, compact and complex analytic, and describe the most important examples of both types and the relations between them.

A. Compact Lie Groups

Example 2. Toruses. In a real n-dimensional vector space L, consider the lattice $C = \mathbb{Z}e_1 + \cdots + \mathbb{Z}e_n$, where $e_1, \ldots, e_n$ is some basis of L. The quotient group $T = L/C$ is compact. It is a Lie group and is called a *torus*. Since $L = \mathbb{R}e_1 + \cdots + \mathbb{R}e_n$, we have

$$T \cong (\mathbb{R}/\mathbb{Z}) \times \cdots \times (\mathbb{R}/\mathbb{Z}).$$

The quotient group $\mathbb{R}/\mathbb{Z}$ is a circle, and an n-dimensional torus is a direct product of n circles. Toruses have an enormous number of applications, and we indicate three of these.

(a) A periodic function of period 2π is a function on the circle $\mathbb{R}/2\pi$. As we will see later, this point of view gives a new way of looking at the theory of Fourier series.

(b) Take $L = \mathbb{C}$ to be the plane of one complex variable; we have already considered this example in § 14, Example 1. If $C \subset \mathbb{C}$ is a lattice then the torus $\mathbb{C}/C$ inherits from $\mathbb{C}$ the structure of a complex analytic manifold. As a complex manifold $\mathbb{C}/C$ is 1-dimensional. It can be shown that these are the unique compact complex analytic Lie groups of complex dimension 1. In a similar way, an arbitrary compact complex analytic Lie group is a torus $\mathbb{C}^n/C$, with $C = \mathbb{Z}e_1 + \cdots + \mathbb{Z}e_{2n}$ a lattice in the $2n$-dimensional real vector space $\mathbb{C}^n$. In particular, such a group is necessarily Abelian.

(c) In Arnol'd's treatment of classical mechanics, Liouville's theorem asserts that given a mechanical system with n degrees of freedom, if we know n independent first integrals $I_1, \ldots, I_n$ in involution (that is, all Poisson brackets vanish, $[I_\alpha, I_\beta] = 0$), then the system can be integrated by quadratures. The proof is based on the fact that in the $2n$-dimensional phase space, the n-dimensional level manifold T_c, where $c = (c_1, \ldots, c_n)$, given by

$$T_c\colon I_\alpha = c_\alpha \quad (\text{for } \alpha = 1, \ldots, n)$$

is a torus. This follows at once from the fact that on the T_c, the functions I_α define n vector fields: these are given by the differential forms dI_α using the symplectic structure defined on phase space. Each vector field defines a 1-parameter group $U_\alpha(t)$ of transformations, and the relations $[I_\alpha, I_\beta] = 0$ mean that the transformations $U_\alpha(t_1)$ and $U_\beta(t_2)$ commute. Thus the Lie group $\mathbb{R}^n$ acts on the manifold T_c, with $(t_1, \ldots, t_n) \in \mathbb{R}^n$ corresponding to the transformation $U_1(t_1), \ldots, U_n(t_n)$. It follows from this that T_c is a quotient group of $\mathbb{R}^n$ by the stabiliser subgroup H of some point $x_0 \in T_c$. Since T_c is n-dimensional and compact (because kinetic energy, which is a positive definite form, is constant on it), $T_c = \mathbb{R}^n/H$ is a torus. Thus the motion of a point corresponding to the system always takes place on a torus, and moreover it can be proved that the point moves in a 1-dimensional subgroup of the torus (to this there corresponds the introduction of the so-called 'action angles').

We move on to describe non-Abelian compact Lie groups. We will describe three series of groups (Examples 3, 4 and 5), usually called the *classical groups*. Each of these groups occurs in several versions, usually as certain matrix groups. For a matrix group G, we write SG for the set of all elements of G with determinant 1 (here S stands for 'special'); the quotient groups of G and SG by their centres are denoted by PG and PSG (where P stands for 'projective').

Example 3. The *orthogonal group* $O(n)$ consists of all orthogonal transformations of n-dimensional Euclidean space. This group acts on the unit sphere S^{n-1} in n-space. If $e \in S^{n-1}$ with $|e| = 1$, the stabiliser subgroup of e acts on the hyperplane orthogonal to e, and is isomorphic to $O(n-1)$. Hence in view of (1), $\dim O(n) = \dim O(n-1) + n - 1$, and therefore

$$\dim O(n) = \binom{n}{2}.$$

The group $O(n)$ is not connected. It has an important subgroup of index 2, denoted by $SO(n)$ and consisting of orthogonal transformations of determinant 1. It is easy to prove that the group $SO(n)$ is connected. If n is odd then the centre of $SO(n)$ consists of E only, and if $n \geqslant 4$ is even, then it consists of E and $-E$. The quotient group of $SO(n)$ by its centre is denoted by $PSO(n)$. The group of motions of 3-space treated in Example 1 is $SO(3)$.

A natural generalisation of $O(n)$ is related to considering an arbitrary non-degenerate quadratic form

$$x_1^2 + \cdots + x_p^2 - x_{p+1}^2 - \cdots - x_{p+q}^2 \quad \text{with} \quad p + q = n.$$

Linear transformations of $\mathbb{R}^n$ preserving this quadratic form form a Lie group denoted by $O(p, q)$. This is a compact group only if $p = 0$ or $q = 0$. We will meet these groups with other values of p and q later.

Example 4. The *unitary group* $U(n)$ consists of the unitary transformations of an n-dimensional Hermitian complex vector space. As in Example 3 it is proved that $\dim U(n) = n^2$. The determinant of a transformation in $U(n)$ is a complex number of absolute value 1; the transformations with determinant 1 form a subgroup $\mathrm{SU}(n) \subset \mathrm{U}(n)$ of dimension $n^2 - 1$. The centre of $\mathrm{SU}(n)$ consists of the transformations εE where $\varepsilon^n = 1$. The quotient group by the centre is denoted by $\mathrm{PSU}(n)$.

Example 5. Consider the n-dimensional vector space $\mathbb{H}^n$ over the algebra of quaternions (§ 8, Example 5). On $\mathbb{H}^n$ we define the scalar product with values in $\mathbb{H}$

$$((x_1, \ldots, x_n), (y_1, \ldots, y_n)) = \sum_{i=1}^{n} x_i \overline{y_i}, \tag{2}$$

where $\overline{y_i}$ are the conjugate quaternions. The group of linear transformations in $\mathrm{Aut}_{\mathbb{H}} \mathbb{H}^n$ that preserves this scalar product is called the *unitary symplectic group* and is denoted by $\mathrm{SpU}(n)$. For $n = 1$ we obtain the group of quaternions q of modulus 1.

In the general case,

$$\dim \mathrm{SpU}(n) = 2n^2 + n.$$

There are relations between the different classical Lie groups that are often useful.

Theorem I can now be rewritten in the form

$$\mathrm{SpU}(1)/\{\pm 1\} \cong \mathrm{SO}(3). \tag{3}$$

As we have seen, it follows from this that $|\pi(\mathrm{SO}(3))| = 2$.

An analogous representation for the groups $\mathrm{SO}(n)$, and even for all of the groups $\mathrm{SO}(p, q)$ (which are in general noncompact) can be obtained using the Clifford algebra (§ 8, Example 10). The fact that for any quaternion q the transformation $x \mapsto qxq^{-1}$ takes the space of purely imaginary quaternions into itself is a special feature of the case $n = 3$. In the general case, consider the Clifford algebra $C(L)$ corresponding to the space L over $\mathbb{R}$ with the metric

$$x_1^2 + \cdots + x_p^2 - x_{p+1}^2 - \cdots - x_{p+q}^2 \quad \text{with} \quad p + q = n.$$

Recall that $L \subset C(L)$. We introduce the group G of invertible elements $a \in C^0(L)$ for which $a^{-1}La \subset L$. Obviously G is a group. It is easy to check that for $a \in G$ the map $x \mapsto a^{-1}xa$ for $x \in L$ preserves the metric of L. Thus we get the homomorphism

$$f\colon G \to \mathrm{O}(p, q).$$

It is easy to see that the kernel of f consists of $a \in \mathbb{R}$ with $a \neq 0$. It can be proved (using the well-known fact that any orthogonal transformation can be expressed as a composite of reflections) that the image of f is $\mathrm{SO}(p, q)$ and that any element of G can be written in the form $a = c_1 \ldots c_r$ with $c_i \in L$ for some even r. It follows from this that for $a \in G$ the element $aa^* \in \mathbb{R}$, where $a \mapsto a^*$ is the involution of $C(L)$ (see § 8, Example 10), and if we set $aa^* = N(a)$ then for a, $b \in G$ we have

$N(ab) = N(a)N(b)$. Hence the elements $a \in G$ with $N(a) = 1$ form a group. This is called the *spinor group* and denoted by $\mathrm{Spin}(p, q)$. When $q = 0$ it is denoted by $\mathrm{Spin}(n)$. It is easy to see that the group $\mathrm{Spin}(p, q)$ is connected. The kernel of the homomorphism $f: \mathrm{Spin}(p, q) \to \mathrm{O}(p, q)$ consists of $\{\pm E\}$. The image depends on the numbers p and q. If $q = 0$ then obviously $\mathrm{O}(p, q) = \mathrm{O}(n)$ and it is easy to check that $f(\mathrm{Spin}(n)) = \mathrm{SO}(n)$. That is,

$$\mathrm{Spin}(n)/\{\pm 1\} = \mathrm{SO}(n). \tag{4}$$

Thus the group $\mathrm{Spin}(n)$ is a double covering of $\mathrm{SO}(n)$. Using an induction, it is easy to prove (starting with $n = 3$) that $\pi(\mathrm{SO}(n))$ has order 1 or 2. But we have constructed a 2-sheeted cover $\mathrm{Spin}(n) \to \mathrm{SO}(n)$, and this proves that $|\pi(\mathrm{SO}(n))| = 2$, and that $\mathrm{Spin}(n)$ is simply connected.

If $p > 0$ and $q > 0$ then G contains both elements with positive and with negative norms, and their images define two distinct components of $\mathrm{SO}(p, q)$. The image of the elements with positive norm form a subgroup $\mathrm{SO}^+(p, q) \subset \mathrm{SO}(p, q)$ of index 2, and $f(\mathrm{Spin}(p, q)) = \mathrm{SO}^+(p, q)$, that is

$$\mathrm{Spin}(p, q)/\{\pm 1\} \cong \mathrm{SO}^+(p, q). \tag{5}$$

As we saw in § 8, Example 6, any quaternion can be written in the form

$$q = z_1 + jz_2 \quad \text{with} \quad z_1, z_2 \in \mathbb{C}; \tag{6}$$

here $j^2 = -1$ and $zj = j\bar{z}$. In the form (6) the quaternions form the 2-dimensional vector space $\mathbb{C}^2$ over $\mathbb{C}$, where multiplication by $z \in \mathbb{C}$ is taken as multiplication on the right. Hence the left regular representation gives a representation of the quaternions by $\mathbb{C}$-linear transformations of $\mathbb{C}^2$, that is by 2×2 matrixes. Taking the basis $\{1, j\}$ sends the quaternion (6) to the matrix $\begin{bmatrix} -z_1 & z_2 \\ -\bar{z}_2 & \bar{z}_1 \end{bmatrix}$.

In the notation (6), the modulus of the quaternion q is $\sqrt{(|z_1|^2 + |z_2|^2)}$. Hence multiplication by a quaternion of modulus 1 gives a unitary transformation of $\mathbb{C}^2$ with the metric $|z_1|^2 + |z_2|^2$. Moreover, the determinant of the above matrix is also $|z_1|^2 + |z_2|^2$, that is, in our case equal to 1. We thus get a homomorphism

$$\mathrm{SpU}(1) \to \mathrm{SU}(2), \tag{7}$$

whose kernel is 1. From considerations of dimension and because $\mathrm{SU}(2)$ is connected, we see that this is an isomorphism, so that $\mathrm{SU}(2)$ is isomorphic to the group of quaternions of modulus 1. Putting together (3) and (7) we get the isomorphism

$$\mathrm{SO}(3) = \mathrm{SU}(2)/\{\pm 1\}. \tag{8}$$

An elementary interpretation of this is as follows: consider the set L of 2×2 Hermitian matrixes of trace 0. The group $\mathrm{SU}(2)$ acts on this by $A \mapsto UAU^{-1}$ for $U \in \mathrm{SU}(2)$. Introducing a metric on L by $|A|^2 = -\det A$ for $A \in L$ makes L into a 3-dimensional Euclidean space. The transformation corresponding to $U \in \mathrm{SU}(2)$ defines a transformation $\gamma \in \mathrm{SO}(3)$. This is the homomorphism

$$\mathrm{SU}(2) \to \mathrm{SO}(3).$$

Let $\mathscr{L}$ be the 4-dimensional real vector space of quaternions, with the metric given by the modulus. We define an action of $\mathrm{SpU}(1) \times \mathrm{SpU}(1)$ on $\mathscr{L}$ by

$$(q_1, q_2)(x) = q_1 x q_2^{-1} \quad \text{for } x \in \mathscr{L} \tag{9}$$

It is easy to see that only the pairs $(1, 1)$ and $(-1, -1)$ act trivially. Our transformations obviously preserve the modulus, and so are orthogonal. The determinant of the transformations (9) is $|q_1| \cdot |q_2|^{-1}$, that is in our case 1. We get a homomorphism

$$\mathrm{SpU}(1) \times \mathrm{SpU}(1) \to \mathrm{SO}(4),$$

the kernel of which we know. By considerations of connectedness and dimension, the image is the whole of SO(4). We have obtained an isomorphism

$$\mathrm{SO}(4) \cong (\mathrm{SpU}(1) \times \mathrm{SpU}(1))/H, \tag{10}$$

where H is a subgroup of the center of order 2. The whole centre of $\mathrm{SpU}(1) \times \mathrm{SpU}(1)$ is the product of two groups of order 2, the centres of SpU(1). Taking the quotient of the left-hand side of (10) by the centre of SO(4), we must on the left-hand side quotient $\mathrm{SpU}(1) \times \mathrm{SpU}(1)$ by the whole of its centre. But the quotient of SpU(1) by its centre is SO(3). Therefore

$$\mathrm{PSO}(4) \cong \mathrm{SO}(3) \times \mathrm{SO}(3). \tag{11}$$

B. Complex Analytic Lie Groups

The three series of complex Lie groups given in the following Examples 6, 7, 8 are also called the *classical groups*.

Example 6. The *general linear group* $\mathrm{GL}(n, \mathbb{C})$ is the group of nondegenerate linear transformations of an n-dimensional complex vector space. The dimension of $\mathrm{GL}(n, \mathbb{C})$ as a complex analytic manifold is obviously n^2. It contains the subgroup $\mathrm{SL}(n, \mathbb{C})$ of linear transformations of determinant 1. The centre of $\mathrm{GL}(n, \mathbb{C})$ consists of scalar multiples of the identity matrix; the quotient of $\mathrm{GL}(n, \mathbb{C})$ by its centre is denoted by $\mathrm{PGL}(n, \mathbb{C})$.

Example 7. The subgroup of $\mathrm{GL}(n, \mathbb{C})$ consisting of transformations that preserve some nondegenerate quadratic form ($x_1^2 + \cdots + x_n^2$ in a suitable coordinate system) is denoted by $\mathrm{O}(n, \mathbb{C})$ and is called the *orthogonal group*. Its dimension as a complex analytic variety is $\binom{n}{2}$.

Example 8. The subgroup of $\mathrm{GL}(2n, \mathbb{C})$ consisting of transformations that preserve some nondegenerate skew-symmetric form (of the form

$$\sum_{i=1}^{n} (x_i y_{n+i} - x_{n+i} y_i)$$

in a suitable coordinate system) is called the *symplectic group* and denoted by $\mathrm{Sp}(2n, \mathbb{C})$.

The compact and complex analytic classical groups are closely related; the simplest example of this is the circle group S^1 viewed as the subgroup $S^1 \subset \mathbb{C}^*$ of complex numbers of absolute value 1:

$$S^1 = \mathrm{SU}(1) \subset \mathrm{GL}(1, \mathbb{C}) = \mathbb{C}^*.$$

More generally, the relation is as follows. Obviously $\mathrm{U}(n) \subset \mathrm{GL}(n, \mathbb{C})$. It can be proved that $\mathrm{U}(n)$ is a *maximal compact subgroup* of $\mathrm{GL}(n, \mathbb{C})$, that is, it is not contained in any bigger compact subgroup. Any other compact subgroup of $\mathrm{GL}(n, \mathbb{C})$ is conjugate to a subgroup of $\mathrm{U}(n)$. The reason for this will be explained in § 17B. Similarly, the group $\mathrm{O}(n) \subset \mathrm{O}(n, \mathbb{C})$ is a maximal compact subgroup, and any compact subgroup of $\mathrm{O}(n, \mathbb{C})$ is conjugate to a subgroup of $\mathrm{O}(n)$. To establish the analogous result for the symplectic groups, we use the expression (6) of the quaternions as a 2-dimensional complex vector space, $\mathbb{H} = \mathbb{C} + j\mathbb{C}$. Then a vector $x = (x_1, \ldots, x_n) \in \mathbb{H}^n$ can be written as $(z_1, \ldots, z_n, z_{n+1}, \ldots, z_{2n})$ where $z_i \in \mathbb{C}$ and $x_k = z_k + jz_{k+n}$. In these coordinates, as one checks easily, the product (2) of Example 5 takes the form

$$(x, y) = \sum_{i=1}^{n} z_i \overline{w_i} + j \sum (z_i w_{i+n} - w_i z_{i+n})$$

(where $y = (y_1, \ldots, y_n)$ and $y_k = w_k + jw_{k+n}$). Thus if $(x, y) = \alpha + j\beta$ then α is a Hermitian scalar product of the complex vectors x and y, and β is the value of the skew-symmetric form $\sum (z_i w_{i+n} - w_i z_{i+n})$. Every $\mathbb{H}$-linear transformation φ of $\mathbb{H}^n$ can be written as a $\mathbb{C}$-linear transformation of $\mathbb{C}^{2n}$ and by what we have said above, the condition $\varphi \in \mathrm{SpU}(n)$ means that $\varphi \in \mathrm{U}(2n)$ and $\varphi \in \mathrm{Sp}(2n, \mathbb{C})$. Thus

$$\mathrm{SpU}(n) = \mathrm{U}(2n) \cap \mathrm{Sp}(2n, \mathbb{C}).$$

In particular, $\mathrm{SpU}(n)$ is a subgroup of $\mathrm{Sp}(2n, \mathbb{C})$. It is a maximal compact subgroup, and any compact subgroup of $\mathrm{Sp}(2n, \mathbb{C})$ is conjugate to a subgroup of $\mathrm{SpU}(n)$.

In all three cases, the dimension of the ambient complex group (as a complex analytic manifold) is equal to the dimension of the compact subgroup (as a differentiable manifold).

To conclude we treat some important Lie groups of small dimensions.

The group $\mathrm{O}(3, 1)$ is called the *Lorentz group*, and $\mathrm{SO}(3, 1)$ the *proper Lorentz group*. If we interpret x_1, x_2, x_3 as space coordinates and x_0 as time, then preserving the form $f = -x_0^2 + x_1^2 + x_2^2 + x_3^2$ is equivalent to preserving the speed of light (which we consider to be equal to 1). The same group has another interpretation, that is just as important. Consider x_0, x_1, x_2, x_3 as homogeneous coordinates in a 3-dimensional projective space $\mathbb{P}^3$. The equation $f = 0$ defines in $\mathbb{P}^3$ a surface of degree 2, which can be written in inhomogeneous coordinates $y_i = x_i/x_0$ for $i = 1, 2, 3$ as $y_1^2 + y_2^2 + y_3^2 = 1$. Thus this is the sphere $S \subset \mathbb{P}^3$. Considering transformations of O(3, 1) in homogeneous coordinates makes them into projective transformations of $\mathbb{P}^3$ preserving S. Of course, multiplying all the coordinates through by -1 gives the identity transformation, so that the group PO(3, 1) acts on $\mathbb{P}^3$. Obviously, this group also preserves the interior of S. But as is well known, in the Cayley-Klein model of non-Euclidean geometry, 3-dimensional Lobachevsky space is represented exactly as the points of the interior of the sphere, and its motions by projective transformations preserving the sphere. This proves the next theorem.

Theorem II. PO(3, 1) *is isomorphic to the group of all motions of* 3*-dimensional Lobachevsky space, and* PSO(3, 1) *to the group of all proper* (*orientation-preserving*) *motions.*

Of course, this assertion is of a general nature: the group PO(n, 1) is isomorphic to the group of motions of n-dimensional Lobachevsky space.

The Lorentz group O(3, 1) has another important interpretation. This is based on considering the spin group Spin(3, 1) (see (5) above). In the course of constructing this group, we saw that for $a \in G$ the norm $N(a) = aa^* \in \mathbb{R}$. But in our particular case this condition is sufficient: if $aa^* = \alpha \in \mathbb{R}$ and $\alpha \neq 0$ then $a^{-1}La \subset L$, that is $\alpha \in G$. In fact from $\alpha^* = \alpha$ we get $a^{-1} = \alpha^{-1}a^*$, and it follows from this that $(a^{-1}xa)^* = a^{-1}xa$ for $x \in L$. On the other hand, $a^{-1}xa \in C^1$, and C^1 consists of linear combinations of the elements e_i and the products $e_ie_je_k$, where the e_i form an orthogonal basis of L. Of these, only the linear combinations of the e_i do not change sign under $x \mapsto x^*$, and hence $a^{-1}xa \in L$. Now we use the fact that we can find an explicit representation of the algebra C^0 as a matrix algebra (see § 8, Example 11 and § 10, Example 6). If e_0, e_1, e_2, e_3 is the basis in which f has the form $-x_0^2 + x_1^2 + x_2^2 + x_3^2$ then $1, e_0e_1, e_0e_2$ and e_1e_2 generate the algebra $M_2(\mathbb{C})$, 1 and $e_0e_1e_2e_3$ the algebra $\mathbb{C}$, and the whole of C^0 is isomorphic to $M_2(\mathbb{C})$. It is easy to check that under this isomorphism the involution $a \mapsto a^*$ corresponds to the map

$$\begin{bmatrix} \alpha & \beta \\ \gamma & \delta \end{bmatrix} \mapsto \begin{bmatrix} \delta & -\beta \\ -\gamma & \alpha \end{bmatrix},$$

and the condition $aa^* \in \mathbb{R}$ means that $\det A \in \mathbb{R}$ and $A \in M_2(\mathbb{C})$. Hence Spin(3, 1) is isomorphic to SL(2, $\mathbb{C}$), and we get a homomorphism

$$\mathrm{SL}(2, \mathbb{C}) \to \mathrm{SO}(3, 1).$$

Its image is $\mathrm{SO}^+(3, 1)$ and the kernel is $\{\pm 1\}$. Thus

$$\mathrm{PSL}(2, \mathbb{C}) \cong \mathrm{SO}^+(3, 1).$$

The homomorphism $\mathrm{SL}(2, \mathbb{C}) \to \mathrm{SO}(3, 1)$ has the following elementary interpretation. Consider the space L of 2×2 Hermitian matrixes, and the action of

$SL(2, \mathbb{C})$ on L by $A \mapsto CAC^*$ for $A \in L$ and $C \in SL(2, \mathbb{C})$. Introduce on L the metric $|A|^2 = -\det A$, which is of the form $-x_0^2 + x_1^2 + x_2^2 + x_3^2$ in some basis. The action of $SL(2, \mathbb{C})$ described above therefore defines a homomorphism $SL(2, \mathbb{C}) \to O(3, 1)$. This is the same as our homomorphism $SL(2, \mathbb{C}) \to SO(3, 1)$.

C. Algebraic Groups

We now consider a class of groups that gives interesting examples of Lie groups, discrete groups and finite groups all at one go.

We treat only one case of these, the *algebraic matrix groups* (also called *linear algebraic groups*). These can be defined over an arbitrary field K as the subgroups of $GL(n, K)$ given by algebraic equations with coefficients in K. Examples are: $SL(n, K)$; $O(f, K)$, the group of matrixes preserving f, for some quadratic form f with coefficients in K; $Sp(2n, K)$; the group of upper-triangular matrixes (a_{ij}) with $a_{ij} = 0$ for $i < j$ and $a_{ii} \neq 0$, or its subgroup in which all $a_{ii} = 1$. In particular, the group consisting of 2×2 matrixes of the form $\begin{bmatrix} 1 & a \\ 0 & 1 \end{bmatrix}$ is isomorphic to the group of elements of K under addition; this is denoted by $\mathbb{G}_a$. The group $GL(1, K)$ is isomorphic to the group of elements of K under multiplication, and is denoted by $\mathbb{G}_m$ or K^*. If the field K is contained in $\mathbb{R}$ or $\mathbb{C}$ and G is an algebraic group defined over K, then the real or complex matrixes in G define a real Lie group $G(\mathbb{R})$ or a complex analytic group $G(\mathbb{C})$. The majority of the Lie groups we have considered are of this type. But the general notion is more flexible, since for example it allows us to consider algebraic groups over the rational number field. Thus considering the group $O(f, \mathbb{Q})$ gives a group-theoretic method of studying arithmetical properties of a rational quadratic form f. Moreover, considering matrixes with integral entries and with determinant ± 1 in an algebraic matrix group G defined over $\mathbb{Q}$, we obtain a discrete subgroup $G(\mathbb{Z})$ of the Lie group $G(\mathbb{R})$. For the groups $G = SL(n)$, $O(f)$, $Sp(n)$ the quotient spaces $G(\mathbb{Z})\backslash G(\mathbb{R})$ are either compact or have finite volume (in the sense of the measure defined by an invariant measure on $G(\mathbb{R})$); for examples, see § 14, Examples 3a–5. Groups of this form, and also their subgroups of finite index are called *arithmetic groups*; to treat this notion in the natural generality would require us to consider as well as $\mathbb{Q}$ any algebraic number field.

Finally, algebraic groups such as $GL(n)$, $O(n)$ and $Sp(n)$ can also be considered over finite fields, and they give interesting examples of finite groups. We have already met the group $GL(n, \mathbb{F}_q)$ in § 13, Example 9.

There exists another completely unexpected way in which discrete groups arise in connection with algebraic groups. If a matrix group G is defined over the rational number field $K = \mathbb{Q}$ then we can consider the groups $G(\mathbb{Q})$ of its rational points, $G(\mathbb{R})$ of its real points or $G(\mathbb{Q}_p)$ of its p-adic points (see the end of § 7). The most invariant way of treating the fields $\mathbb{R}$ and $\mathbb{Q}_p$ on an equal footing is to consider the 'infinite product' of $G(\mathbb{R})$ and all of the $G(\mathbb{Q}_p)$. We will not give a precise definition of this product, which is called the *adèle group* of G and denoted by $G_{\mathbb{A}}$. Since $G(\mathbb{Q}) \subset G(\mathbb{R})$ and $G(\mathbb{Q}) \subset G(\mathbb{Q}_p)$ for all p we have a diagonal inclusion $G(\mathbb{Q}) \subset G_{\mathbb{A}}$. It turns out that $G(\mathbb{Q})$ is a discrete subgroup of $G_{\mathbb{A}}$. The idea that matrixes with rational entries should form a discrete subgroup is a very unfamiliar one, although the principle is easy to understand in the example $G = \mathbb{G}_a$, the group of all numbers under addition. If $x \in G(\mathbb{Q})$ is a rational number then the condition $\varphi_p(x) \leqslant 1$ for all p (see §7 for the definition of the valuation φ_p) means that x is an integer, and $\varphi_R(x) < 1$ then implies that $x = 0$. In a number of cases (such as, for example, $G = \mathrm{SL}(n)$, $\mathrm{O}(f)$, $\mathrm{Sp}(n)$) the quotient space $G_{\mathbb{Q}} \backslash G_{\mathbb{A}}$ is of finite volume. It can be shown that this volume is uniquely determined by the group G only. This volume is the so-called *Tamagawa number* $\tau(G)$ of G, and is a very important arithmetic invariant. For example, if f is a rational positive definite quadratic form then it follows from the Minkowski-Hasse theorem (§7, Theorem IV) that the equation $f(x) = a$ for rational x and a is solvable if and only if it is solvable in $\mathbb{R}$ (that is, $a > 0$) and in all $\mathbb{Q}_p$ (that is the congruences $f(x) \equiv a \bmod p^n$ are all solvable). But if these conditions are satisfied, we can give numerical characterisations of the number of integral solutions in terms of the number of solutions of the congruences $f(x) \equiv a \bmod p^n$; this turns out to be equivalent to finding the Tamagawa number $\tau(G)$ (which is equal to 2: this is a reflection of the fact that $|\pi(SO(n)| = 2)$.

§16. General Results of Group Theory

The ideal result of 'abstract' group theory would be a description of all possible groups up to isomorphism, completely independently of concrete realisations of the groups. In this generality, the problem is of course entirely impracticable. More concretely one could envisage the problem (which is still very wide) of classifying all finite groups. Since only a finite number of Cayley tables (multiplication tables) can be made up from a finite number of elements, there are only finitely many nonisomorphic groups of a given order; ideally one would like a rule specifying all finite groups of given order. For fairly small orders this can be done without much difficulty, and we run through the groups which arise.

We should recall that for finite Abelian groups the answer is provided by the basic theorem on finitely generated modules over a principal ideal ring (see §5, Example 6 and §6, Theorem II): this says that a finite Abelian group (written additively) can be represented as a direct sum of groups $\mathbb{Z}/(p^k)$, where the p are prime numbers, and such a representation is unique. Thus only non-Abelian groups cause any difficulty.

Example 1. We now list all groups of order $\leqslant 10$ up to isomorphism:

$$|G| = 2: \quad G \cong \mathbb{Z}/(2).$$

$$|G| = 3: \quad G \cong \mathbb{Z}/(3).$$

$$|G| = 4: \quad G \cong \mathbb{Z}/(4) \text{ or } \mathbb{Z}/(2) \oplus \mathbb{Z}/(2).$$

$$|G| = 5: \quad G \cong \mathbb{Z}/(5).$$

For $|G| = 6$, a non-Abelian group appears for the first time, namely the group isomorphic to $\mathfrak{S}_3$ or to the symmetry group of an equilateral triangle; defining relations for it are given in § 12, (6) and (7). Thus:

$$|G| = 6: \quad G \cong \mathfrak{S}_3 \text{ or } \mathbb{Z}/(2) \oplus \mathbb{Z}/(3).$$

$$|G| = 7: \quad G \cong \mathbb{Z}/(7).$$

For $|G| = 8$ there are already two nonisomorphic non-Abelian groups. One of these is the group D_4 of symmetries of the square; this is also the group generated by two elements s and t with defining relations $s^2 = e$, $t^4 = e$, $(st)^2 = e$ (here s is a reflection in one of the medians and t is a rotation through 90°). The other non-Abelian group H_8 can be described in terms of the quaternion algebra (§ 8, Example 5); it consists of $1, i, j, k, -1, -i, -j, -k$, multiplying together as quaternions. Thus:

$$|G| = 8: \quad G \cong D_4, H_8, \mathbb{Z}/(8), \mathbb{Z}/(4) \oplus \mathbb{Z}/(2) \text{ or } (\mathbb{Z}/(2))^{\oplus 3}.$$

$$|G| = 9: \quad G \cong \mathbb{Z}/(9) \text{ or } \mathbb{Z}/(3) \oplus \mathbb{Z}/(3).$$

For $|G| = 10$ a non-Abelian group again arises, isomorphic to the group D_5 of symmetries of a regular pentagon, generated by elements s and t with defining relations $s^2 = e$, $t^5 = e$, $(st)^2 = e$ (where s is a reflection in an axis and t a rotation through $2\pi/5$). Thus:

$$|G| = 10: \quad G \cong D_5 \text{ or } \mathbb{Z}/(5) \oplus \mathbb{Z}/(2).$$

We include a table giving the number of groups of given order $\leqslant 32$:

$\lvert G\rvert$	2	3	4	5	6	7	8	9	10	11	12	13	14	15	16	17
number of groups	1	1	2	1	2	1	5	2	2	1	5	1	2	1	14	1

18	19	20	21	22	23	24	25	26	27	28	29	30	31	32
5	1	5	2	2	1	15	2	2	5	4	1	4	1	51

Of course, in fact the study of the structure of finite groups uses not just their order, but also other more precise invariants, and moreover uses methods of constructing more complicated groups from simpler ones.

We return to the general notion of group and describe the basic methods of constructing groups. One of these we have already used, namely direct product. By analogy with the case of two factors, we define the *direct product* $G_1 \times \cdots \times G_m$ of any finite number of groups $G_1, \ldots, G_m$: this consists of sequences $(g_1, \ldots, g_m)$ with $g_i \in G_i$, with multiplication defined component-by-component. The groups G_i can be viewed as subgroups of the product $G_1 \times \cdots \times G_m$, by identifying $g \in G_i$ with the sequence $(e, \ldots, g, \ldots, e)$ with g in the ith place. Under this the G_i are normal subgroups of the product $G_1 \times \cdots \times G_m$, generating it and satisfying $G_i \cap (G_1 \times \cdots \times G_{i-1} \times e \times G_{i+1} \times \cdots \times G_m) = e$. It is not hard to show that conversely, if a group G is generated by normal subgroups G_i with the property that $G_i \cap (G_1 \ldots G_{i-1} G_{i+1} \ldots G_m) = e$ then G is isomorphic to the direct product of the G_i.

We have seen that for finite Abelian groups, or even for finitely generated Abelian groups, direct sum turns out to be a powerful construction, sufficient for a complete classification of these groups. However, for this it is essential that the direct sum decomposition appearing in the classification theorem (§6, Theorem II) is unique. It is natural to ask whether for non-Abelian groups also, decomposition as a direct product of groups which cannot be decomposed any further is unique. We give the answer to this in the simplest case, which will however be sufficient in most applications. By analogy with modules, we consider chains of subgroups

$$G \supseteq H_1 \supseteq H_2 \supseteq \cdots \supseteq H_k. \tag{1}$$

If the length of all such chains is bounded then G is a *group of finite length*. Finite groups are of this kind. If G is a Lie group or an algebraic group, then it is natural to consider in the definition only chains (1) with H_i connected closed Lie subgroups or algebraic subgroups. Then the dimension of subgroups in a chain (1) must decrease, so that the length of the chain is bounded by the dimension of G.

I. The Wedderburn-Remak-Shmidt Theorem. *A group of finite length has one and only one decomposition as a direct sum of normal subgroups which cannot be decomposed any further. More precisely, any two such decompositions must have the same number of factors and the factors must be isomorphic in pairs.*

However, a non-Abelian group (for example a finite group or a Lie group) is only in exceptional cases decomposable as a direct sum: the great majority of them are indecomposable. A more universal method of reducing groups to simpler component parts is provided by the notion of homomorphism. If $G' = G/N$ then the homomorphism $G \to G'$ allows us to view G' as a kind of simplified version of G, obtained by considering G 'up to elements of N'. Now, what is the extreme nontrivial extent to which a group can be 'simplified' in this way? We could consider a group homomorphism $G' \to G''$, and so on. If G is of finite length then this process must stop, which happens when we arrive at a group $\bar{G}$ not having any nontrivial homomorphisms; this means that $\bar{G}$ does not have any normal subgroups other than $\{e\}$ and $\bar{G}$ itself.

We say that a group G without any normal subgroups other than $\{e\}$ and G itself is *simple*. In the case of Lie groups or algebraic groups it is natural to talk of connected normal Lie subgroups or algebraic subgroups. Thus a Lie group which is simple according to our definition may fail to be simple as an abstract group if it contains a discrete normal subgroup. An example is $\mathbb{R}$, containing the subgroup $\mathbb{Z}$. We have seen that every group of finite length has a homomorphism to a simple group. Let $G \to G'$ be one such homomorphism and N_1 its kernel. It is then natural to apply the same construction to N_1. We obtain a homomorphism $N_1 \to G''$ to a simple group G'' with kernel N_2 which is a normal subgroup in N_1 (but not necessarily in G). Continuing this procedure, we obtain a chain $G = N_0 \lhd N_1 \lhd \cdots \lhd N_k \lhd N_{k+1} = \{e\}$, in which the quotient groups N_i/N_{i+1} are simple. Such a chain is called a *composition series* for G, and the quotient groups N_i/N_{i+1} the *quotients* of the composition series. Of course, the same group may have different composition series, so that the following result is very important.

II. The Jordan-Hölder Theorem. *Two composition series of the same group have the same length, and their quotients are isomorphic in pairs (but possibly occur in a different order).*

The proofs of the Jordan-Hölder and the Wedderburn-Remak-Shmidt theorems make very little use of properties of groups. The basic fact which they use is the following.

Lemma. *If H is a subgroup of G and N a normal subgroup of G then $H \cap N$ is normal in H and*

$$H/H \cap N \cong HN/N. \tag{2}$$

Here HN is the subgroup of G consisting of all products of the form hn with $h \in H$ and $n \in N$.

For consider the homomorphism $f \colon G \to G/N$; the restriction of f to H defines a homomorphism $f_1 \colon H \to G/N$ with kernel $H \cap N$ and image $H_1 \cong H/H \cap N$. The inverse image of H_1 under f is HN, so that $H_1 \cong HN/N$, and (2) follows from this.

The proof of both theorems is based on the idea that replacing the pair $(H, H \cap N)$ by (HN, N) for various choices of H and N, we can pass from one decomposition of a group as a direct sum of indecomposable subgroups to another, or from one composition series to another. In essence these arguments use only the properties of the partially ordered set of subgroups of G, and in this form they can be axiomatised. This treatment is useful, in that it applies also to modules of finite length, and gives analogues of these two theorems for these (compare §9, Theorem II).

Thus the problem of describing groups (finite groups, Lie groups, or algebraic groups) reduces to the following two questions:

(1) Describe all groups having a given collection of groups as quotients of a composition series.

(2) Which groups can be the quotients of a composition series?

The first question can be studied inductively, and we then arrive at the following question: for given N and F, describe all groups G having a normal subgroup isomorphic to N with quotient group isomorphic to F; in this case we say that G is an *extension* of F by N. For example, a crystallographic group G (§14, Example 2) is an extension of F by A, where A is the group of translations contained in G, consisting of translations in the vectors of a certain lattice C, and F is a symmetry group of C.

Although in this generality the question is unlikely to have a complete answer, various approaches to it are known which in concrete situations lead to a more-or-less satisfactory picture (concerning this, see §21, Example 4).

Much more intriguing is the second of the above questions. Since the quotients of composition series are simple groups, and any set of simple groups is a factor of a composition series of some group (for example, their direct sum), our question is equivalent to the following:

WHAT ARE THE SIMPLE GROUPS?

At the present time, the answer to this is known for the most important types of groups: finite groups, Lie groups and algebraic groups.

We start with a case which would seem to be trivial, but which has a large number of applications, the simple Abelian groups. From the point of view of abstract group theory, the answer is obvious: the simple Abelian groups are just the cyclic groups of prime order. In the theory of Lie groups, we gave the definition of a simple group in terms of connected normal subgroups. Hence in the theory of connected differentiable Lie groups there are two further examples: the additive group $\mathbb{R}$ of real numbers, and the circle group. We will not give the answer for complex-analytic Lie groups, which is more complicated. Similarly, for algebraic matrix groups (or linear algebraic groups) over an algebraically closed field there are two new example: $\mathbb{G}_a$, the additive group of elements of the ground field, and $\mathbb{G}_m$, the multiplicative group.

This meagre collection of simple Abelian groups leads to a quite extensive class of groups when used as quotients in composition series, following the ideas described above. A group having a composition series with Abelian quotients is said to be *solvable*. It is easy to check the following properties:

Theorem III. *A subgroup or homomorphic image of a solvable group is solvable; if a group has a solvable normal subgroup with solvable quotient then it is solvable. If a group is solvable then it has a normal subgroup with Abelian quotient, that is, a nontrivial homomorphism to an Abelian group.*

For any group G, the intersection of the kernels of all the homomorphisms $f: G \to A$ of G to Abelian groups is called the *commutator subgroup* of G, and written G'.

Obviously, for any elements $g_1, g_2 \in G$, the products $g_1 g_2$ and $g_2 g_1$ go into the same element under any homomorphism to an Abelian group, and hence

$g_1 g_2 (g_2 g_1)^{-1} = g_1 g_2 g_1^{-1} g_2^{-1}$ goes to the identity. An element of the form $g_1 g_2 g_1^{-1} g_2^{-1}$ is called a *commutator*. We see that all commutators are contained in the commutator subgroup; it is not hard to prove that they generate it, so that

$$G' = \langle g_1 g_2 g_1^{-1} g_2^{-1} | g_1, g_2 \in G \rangle.$$

If G is solvable, then $G' \neq G$ (provided that $G \neq \{e\}$). But as a subgroup of a solvable group, G' is again solvable, so that either $G' = \{e\}$ or $G'' = (G') \neq G'$. Continuing this procedure shows that taking successive commutator subgroups of a solvable group, we eventually arrive at $\{e\}$; that is, if we set $G^{(i)} = (G^{(i-1)})'$, then $G^{(n)} = \{e\}$ for some n. It is easy to see that this is also a sufficient condition for a group of finite length to be solvable. Abelian groups are characterised by the fact that $G' = \{e\}$. In this sense, solvable groups are a natural generalisation of Abelian groups.

Example 2. Among the finite groups we have met, the following are solvable (in addition to Abelian groups):

$\mathfrak{S}_3$, with the composition series $\mathfrak{S}_3 \supset \mathfrak{A}_3 \supset \{e\}$.

$\mathfrak{S}_4$, with the composition series $\mathfrak{S}_4 \supset \mathfrak{A}_4 \supset \mathfrak{B}_4 \supset \{e\}$, where $\mathfrak{B}_4$ is the subgroup consisting of e and all the elements of cycle-type $(2, 2)$.

$\mathrm{GL}(2, \mathbb{F}_2)$ and $\mathrm{GL}(2, \mathbb{F}_3)$.

We say that a finite group is a *p-group* if its order is a power of a prime number p.

Theorem IV. *A finite p-group is solvable.*

In fact, consider the adjoint action of G on itself; its orbits are the conjugacy classes of elements of G, say $C_1, \ldots, C_h$. Suppose that C_i has k_i elements, so that $|G| = k_1 + \cdots + k_h$. Then as we have seen (§ 12, (10)), $k_i = (G : S_i)$ where S_i is the stabiliser of some element $g_i \in C_i$. The stabiliser S_i is a subgroup of G, so that $(G : S_i)$ divides the order of G, and hence is a power of p. In particular, $k_i = 1$ if and only if C_i consists of one element only, contained in the centre of G. In the equation $|G| = k_1 + \cdots + k_h$, the left-hand side is a power of p, and the right-hand side is a sum of terms which are also powers of p (possibly equal to 1). It follows from this that the number of k_i which are equal to 1 must be divisible by p. This proves that a finite p-group has a nontrivial centre Z. Since Z is Abelian, it is solvable, and by induction, we can assume that the quotient G/Z is also solvable. Hence G is solvable.

We now give some examples of solvable Lie groups.

Example 3. $E(2)$, the group of all orientation-preserving motions of the Euclidean plane; this has the composition series $E(2) \supset T \supset T' \supset \{e\}$, where T is the group of all translations, and T' the subgroup of all translations in some given direction. The quotients of this composition series are $E(2)/T \cong \mathrm{SO}(1) \cong \mathbb{R}/\mathbb{Z}$, $T/T' \cong \mathbb{R}$ and $T' \cong \mathbb{R}$.

Example 4. The group of all upper-triangular matrixes of the form

$$\begin{bmatrix} a_{11} & a_{12} & \cdots & a_{1n} \\ 0 & a_{22} & \cdots & a_{2n} \\ \cdot & \cdot & \cdots & \cdot \\ 0 & \cdot & \cdots & a_{nn} \end{bmatrix}, \quad \text{with } a_{11}a_{22}\ldots a_{nn} \neq 0,$$

where a_{ij} belongs to some field K. This is an algebraic group; for $K = \mathbb{R}$ or $\mathbb{C}$ it is a Lie group; and for K a finite field it is a finite group.

We return to the original question on the structure of simple groups and restrict ourselves to the nontrivial case of non-Abelian simple groups. For completely arbitrary groups there is of course no precise answer to our question. We now run through the non-Abelian groups we have met and say which of them are simple.

Theorem V. *The following series of groups are simple.*

(a) *Finite groups*:

$\mathfrak{A}_n$, *the alternating group, for* $n \geqslant 5$.

$\mathrm{PSL}(n, \mathbb{F}_q)$, *except for the case* $n = 2$, $q = 2$ *or* 3.

(b) *The series* $\mathscr{C}\!\mathit{omp}$ *of classical compact Lie groups*:

$\mathrm{SU}(n)$ *for* $n > 1$;

$\mathrm{SO}(n)$ *for* $n \neq 1, 2, 4$;

$\mathrm{SpU}(n)$ *for* $n \geqslant 1$.

(c) *The series* $\mathscr{L}\!\mathit{ie}_{\mathbb{C}}$ *of classical complex analytic Lie groups*:

$\mathrm{SL}(n, \mathbb{C})$ *for* $n > 1$;

$\mathrm{SO}(n, \mathbb{C})$ *for* $n \neq 1, 2, 4$;

$\mathrm{Sp}(2n, \mathbb{C})$ *for* $n \geqslant 1$.

(d) *The series* $\mathscr{A}\!\mathit{lg}_K$ *of classical algebraic matrix groups (over an arbitrary algebraically closed field* K*).*

$\mathrm{SL}(n, K)$ *for* $n > 1$;

$\mathrm{SO}(n, K)$ *for* $n \neq 1, 2, 4$;

$\mathrm{Sp}(2n, K)$ *for* $n \geqslant 1$.

As we have already observed, groups of the series (b), (c) and (d) are not simple as abstract groups. They contain a nontrivial centre Z, and if $Z_0 \subset Z$ is any subgroup in the centre of one of the groups G listed then G/Z_0 will also be a simple group in the sense of our definition; for example $\mathrm{PSU}(n) = \mathrm{SU}(n)/Z_0$ for

$Z = Z_0$. In what follows we will without further mention include these trivial modifications in the same series $\mathscr{C}omp$, $\mathscr{L}ie_{\mathbb{C}}$ and $\mathscr{A}lg_K$.

One of the greatest achievements of mathematics of modern times has been the proof that in the final three cases the examples given almost exhaust all simple groups. In various formulations of the problem this discovery, first made in the 19th century, has been extended and made more precise, to cover all the cases considered (and also arbitrary differentiable simple Lie groups, not necessarily compact). This theory has an enormous number of applications. The discovery of the regular polyhedrons, corresponding to the finite subgroups of motions of space, is considered as the highest achievement of mathematics in antiquity—Euclid's Elements ends with the description of the regular polyhedrons. These were the most profound symmetries discovered by mathematics in antiquity. The discovery and classification of the simple Lie groups occupies the same position in the mathematics of modern times: these are the most delicate symmetries accessible to the understanding of modern mathematics. And just as Plato considered the tetrahedron, the octahedron, the cube and the icosahedron to be forms of the four elements—fire, air, earth and water (leaving the dodecahedron as a symbol of the cosmos), so modern physicists attempt to find general laws governing the variety of elementary particles in terms of properties of various simple groups SU(2), SU(3), SU(4), SU(6) and others.

We will not give a full statement of the result. It turns out that there exist exactly another 5 groups, the *exceptional simple groups*, denoted by E_6, E_7, E_8, G_2 and F_4, of dimensions 78, 133, 248, 14 and 52, which need to be added to the three series indicated above to provide a list of all simple groups (in each of the three series (b), (c) and (d)). The relation between the three resulting types of simple groups, the compact, complex analytic and matrix algebraic groups over a field K, is very simple for the latter two types: the complex groups arise from the algebraic groups when $K = \mathbb{C}$. The relation between the complex and compact groups has in fact already been indicated in § 15: the compact groups are maximal compact subgroups of the corresponding complex analytic ones, and all maximal compact subgroups are conjugate.

The classification of differentiable simple Lie groups is a little more complicated than the classification of just the compact ones, but conceptually it is just as clear. Each type of compact simple groups has a number of analogues in the noncompact case. Let us describe for example the analogues of the compact groups SU(n): these are the groups SU(p, q) where U(p, q) is the group of complex linear transformations preserving the form $|z_1|^2 + \cdots + |z_p|^2 - |z_{p+1}|^2 - \cdots - |z_{p+q}|^2$, and the groups SL($n$, $\mathbb{R}$), SU(n, $\mathbb{C}$) and SL(n/2, $\mathbb{H}$) (if n is even), considered as differentiable Lie groups. The final group SL(n, $\mathbb{H}$) requires some explanation. Since the usual definition of determinant is not applicable to matrixes over a noncommutative ring, it is not clear what the notation S means. The answer consists of noting that SL(n, $\mathbb{R}$) and SL(n, $\mathbb{C}$) can also be defined in another way. That is, it is not hard to prove that SL(n, $\mathbb{R}$) coincides with the commutator subgroup of GL(n, $\mathbb{R}$), and SL(n, $\mathbb{C}$) is obtained in the same way from

$\mathrm{GL}(n, \mathbb{C})$. Thus for $\mathrm{SL}(n, \mathbb{H})$ we take by definition the commutator subgroup of $\mathrm{GL}(n, \mathbb{H})$.

We have not said anything so far about *finite simple groups*. Their classification is a very extensive problem. The brave conjecture that they should in some sense be analogous to the simple Lie groups had already been made in the 19th century. This relation is hinted at by the example of $\mathrm{PSL}(n, \mathbb{F}_q)$ which we have met. More concretely, the following approach is possible, which can at least provide many examples. For each simple algebraic matrix groups over the finite fields $\mathbb{F}_q$, we need to consider the group $G(\mathbb{F}_q)$ consisting of matrixes with entries in $\mathbb{F}_q$. The simple algebraic groups G over $\mathbb{F}_q$ are practically already known; if we replace $\mathbb{F}_q$ by its algebraic closure K then the simple algebraic groups over K are provided by the theory we have just described: the list of groups $\mathscr{A}\ell g_K$ and the 5 exceptional groups. However, it may happen that two groups defined over $\mathbb{F}_q$ and not isomorphic over $\mathbb{F}_q$ turn out to be isomorphic over K (or might no longer be simple over K). In essence, the same phenomenon has already appeared in the theory of real Lie groups, where the analogue of $\mathbb{F}_q$ is $\mathbb{R}$ and of K is $\mathbb{C}$. For example, all the groups $\mathrm{SU}(p, q)$ with $p + q = n$ are algebraic groups over $\mathbb{R}$ (because each complex entry of the matrix can be given by its real and complex parts). But one can show that they all become isomorphic to each other and to $\mathrm{SL}(n, \mathbb{C})$ when considered over $\mathbb{C}$. A similar situation occurs also for the fields $\mathbb{F}_q$. The question of classifying all algebraic groups over $\mathbb{F}_q$, assuming them known over the algebraic closure of $\mathbb{F}_q$, is essentially a matter of overcoming technical difficulties, and the answer to it is known. This leads to a list of simple algebraic groups G defined over $\mathbb{F}_q$, and for each of these we can construct a finite group $G(\mathbb{F}_q)$. If we carry out this construction with the appropriate amount of care (for example, we should consider the group $\mathrm{PSL}(n, \mathbb{F}_q)$ rather than $\mathrm{SL}(n, \mathbb{F}_q)$), then it turns out that all of these groups are simple, now as finite groups rather than as algebraic groups. There exist just a few exceptions corresponding to groups G of small dimension and small values of q (we have already seen this effect in the example of $\mathrm{PSL}(n, \mathbb{F}_q)$). In this way one arrives at a number of series of simple finite groups, called *groups of algebraic type*.

To the groups of algebraic type we must add one more series, the alternating groups $\mathfrak{A}_n$ for $n \geqslant 5$. Already in the 19th century, however, examples of groups began to appear which cannot be fitted into any of these series. But such examples always turned up as individuals, and not as infinite series. Up to now 26 such finite simple groups have been discovered, which are neither groups of algebraic type nor alternating groups $\mathfrak{A}_n$. These are called *sporadic simple groups*. The biggest of these is of order

$$2^{46} \cdot 3^{20} \cdot 5^9 \cdot 7^6 \cdot 11^2 \cdot 13^3 \cdot 17 \cdot 19 \cdot 23 \cdot 29 \cdot 31 \cdot 41 \cdot 47 \cdot 59 \cdot 71$$

(not for nothing is it called the *Monster*). At the present time, it seems to have been proved that the groups of algebraic type, the alternating groups and the 26 sporadic groups exhaust all finite simple groups. This is without doubt a result of the first importance. Unfortunately, it has been obtained as the result of many years of effort on the part of several dozen mathematicians, and the proof is scattered over hundreds of articles, adding up to tens of thousands of pages. Hence a certain time probably has to elapse before this achievement has been accepted and digested by mathematicians to the same extent as the analogous classification of simple Lie groups and algebraic groups.

§ 17. Group Representations

We recall that a *representation* of a group G is a homomorphism of G to the group $\operatorname{Aut} L$ of linear transformations of some vector space L (see § 9); this notion is closely related to the idea of 'coordinatisation'. The meaning of coordinatisation is to specify objects forming a homogeneous set X by assigning individually distinguishable quantities to them. Of course such a specification is in principle impossible: considering the inverse map would then make the objects of X themselves individually distinguishable. The resolution of this contradiction is that, in the process of coordinatisation, apart from the objects and the quantities, there is in fact always a third ingredient, the coordinate system (in one or other sense of the world), which is like a kind of physical measuring instrument. Only after fixing a coordinate system S can one assign to a given object $x \in X$ a definite quantity, its 'generalised coordinate'. But then the fundamental problem arises: how to distinguish the properties of the quantities that reflect properties of the objects themselves from those introduced by the choice of the coordinate system? This is the *problem of invariance* of various relations arising in theories of this kind. In spirit, it is entirely analogous to the *problem of the observer* in theoretical physics.

If we have two coordinate systems S and S', then usually one can define an automorphism g of X (that is, a transformation of X preserving all notions defined in X) that takes S into S': that is, g is defined by the fact that it takes each object x into an object x' whose coordinate with respect to S' equals that of x with respect to S. Thus all the admissible coordinate systems of our theory correspond to certain automorphisms of X, and it is easy to see that the automorphisms obtained in this way form a group G. This group acts naturally on the set of quantities: if $g \in G$ and $gS = S'$, then g takes the coordinate of each object in the coordinate system S into that of the same object in the coordinate system S'. If the set of quantities forms a vector space, then this action defines a representation of G.

Let's explain all this by an example. Consider an n-dimensional vector space L over a field K. Choosing a coordinate system S in L (that is, a basis), we can specify a vector by a set of n numbers. Passing to a different coordinate system is given by a linear transformation $g \in \mathrm{GL}(n, K)$, which at the same time transforms the n coordinates by means of the matrix of the linear transformation. We obtain the *tautological representation* of $\mathrm{GL}(n, K)$ by matrixes. But if instead we take as our objects the quadratic forms, given in each coordinate system by a symmetric matrix, then passing to a different coordinate system involves passing from A to CAC^*, and we obtain a representation of the group $\mathrm{GL}(n, K)$ in the space of symmetric matrixes, taking $C \in \mathrm{GL}(n, K)$ into the linear transformation $A \mapsto CAC^*$. In exactly the same way, considering linear transformations instead of quadratic forms, we get another representation of $\mathrm{GL}(n, K)$, this time in the space of all matrixes, taking $C \in \mathrm{GL}(n, K)$ into the linear transformation $A \mapsto CAC^{-1}$. Clearly, the same ideas apply to any tensors. In either case, quadratic forms or linear transformations, we are usually interested in properties of the corresponding matrixes that are independent of the choice of a coordinate system, that is, are preserved under the substitution $A \mapsto C^*AC$ in the first case, or $A \mapsto C^{-1}AC$ in the second. Examples of such properties are the rank of a matrix in the first case, and the coefficients of the characteristic polynomial in the second.

A similar situation arises if the conditions of a problem admit a given symmetry (that is, they are preserved by some transformation g). Then the set X of all solutions of the problem should be taken to itself under the same transformation g; that is, the symmetry group of the problem acts on the set X, and this usually leads to a representation of the symmetry group.

A beautiful example of this situation in given in [Michel **84** (1980)]. Consider the problem of how to construct a road network taking us from any vertex of a square $ABCD$ into any other vertex, and of the shortest possible length. It is not hard to prove that the answer is given by the network shown in Figure 38, (a), in which the angles AED and BFC are equal to 120°. The square $ABCD$ has the symmetry group D_4 (§ 13, Theorem III), but Figure 38, (a) obviously does not go into itself under this group! The explanation is that the given problem has two solutions, shown in Figure 38, (a) and (b), and that together these have the full D_4 symmetry: D_4 acts on the set of two figures (a) and (b).

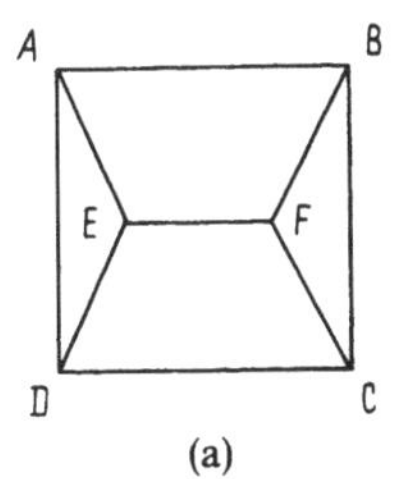

(a)

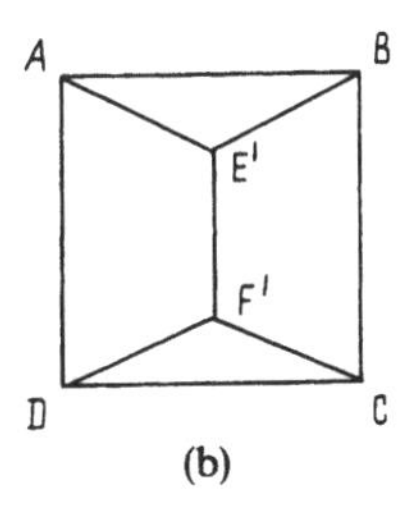

(b)

Fig. 38

Here is another example which leads to a representation of a symmetry group. Suppose given a linear differential equation of the form

$$\sum_{i=0}^{n} a_i(t)\frac{d^{n-i}f}{dt^{n-i}} = 0, \tag{1}$$

with coefficients periodic functions of period 2π. Then together with any solution $f(t)$, the function $f(t + 2\pi k)$ is also a solution for $k \in \mathbb{Z}$, and the map $f(t) \mapsto f(t + 2\pi k)$ defines a linear transformation u_k of the n-dimensional space of solutions. We obtain a representation of $\mathbb{Z}$ given by $k \mapsto u_k$, with $u_{k+l} = u_k u_l$, and hence $u_k = u_1^k$. A more complicated version of the same phenomenon involves considering equation (1) in the complex domain. Suppose that the $a_i(t)$ are rational functions of a complex variable t. If t_0 is not a pole of any of the functions $a_i(t)$ then near t_0, (1) has n linearly independent solutions, which are holomorphic in t. Analytically continuing these solutions along a closed curve s not passing through any poles of the $a_i(t)$, and with starting point and end point at t_0, we return to the same space of solutions; thus, as above, we obtain a linear transformation $u(s)$, which defines an n-dimensional representation of the fundamental group $\pi(\mathbb{C} \setminus P_1, \ldots, P_m)$, where $P_1, \ldots, P_m$ are the poles of the $a_i(t)$. This representation is called the *monodromy* of equation (1).

Another example. Suppose that a linear differential equation $L(x_1, \ldots, x_n, F) = 0$ has coefficients which depend symmetrically on $x_1, \ldots, x_n$. Then permutations of $x_1, \ldots, x_n$ define a representation of the group $\mathfrak{S}_n$ in the space of solutions. This situation occurs in the quantum mechanical description of a system consisting of n identical particles. The state of such a system is given by a wave function $\psi(q_1, \ldots, q_n)$, where q_i is the set of coordinates of the ith particle, and ψ is determined up to a scalar multiple λ with $|\lambda| = 1$. A permutation σ of the particles does not change the state, that is, it must multiply ψ by a constant. We get the relation $\psi(q_{\sigma(1)}, \ldots, q_{\sigma(n)}) = \lambda(\sigma)\psi(q_1, \ldots, q_n)$, from which it follows that $\lambda(\sigma_1\sigma_2) = \lambda(\sigma_1)\lambda(\sigma_2)$, that is, $\sigma \mapsto \lambda(\sigma)$ is a 1-dimensional representation of the group $\mathfrak{S}_n$. We know two such representations, the identity $\varepsilon(\sigma) = 1$, and the parity representation η given by $\eta(\sigma) = 1$ for even and -1 for odd permutations. It is easy to prove that there are no other 1-dimensional representations of $\mathfrak{S}_n$. Thus

$$\text{either} \quad \psi(q_{\sigma(1)}, \ldots, q_{\sigma(n)}) = \psi(q_1, \ldots, q_n) \text{ for all } \sigma,$$

$$\text{or} \quad \psi(q_{\sigma(1)}, \ldots, q_{\sigma(n)}) = \eta(\sigma)\psi(q_1, \ldots, q_n) \text{ for all } \sigma.$$

Which of these two cases occurs depends on the nature of the particles. We say that the particles are governed by *Bose-Einstein statistics* in the first case (for example, photons), and by *Fermi-Dirac statistics* in the second (for example, electrons, protons and neutrons).

In § 9 we defined the main notions of group representation theory: invariant subspace, irreducible representation, direct sum of representations, regular representation, the character of a representation. In the following, we consider representations over the complex number field of some of the main types of groups we have met: finite groups, compact Lie groups and complex Lie groups.

A. Representations of Finite Groups

In § 10 we obtained a series of theorems for finite groups as a particular case of the theory of semisimple algebras: the finiteness of the number of irreducible representations; the theorem that the regular representation decomposes into irreducible factors, among which each irreducible representation occurs; Burnside's theorem $|G| = \sum_{i=1}^{h} n_i^2$ (where n_i are the ranks of the irreducible representations of G); the theorem that the number h of irreducible representations is equal to the rank of the centre $Z(\mathbb{C}[G])$ of the group algebra of G; the fact that an irreducible representation is uniquely determined by its character.

Consider first the case of Abelian groups. Let $\rho\colon G \to \mathrm{Aut}_{\mathbb{C}} L$ be an irreducible representation of such a group (whether G is finite or not plays no role). Then the transformations $\{\sum \alpha_g \rho(g) | g \in G \text{ and } \alpha_g \in \mathbb{C}\}$ form an irreducible algebra in $\mathrm{End}_{\mathbb{C}} L$, which by Burnside's theorem (§ 10, Theorem XVII) must be the whole of $\mathrm{End}_{\mathbb{C}} L$. Since G is Abelian, the algebra $\mathrm{End}_{\mathbb{C}} L$ is commutative, and this is only possible if the rank of the representation is 1. Thus, an irreducible representation of an Abelian group has rank 1.

For finite Abelian groups, the same result is also clear from other considerations. The group algebra $\mathbb{C}[G]$ is commutative and hence by § 10, Theorem V, is isomorphic to a direct sum of fields $\mathbb{C}[G] \cong \mathbb{C}^n$. Its irreducible representations come from the projection onto the factors $\mathbb{C}$ of this decomposition, that is, if $x = (z_1, \ldots, z_n)$ then $\chi_i(x) = z_i$. We therefore have the following result.

Theorem I. *All irreducible representations of a finite Abelian group are* 1-*dimensional, and their number is equal to the order of the group.*

Thus irreducible representations of a finite Abelian group G are the same thing as homomorphisms $\chi\colon G \to \mathbb{C}^* = \mathrm{GL}(1, \mathbb{C})$ to the multiplicative group of complex numbers; χ coincides with its trace, and hence is called a *character*.

Homomorphisms of any group G (not necessarily Abelian) into $\mathbb{C}^*$ (or into any Abelian group) can be multiplied element-by-element: by definition, the *product* of characters χ_1 and χ_2 is the character $\chi = \chi_1 \chi_2$ defined by

$$\chi(g) = \chi_1(g)\chi_2(g). \tag{2}$$

It is easy to see that under this definition of product, the characters of an Abelian group G themselves form a group, the *character group* of G, which is denoted by $\hat{G}$. The identity is the character ε with $\varepsilon(g) = 1$ for $g \in G$; the inverse of χ is the character $\chi^{-1}(g) = \chi(g)^{-1}$. It can be shown that for an Abelian group G, the character group $\hat{G}$ not only has the same order as G, but is isomorphic to G as an abstract group. However, there does not exist any 'natural' isomorphism between these groups. But the character group $\hat{\hat{G}}$ of $\hat{G}$ is naturally isomorphic to G: the formula (2) shows that the map $G \to \hat{\hat{G}}$ given by $g(\chi) = \chi(g)$ is a homomorphism, which is an isomorphism, as one checks easily. The situation here is similar to the notion of the dual of a vector space.

The relation between G and its character group extends also to subgroups. Sending a subgroup $H \subset G$ to the subgroup $H^* \subset \hat{G}$ of characters taking the value 1 on all the elements of H, we get a 1-to-1 correspondence between subgroups of G and those of its character group $\hat{G}$. This correspondence is order-reversing: if $H_1 \subset H_2$ then $H_1^* \supset H_2^*$. Moreover, $H^* = (G/H)^\wedge$.

Characters satisfy a number of important relations. First of all, if $\chi \neq \varepsilon$ (the identity character) then

$$\sum_{g \in G} \chi(g) = 0. \tag{3}$$

Indeed, by assumption there exist $g_0 \in G$ such that $\chi(g_0) \neq 1$. If we substitute $g_0 g$ for g in the left-hand side of (3), we see easily that on the one hand it remains unchanged, and on the other, it is multiplied through by $\chi(g_0)$; this proves (3). For $\chi = \varepsilon$, $\varepsilon(g) = 1$ for all $g \in G$, and hence

$$\sum_{g \in G} \varepsilon(g) = |G|. \tag{4}$$

Substituting $\chi = \chi_1 \chi_2^{-1}$ in (3), where χ_1 and χ_2 are two characters, we get from (3) and (4) that

$$\sum_{g \in G} \chi_1(g)\chi_2(g)^{-1} = \begin{cases} 0 & \text{if } \chi_1 \neq \chi_2, \\ |G| & \text{if } \chi_1 = \chi_2. \end{cases} \tag{5}$$

Since each element g is of finite order, the numbers $\chi(G)$ are roots of unity, and so have absolute value 1. Hence $\chi(g)^{-1} = \overline{\chi(g)}$, which allows us to interpret (5) as saying that the characters are orthonormal in the space of complex-valued functions on G, where we give this space the scalar product defined by

$$(f_1, f_2) = \frac{1}{|G|} \sum_{g \in G} f_1(g)\overline{f_2(g)}.$$

Thus the characters form an orthonormal basis, and any function on G can be written as a combination of them:

$$f = \sum_{\chi \in \hat{G}} c_\chi \cdot \chi,$$

where the 'Fourier coefficients' c_χ are determined by the formula

$$c_\chi = \frac{1}{|G|} \sum_{g \in G} f(g)\overline{\chi(g)}.$$

Using the symmetry between a group and the character group ($\hat{\hat{G}} = G$), from (3) and (5) we get the relations

$$\sum_{\chi \in \hat{G}} \chi(g) = 0 \qquad \text{for } g \neq e, \tag{4'}$$

and

$$\sum_{\chi \in \hat{G}} \chi(g_1)\chi(g_2^{-1}) = \begin{cases} 0 & \text{if } g_1 \neq g_2, \\ |G| & \text{if } g_1 = g_2. \end{cases} \tag{5'}$$

One of the important applications of the character theory of finite Abelian groups relates to number theory. For G we take the multiplicative group $(\mathbb{Z}/(m))^*$ of invertible elements of the ring $\mathbb{Z}/(m)$, that is, the group of residue classes $a + m\mathbb{Z}$, consisting of numbers coprime to m. A character χ of $(\mathbb{Z}/(m))^*$, with its definition extended to be 0 on noninvertible elements, can be viewed as a periodic functions on $\mathbb{Z}$ with period m. Such a function is called a *Dirichlet character*. The *Dirichlet series*

$$L(s, \chi) = \sum_{n>0} \frac{\chi(n)}{n^s}$$

associated with these are one of the basic instruments of number theory. They form, for example, the basis of the proof of Dirichlet's theorem, that if a and m are coprime, the residue class $a + m\mathbb{Z}$ contains an infinite number of prime numbers. In the course of the proof it becomes necessary to separate off the partial sum $\sum' \frac{1}{n^s}$ taken over the residue class $a + m\mathbb{Z}$. Here we apply the relations (5′), from which it follows that this sum can be expressed as $\frac{1}{\varphi(m)} \sum_{\chi} \overline{\chi(a)} L(s, \chi)$, where $\varphi(m)$ is the Euler function, the order of $(\mathbb{Z}/(m))^*$, and the sum takes place over all characters of the group.

Proceeding to non-Abelian finite groups, we start with the problem of the number of their irreducible representations. According to §10, Theorem XIII, this equals the rank of the centre $Z(\mathbb{C}[G])$ of the group algebra $\mathbb{C}[G]$ of G. It is easy to see that an element $x = \sum_{g \in G} f(g)g \in \mathbb{C}[G]$ belongs to the centre of $\mathbb{C}[G]$ (that is, x commutes with all $u \in G$) if and only if $f(ugu^{-1}) = f(g)$ for all $u, g \in G$. In other words, the function $f(g)$ is constant on conjugacy classes of elements of G. Therefore, a basis of the centre is formed by the elements $z_C = \sum_{g \in C} g$, where the C are the different conjugacy classes. In particular, we obtain the result:

Theorem II. *The number of irreducible representations of a finite group G equals the number of conjugacy classes of elements of G.*

Can we find in the general case an analogue of the fact that the characters of an Abelian group themselves form a group? We have an analogue of the identity character, the identity 1-dimensional representation $\varepsilon(g) = 1$ for $g \in G$. One can also propose an analogue of the inverse element for a representation ρ, the so-called *contragredient representation* $\hat{\rho}(g) = \rho(g^{-1})^*$, where $*$ denotes the adjoint operator, acting on the dual space L^* to the space L on which ρ acts. If ρ is a unitary representation (that is, all the $\rho(g)$ are unitary with respect to some Hermitian metric; we saw in §10 that such a metric always exists) then the matrix of the transformation $\hat{\rho}(g)$ is simply the complex conjugate of that of $\rho(g)$.

Finally, there also exists an analogue of the product of characters. We start with the case that two groups G_1 and G_2 are given, with two representations $\rho_1: G_1 \to \operatorname{Aut} L_1$ and $\rho_2: G_2 \to \operatorname{Aut} L_2$ of them on vector spaces L_1 and L_2. Consider the tensor product $L = L_1 \otimes_{\mathbb{C}} L_2$ of these vector spaces (see § 5) and the map ρ of the direct product $G_1 \times G_2$ of G_1 and G_2 defined by $\rho(g_1, g_2)(x_1 \otimes x_2) = \rho_1(g_1)(x_1) \otimes \rho_2(g_2)(x_2)$. It is easy to see that this defines a map $\rho: G_1 \times G_2 \to \operatorname{Aut}(L_1 \otimes_{\mathbb{C}} L_2)$ which is a representation of $G_1 \times G_2$; it is called the *tensor product* of the representations ρ_1 and ρ_2, and denoted by $\rho_1 \otimes \rho_2$.

For example, if G_1 and G_2 act on sets X_1, X_2, and L_1, L_2 are certain spaces of functions on X_1, X_2 preserved by these actions, with ρ_1, ρ_2 the actions of G_1, G_2 in L_1, L_2, then the representation $\rho_1 \otimes \rho_2$ acts on the space of functions on $X_1 \times X_2$ spanned by $f_1(x_1)f_2(x_2)$ with $f_1 \in L_1$ and $f_2 \in L_2$. It is not hard to see that all irreducible representations of $G_1 \times G_2$ are of the form $\rho_1 \otimes \rho_2$, where ρ_1 is an irreducible representation of G_1 and ρ_2 of G_2. All of these arguments carry over to representations of any semisimple algebras (and not just group algebras $\mathbb{C}[G]$).

Suppose now that $G_1 = G_2 = G$. Then we can construct a diagonal embedding $\varphi: G \to G \times G$ defined by $\varphi(g) = (g, g)$. (Here the construction uses the specific group situation in an essential way, and is not meaningful for algebras.) The composite $(\rho_1 \otimes \rho_2) \circ \varphi$ defines a representation $G \to \operatorname{Aut}(L_1 \otimes_{\mathbb{C}} L_2)$ called the *tensor product* of the representations ρ_1 and ρ_2.

The essential difference from the Abelian case is the fact that for two irreducible representations ρ_1 and ρ_2 of G, the product $\rho_1 \otimes \rho_2$ may turns out to be reducible. Thus irreducible representations do not form a group: the product of two of them is a linear combination of the remainder. For example, the representations ρ and $\hat{\rho}$ acting on L and L^* define a representation $\rho \otimes \hat{\rho}$ in $L \otimes_{\mathbb{C}} L^*$. It is well known from linear algebra that $L \otimes_{\mathbb{C}} L^*$ is isomorphic to the space of linear transformations $\operatorname{End} L$ (the isomorphism takes a vector $a \otimes \varphi \in L \otimes_{\mathbb{C}} L^*$ into the linear transformation of rank 1 $x \mapsto \varphi(x)a$). It is easy to see that in $\operatorname{End} L$ the representation $\rho \otimes \hat{\rho}$ is written as $\alpha \mapsto \rho(g)\alpha\rho(g)^{-1}$ for $\alpha \in \operatorname{End} L$. But transformations which are multiples of the identity are invariant under this representation, and hence the representation $\rho \otimes \hat{\rho}$ is reducible, having the identity representation among its irreducible factors. (It can be shown that for an irreducible representation ρ the representation $\hat{\rho}$ is the unique irreducible representation σ such that the identity representation appears in the decomposition of $\rho \otimes \sigma$ into irreducible representations—in this very weak sense $\hat{\rho}$ is still an inverse of ρ.) It is easy to check that if χ_1 and χ_2 are characters of representations ρ_1 and ρ_2 then the character of $\rho_1 \otimes \rho_2$ is $\chi_1 \chi_2$.

Iterating this construction, for a group G we can consider the representation $\rho \otimes \rho \otimes \cdots \otimes \rho$ in the space $L \otimes L \otimes \cdots \otimes L$. This is called the *pth tensor power* of ρ, and denoted by $T^p(\rho)$, where p is the number of factors. From this by factorisation we obtain the representation $S^p(\rho)$ in the space $S^p L$ and $\bigwedge^p(\rho)$ in $\bigwedge^p L$.

For each irreducible representation ρ_i, we choose a basis in which $\rho_i(g)$ can be written as unitary matrixes $r^i_{jk}(g)$ (such a basis exists by § 10, Example 3).

We introduce the scalar product

$$(f_1, f_2) = \frac{1}{|G|} \sum_{g \in G} f_1(g)\overline{f_2(g)}$$

on the set of functions on G. In approximately the same way as for relation (5), one proves the result:

Theorem III. *The functions $r^i_{jk}(g)$ are mutually orthogonal and the square of the modulus of r^i_{jk} equals $\frac{1}{n_i}$, where n_i is the rank of the representation ρ_i.*

In particular, the characters form an orthonormal system of functions.

Example 1. The group $\mathfrak{S}_3$ has two 1-dimensional representations, the identity representation $\varepsilon(\sigma)$ and the representation $\eta(\sigma)$ which is ± 1 depending on the parity of σ. Realising $\mathfrak{S}_3$ as the permutations of the set $X = \{x_1, x_2, x_3\}$ we get a rank 2 representation ρ_2 of $\mathfrak{S}_3$ on the space of functions on X with $\sum_{x \in X} f(x) = 0$, which is also irreducible. Since $|\mathfrak{S}_3| = 6 = 1^2 + 1^2 + 2^2$, it follows from Burnside's theorem that these are all the irreducible representations of $\mathfrak{S}_3$.

Example 2. The octahedral group O (§ 13, Example 4). The group O permutes the pairs of opposite vertexes of the octahedron, which defines a homomorphism $O \to \mathfrak{S}_3$. Hence the representations of $\mathfrak{S}_3$ found in Example 1 give us certain irreducible representations of O. From the point of view of the geometry of the octahedron, these have the following meaning. We saw in § 13, Example 4 that O contains the tetrahedral group T as a subgroup and $(O : T) = 2$; then $\eta(g) = 1$ for $g \in T$ and -1 for $g \notin T$. The representation ρ_2 is realised in the space of functions on the vertexes of the octahedron taking the same value at opposite vertexes and with the sum of all values equal to 0. Furthermore, the inclusion $O \to \mathrm{SO}(3)$ defines a 3-dimensional tautological representation ρ_3 of O. Finally, the tensor product $\rho_3 \otimes \eta$ (which in the present case reduces just to multiplying the transformation $\rho_3(G)$ by the number $\eta(g)$) defines another representation ρ_3'. From the point of view of the geometry of the octahedron, it has the following meaning. We saw in § 13, Theorem V that the group $\mathrm{O}(3)$ contains a subgroup OT isomorphic to the group of the octahedron, but not contained in $\mathrm{SO}(3)$. The composite $O \cong OT \to \mathrm{O}(3)$ defines ρ_3'. Since $|O| = 24 = 1^2 + 1^2 + 2^2 + 3^2 + 3^2$, we have found all the irreducible representations of the octahedral group.

B. Representations of Compact Lie Groups

Representations of compact Lie groups enjoy almost all the properties present for finite groups. At the root of all the properties of representations of finite groups is the fact that they are semisimple, and as we have seen in § 10, in its various forms (for representations over $\mathbb{C}$ or over arbitrary fields) this can be deduced by one and the same idea: considering sums of the form $\sum_{g \in G} F(g)$ for various quantities $F(g)$ related to elements of the group; that is, the possibility of summing or averaging over the group. Now this idea has an analogue in the theory of compact Lie groups. The corresponding expression is called the integral over the group. It takes any continuous function $f(g)$ on a compact Lie group G into a number $I(f) \in \mathbb{R}$ called the *integral over the group*, and satisfies the following properties:

$$I(f_1 + f_2) = I(f_1) + I(f_2),$$

$$I(\alpha f) = \alpha I(f) \qquad \text{for } \alpha \in \mathbb{C},$$

$$I(f) = 1 \qquad \text{if } f \equiv 1$$

$$I(|f|^2) > 0 \qquad \text{if } f \not\equiv 0,$$

$$I(f_i) = I(f) \qquad \text{for } i = 1, 2, 3, \text{ where } f_1(g) = f(g^{-1}), f_2(g) = f(ug) \text{ and } f_3(g) = f(gu) \text{ for } u \in G.$$

The proof of the existence of I is based on constructing an invariant differential n-form ω on G, where $n = \dim G$. We then set

$$I(f) = c \int_G f\omega,$$

where c is chosen such that $I(f) = 1$ for $f \equiv 1$. A form ω is *invariant* if $\tau_u^* \omega = \omega$ for all $u \in G$, where τ_u is the transformation $g \mapsto gu$ of G. An invariant form is constructed by the method described at the beginning of § 15: we need to choose an n-form $\omega_e \in \bigwedge^n T_e$ on the tangent space T_e to G at the identity e, and define the value of ω_g on T_g as $(\tau_g^*)^{-1} \omega_e$.

The existence of integration on the group $I(f)$ allows us to carry over word-for-word the argument of § 10, Example 3 to compact groups, and to prove that a compact subgroup $G \subset \mathrm{GL}(n, \mathbb{C})$ leaves invariant some Hermitian positive definite form φ. Since this form is equivalent to the standard form $\sum |z_i|^2$, we have $G \subset C\mathrm{U}(n)C^{-1}$ for some $C \in \mathrm{GL}(n, \mathbb{C})$, that is, G is conjugate to a subgroup of the unitary group. This result on compact subgroups of $\mathrm{GL}(n, \mathbb{C})$ was discussed without proof in § 15, Example 8. In a similar way, a compact subgroup of $\mathrm{GL}(n, \mathbb{R})$ is conjugate to a subgroup of $\mathrm{O}(n)$. Here is one famous application of the same ideas.

Example 3. The Helmholtz-Lie Theorem. A *flag* F in an n-dimensional real vector space L is a sequence of oriented embedded subspaces $L_1 \subset L_2 \subset \cdots \subset L_{n-1} \subset L$, with $\dim L_i = i$. If we introduce a Euclidean metric in L, then a flag corresponds uniquely to an orthonormal basis $e_1, \ldots, e_n$, with $L_i = \{e_1, \ldots, e_i\}$ (as oriented subspace). It follows from this that the manifold $\mathscr{F}$ of all flags is compact. The group $\mathrm{GL}(n, \mathbb{R})$ acts on $\mathscr{F}$: $g(L_1, L_2, \ldots, L_{n-1}) = (g(L_1), g(L_2), \ldots, g(L_{n-1}))$, for $g \in \mathrm{GL}(n, \mathbb{R})$ and $(L_1, L_2, \ldots, L_{n-1}) \in \mathscr{F}$.

Theorem. *Suppose that a subgroup $G \subset \mathrm{GL}(n, \mathbb{R})$ acts simply transitively on $\mathscr{F}$ (that is, for any two flags F_1 and F_2 there exists a unique transformation $g \in G$ for which $g(F_1) = F_2$). Then L can be given a Euclidean metric so that G becomes the group of orthogonal transformations.*

Indeed, since the stabiliser subgroup of a point of $\mathscr{F}$ under the action of G is trivial, G can be identified with the orbit of any point, which by the transitivity of the action is the whole of $\mathscr{F}$. Hence since $\mathscr{F}$ is compact, so is G. From this, as we have seen, it already follows that there exists a Euclidean metric invariant under G. The fact that G is the whole of the orthogonal group of this metric follows easily from the fact that it acts transitively on $\mathscr{F}$.

The assertion we have proved is a local analogue and a first step in the proof of the famous Helmholtz-Lie Theorem, which gives an intrinsic characterisation of Riemannian manifolds of constant curvature. Namely, suppose that X is a differentiable manifold and G a group of diffeomorphisms which acts simply transitively on the set of points $x \in X$ and flags in the tangent spaces T_x (that is, for any two points $x, y \in X$ and flags F_x in T_x, F_y in T_y, there exists a unique transformation $g \in G$ such that $g(x) = y$, $g(F_x) = F_y$). Then X can be given a metric which turns it into one of the spaces of constant curvature, Euclidean, Lobachevsky, spherical or Riemannian (the quotient of the n-dimensional sphere by a central reflection), and G into the group of motions of the geometry. The assertion we have proved allows us to introduce on X a Riemannian metric, and then apply the technique of Riemannian geometry.

The transitivity of the action of a group of motions on the set of flags is called the *complete isotropy axiom* of a Riemannian manifold. Thus Riemannian manifolds satisfying this axiom are analogues of the regular polyhedrons (see § 13, Example 4) and conversely, regular polyhedrons are finite models of Riemannian spaces of constant curvature.

It was apparently in the proof of the Helmholtz-Lie theorem that the role of the flag manifold $\mathscr{F}$ was first understood. Subsequently it appears repeatedly: in topology, in the representation theory of Lie groups and in the theory of algebraic groups; and it always reflects the property we met above: it is the 'best' compact manifold on which the group $\mathrm{GL}(n, \mathbb{R})$ acts transitively.

We now return to the representation theory of compact groups, which is also based on the existence of integration over the group.

Theorem IV. (1) *A finite-dimensional representation of a compact Lie group G is equivalent to a unitary representation and is semisimple.*

(2) *In the space $L^2(G)$ of square-integrable functions f (for which $I(|f|^2) < \infty$), we introduce the inner product*

$$(f, g) = I(f\bar{g}).$$

Then the analogue of the orthogonality relations holds word-for-word for the matrixes of the irreducible finite-dimensional representations (see Theorem III).

(3) *Define the* regular representation *of G in the space $L^2(G)$ by the condition $T_g(f)(u) = f(ug)$. Then the regular representation decomposes as a direct sum of a countable number of finite-dimensional irreducible representations, and each irreducible representation appears in this decomposition the same number of times as its rank. A finite-dimensional representation is uniquely determined by its traces.*

(4) *If a group G admits an embedding $\rho: G \to \mathrm{Aut}(V)$ then all its irreducible representations are contained among those appearing in the decomposition of representations of the form $T^p(\rho) \otimes T^q(\bar{\rho})$.*

(5) *The irreducible representations of a direct product $G_1 \times G_2$ are of the form $\rho_1 \otimes \rho_2$ where ρ_1 and ρ_2 are irreducible representations of G_1 and G_2.*

The first and second assertions are proved word-for-word in the same way as for finite groups. The idea of the proof of the third can be illustrated if we assume that G is a closed subgroup of the group of linear transformations $\mathrm{Aut}(V)$ of a finite-dimensional vector space V (in fact such a representation of G is always possible, and for important examples of groups such as the classical groups, it is part of the definition). Then we have a tautological representation $\rho: G \to \mathrm{Aut}(V)$ or $G \to \mathrm{GL}(n, \mathbb{C})$, where $n = \dim V$. If $\rho(g)$ is the matrix $(r_{jk}(g))$ then the values of the $2n^2$ functions $x_{jk} = \mathrm{Re}\, r_{jk}(g)$ and $y_{jk} = \mathrm{Im}\, r_{jk}(g)$ determine the element g uniquely. By Weierstrass' approximation theorem, any continuous function f on G can be approximated by polynomials in x_{jk} and y_{jk}. But one sees easily that polynomials in x_{jk} and y_{jk} coincide with linear combinations of the matrix elements of all possible representations $T^p(\rho) \otimes T^q(\bar{\rho})$ or of their irreducible components. Therefore any continuous function on G can be approximated by linear combinations of functions $r^i_{jk}(g)$ corresponding to irreducible finite-dimensional representations ρ_i, from which it follows easily that any function $f \in L^2(G)$ can be expanded as a series in this orthogonal system. (3) follows easily from this. The proof gives even more, the information about irreducible representations of a group G contained in (4), (5).

Note finally that the same properties (1)–(3) hold for any compact topological group. In this case, the 'integral over the group' $I(f)$ is defined in a different way, by a beautiful construction of set theory.

Example 4. Compact Abelian Lie Groups. Here all irreducible representations are 1-dimensional and we have a complete analogue of the character theory of finite Abelian groups. A compact Abelian group G has a countable number of characters, or homomorphisms $\chi: G \to \mathbb{C}^*$. They are related by orthogonality relations analogous to (5), and any function $f \in L^2(G)$ can be expanded as a series $f = \sum_{\chi} c_\chi \cdot \chi$ in them. Characters form a (discrete) group $\hat{G}$, and we have the same relation between subgroups of G and $\hat{G}$ as in the case of finite groups. In the particular case of the circle group $G = \mathbb{R}/\mathbb{Z}$, the group $\hat{G}$ is isomorphic to $\mathbb{Z}$, since all characters of G are of the form $\chi_n(\varphi) = e^{2\pi i n \varphi}$. The expansion $f = \sum c_n \chi_n$ is the *Fourier series* of f. This explains the role of the functions $e^{2\pi i n \varphi}$ (or $\sin 2\pi n \varphi$ and $\cos 2\pi n \varphi$) in the theory of Fourier series, as characters of G.

The theory takes on the most complete appearance if we extend it to the class of locally compact Abelian group. The relation between G and its character group $\hat{G}$ (which is also locally compact) is described by *Pontryagin duality.*

Example 5. Let $\rho\colon \mathrm{SO}(n) \to L$ be the tautological representation of the group $\mathrm{SO}(n)$ as the orthogonal transformations of an n-dimensional Euclidean space L. The representation $S^2\rho$ can be realised in the space of symmetric bilinear forms on L (identifying L with L^* using the Euclidean structure of L). Then $u \in \mathrm{SO}(n)$ acts on a function $\varphi(a, b)$ (for a, $b \in L$) by taking it to $\varphi(\rho(u)^{-1}a, \rho(u)^{-1}b)$. In an orthonormal basis, φ can be written as a symmetric matrix A, and the transformation law has the usual form $A \mapsto \rho(u)A\rho(u)^{-1}$ (since $\rho(u)^* = \rho(u)^{-1}$). Obviously, under this the identity matrix remains invariant. Hence $S^2\rho = I \oplus S_0^2\rho$, where I is the identity representation and $S_0^2\rho$ is the representation in the space of matrixes with trace 0. It is easy to see that $S_0^2\rho$ is irreducible.

The following Examples 6–8 play a role in the theory of 4-dimensional Riemannian manifolds.

Example 6. Consider in particular the case $n = 4$. Then $\mathrm{SO}(4) \cong (\mathrm{SpU}(1) \times \mathrm{SpU}(1))/\{\pm(1, 1)\}$ (by §15, Formula (10)). Hence $S_0^2\rho$ can be viewed as a representation of $\mathrm{SpU}(1) \times \mathrm{SpU}(1)$ and hence, in view of Theorem IV, (5), is of the form $\rho_1 \otimes \rho_2$ where ρ_1 and ρ_2 are representations of $\mathrm{SpU}(1)$; let's find these representations. The representation $S_0^2\rho$ acts, as in the preceding example, on the 9-dimensional space of symmetric bilinear forms $\varphi(a, b)$ of trace 0, where now we can assume that a, $b \in \mathbb{H}$ are quaternions. Now $u \in \mathrm{SpU}(1) \times \mathrm{SpU}(1)$ is of the form $u = (q_1, q_2)$ where $q_1, q_2 \in \mathbb{H}$ with $|q_1| = |q_2| = 1$ and $\rho(u)a = q_1 a q_2^{-1}$ (§15, Formula (10)). Consider the action ρ_1 of $\mathrm{SpU}(1)$ on the 3-dimensional space $\mathbb{H}^-$ of purely imaginary quaternions given by $\rho_1(q)\colon x \mapsto qxq^{-1}$, for $x \in \mathbb{H}^-$ and $q \in \mathbb{H}$ with $|q| = 1$. Starting from two elements x, $y \in \mathbb{H}^-$ construct the function $\varphi_{x,y}(a, b) = \mathrm{Re}(xay\bar{b})$ for a, $b \in \mathbb{H}$. It is easy to see that $\varphi_{x,y}(a, b) = \varphi_{x,y}(b, a)$ and that the action $x \mapsto q_1 x q_1^{-1}$, $y \mapsto q_2 y q_2^{-1}$ is equivalent to the transformation of the function $\varphi_{x,y}(a, b)$ under the representation $S^2\rho$ (we need to use $\mathrm{Re}(\bar{\xi}) = \mathrm{Re}(\xi)$, $\mathrm{Re}(\xi\eta) = \mathrm{Re}(\eta\xi)$, $\bar{x} = -x$, $\bar{y} = -y$ and $\bar{q} = q^{-1}$ if $|q| = 1$). Thus we have a homomorphism $\rho_1 \otimes \rho_1 \to S^2\rho$ given by $x \otimes y \mapsto \varphi_{x,y}$, where ρ_1 is the standard representation of $\mathrm{SpU}(1)$ (or even of $\mathrm{SO}(3)$) in $\mathbb{H}^-$. It is easy to see that its kernel is 0. The image can only be $S_0^2\rho$. Hence $S_0^2\rho \cong \rho_1 \otimes \rho_1$.

Example 7. Again for $n = 4$, consider the representation $\bigwedge^2\rho$ (where ρ is as in Examples 5–6), which we realise in the space of skew-symmetric bilinear forms on $\mathbb{H}$. For $x \in \mathbb{H}^-$ consider the bilinear form $\psi_x(a, b) = \mathrm{Re}(ax\bar{b})$. Then $\psi_x(b, a) = -\psi_x(a, b)$ and $x \mapsto \psi_x$ defines a homomorphism of representations $(1 \otimes \rho_1) \to \bigwedge^2\rho$. Similarly, sending $x \in \mathbb{H}^-$ into the form $\xi_x(a, b) = \mathrm{Re}(\bar{a}xb)$ defines a homomorphism $(\rho_1 \otimes 1) \to \bigwedge^2\rho$. Adding these, we get a homomorphism $(\rho_1 \otimes 1) \oplus (1 \otimes \rho_1) \to \bigwedge^2\rho$ which one checks easily is an isomorphism, and defines a decomposition of $\bigwedge^2\rho$ into irreducible summands.

Example 8. Consider finally the representation $S^2\bigwedge^2\rho$. This is a representation of SO(4) in tensors with the same symmetry conditions as those satisfied by the curvature tensor of a 4-dimensional Riemannian manifold. According to Example 7, $\bigwedge^2\rho \cong (\rho_1 \otimes 1) \oplus (1 \otimes \rho_1)$ where ρ_1 is the tautological representation of SO(3) in $\mathbb{R}^3$. It is easy to see that for any representations ξ and η of any group, $S^2(\xi \oplus \eta) = S^2\xi \oplus S^2\eta \oplus (\xi \otimes \eta)$. In particular, $S^2\bigwedge^2\rho \cong (S^2\rho_1 \otimes 1) \oplus (1 \otimes S^2\rho_1) \oplus (\rho_1 \otimes \rho_1)$. By the result of Examples 5 and 6 we can write

$$S^2\bigwedge^2\rho \cong (I \otimes 1) \oplus (S_0^2\rho_1 \otimes 1) \oplus (1 \otimes I) \oplus (1 \otimes S_0^2\rho_1) \oplus S_0^2\rho, \tag{6}$$

which gives the decomposition of $S^2\bigwedge^2\rho$ into irreducible representations. This shows which groups of the component of the curvature tensor can be picked out in an invariant way, so that they have geometric meaning.

According to (6), we write an element $\xi \in S^2\bigwedge^2\rho$ as

$$\xi = \alpha_0 + \alpha_1 + \beta_0 + \beta_1 + \gamma$$

with $\alpha_0 \in I \otimes 1, \alpha_1 \in S_0^2\rho_1 \otimes 1, \beta_0 \in 1 \otimes I, \beta_1 \in 1 \otimes S_0^2\rho_1$ and $\gamma \in S_0^2\rho$. Since $I \otimes 1$ and $1 \otimes I$ are 1-dimensional representations, α_0 and β_0 are given by numbers a, b; the so-called Bianchi identity shows that for the curvature tensor of a Riemannian manifold we always have $a = b$. The number $12a = 12b$ is called the *scalar curvature*, the symmetric matrix γ (of trace 0) the *trace-free Ricci tensor*, and α_1 and β_1 the *positive and negative Weyl tensors*.

Example 9. Irreducible representations of SU(2). SU(2) has the tautological representation ρ: SU(2) $\to \operatorname{Aut}\mathbb{C}^2 = \operatorname{Aut} L$. According to Theorem IV, (4), it follows from this that all irreducible representations of SU(2) are obtained among the tensor product of any number of copies of ρ and $\hat{\rho}$. The representation $\hat{\rho}$ is equivalent simply to the complex conjugate of ρ, and in the present case, to ρ itself. In fact when $\dim L = 2$ the space $\bigwedge^2 L$ is 1-dimensional; choosing a basis ω_0 of $\bigwedge^2 L$ gives a bilinear form φ on L, with $x \wedge y = \varphi(x, y)\omega_0$, and thus establishes an isomorphism between L and the dual space L^*. On the other hand, the Hermitian structure on L (contained in the definition of SU(2)) gives an isomorphism of L with its Hermitian conjugate $\overline{L^*}$. From these two isomorphisms it follows that $\overline{L^*}$ and L^* are isomorphic, hence L and $\bar{L}$, and hence the equivalence of ρ and $\hat{\rho}$.

Thus all irreducible representations of SU(2) are obtained from decompositions of the representations $T^p(\rho)$ only. One set of representations is immediately obvious. The 2×2 unitary matrixes (or indeed any 2×2 matrixes) can be viewed as transformation matrixes of two variables x, y. As such they act on the space of homogeneous polynomials (or *forms*) of degree n in x and y

$$\begin{bmatrix} \alpha & \beta \\ \gamma & \delta \end{bmatrix}: F(x, y) \to F(\alpha x + \gamma y, \beta x + \delta y).$$

This representation of rank $n + 1$ is denoted by ρ_j, where $j = n/2$, following a tradition which arises in quantum mechanics (the theory of *isotopic spin*, compare

§ 18.E below). Obviously, $\rho_j = S^{2j}\rho$. If we take the homogeneous polynomial $F(x, y)$ into the inhomogeneous polynomial $f(z) = F(z, 1)$ then ρ_j can be written

$$f(z) \mapsto (\beta z + \delta)^n f\left(\frac{\alpha z + \gamma}{\beta z + \delta}\right). \tag{7}$$

It is not hard to check that the ρ_i are irreducible (this will become completely obvious in § 17.C below). By what we have said above, to prove that the ρ_j for $j = 1/2, 1, 3/2, \ldots$ are all the irreducible representations of SU(2), we need only check that the representations $T^p(\rho)$ decompose as a sum of the representations ρ_j; since ρ itself is $\rho_{1/2}$, by induction it is enough to prove that $\rho_j \otimes \rho_{j'}$ decomposes as a sum of representations ρ_k. We can guess the rule for this decomposition if we consider the subgroup H of SU(2) consisting of diagonal matrixes. This is the group

$$g_\alpha = \begin{bmatrix} \alpha & 0 \\ 0 & \alpha^{-1} \end{bmatrix}, \quad \text{with } |\alpha| = 1. \tag{8}$$

This is Abelian, and hence when we restrict to it, ρ_j decomposes into 1-dimensional representations. Indeed, a basis of the invariant 1-dimensional subspaces (realised in the space of forms) is given by the monomials

$$x^n, x^{n-1}y, \ldots, y^n, \tag{9}$$

on which H acts via the characters $\chi_n(g_\alpha) = \alpha^n$, $\chi_{n-2}(g_\alpha) = \alpha^{n-2}, \ldots, \chi_{-n}(g_\alpha) = \alpha^{-n}$. On decomposing the restriction of $\rho_j \otimes \rho_{j'}$ to H, we get the products in pairs of the characters occuring in the restriction of ρ_j and $\rho_{j'}$ to H, that is, the character $\chi_{n+n'}$ once, (where $n = 2j$, $n' = 2j'$), $\chi_{n+n'-2}$ twice, $\chi_{n+n'-4}$ three times, and so on. From this one can easily guess that if $\rho_j \otimes \rho_{j'}$ decomposes as a sum of representations ρ_k, then this decomposition can only be of the form

$$\rho_j \otimes \rho_{j'} = \rho_{j+j'} \oplus \rho_{j+j'-2} \oplus \cdots \oplus \rho_{|j-j'|}. \tag{10}$$

To prove relation (10), we can use Theorem IV, (3), that is, the fact that a representation is uniquely determined by its trace. It is enough to show that a unitary matrix is always diagonalisable, and hence conjugate to a matrix of the form (8), so that the character of the representation is determined by specifying it on such matrixes. In particular, for the representations ρ_j we can find it easily (using the description of the action of g_α in the basis (9)):

$$\chi_j(g_\alpha) = \chi_j(\alpha) = \frac{\alpha^{2j+1} - \alpha^{-(2j+1)}}{\alpha - \alpha^{-1}}. \tag{11}$$

Now (10) reduces to the simple formula

$$\chi_j(\alpha)\chi_{j'}(\alpha) = \chi_{j+j'}(\alpha) + \chi_{j+j'-2}(\alpha) + \cdots + \chi_{|j-j'|}(\alpha),$$

which can easily be checked. Thus we have found the irreducible representations of SU(2) and have proved (10): this is called the *Clebsch-Gordan formula.*

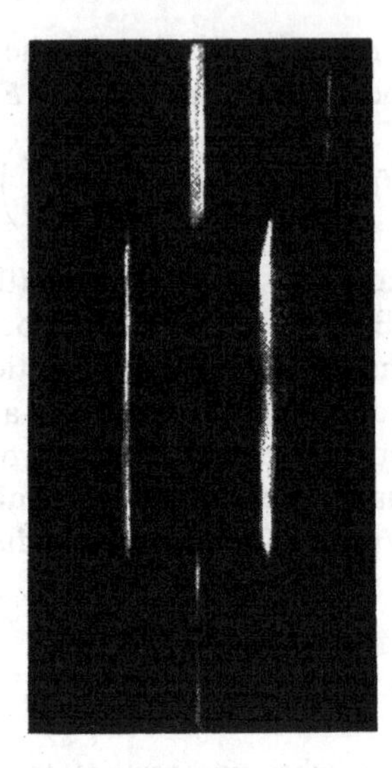

Fig. 39

Since $SO(3) \cong SU(2)/(+1, -1)$, the irreducible representations of SO(3) are contained among those of SU(2); they are just those on which the matrix $-E$ acts trivially. Obviously this happens exactly when j is an integer. From (11) we get a formula for the characters of these representations: for a rotation g_φ through an angle φ,

$$\chi_j(g_\varphi) = \frac{\sin(2j + 1)\varphi}{\sin \varphi}.$$

It is interesting that the method of restricting to an Abelian subgroup we have used (in the case of SO(3), this was the group of rotations around an axis) has a quantum-mechanical interpretation. Suppose that we have an electron in a centrally symmetric field. This symmetry should be reflected in the Hamiltonian, which must be SO(3)-invariant. Then the state space must be SO(3)-invariant, and hence must decompose as a direct sum of irreducible representations of SO(3). An irreducible subspace which occurs in the state space is defined by two numbers, of which one (the azimuthal quantum number) is j, and determines the type of the corresponding irreducible representation, and the other (the principal quantum number) distinguishes the different invariant subspaces corresponding to equivalent representations. All states occuring in one irreducible subspace must have the same energy level.

If we switch on a magnetic field having rotational symmetry about an axis, then each irreducible representation of SO(3) restricts to the subgroup $H \subset SO(3)$ of rotations about the axis of symmetry. The restriction to H of an irreducible representation of SO(3) decomposes, as we have seen, into 1-dimensional irreducible representations, and the states corresponding to different invariant subspaces with respect to the subgroup H already have different energy levels. This described the splitting of spectral lines in a magnetic field, the *Zeeman effect*. For example, Figure 39, which is taken from the article [F.J. Dyson, **37** (1964)], gives spectrograms showing that the state space of an atom of the metal niobium transforms according to the representation ρ_1 of SO(3), which breaks up into 3 representations of H after switching on the magnetic field.

C. Representations of the Classical Complex Lie Groups

In the following, we consider analytic representations of the classical groups, that is, we assume that the matrix of the linear transformation $\rho(g)$ has entries which are complex-analytic functions of the entries of the matrix $g \in G$. We use the relation between the classical complex and compact Lie groups, as described in § 15. Each classical compact group is contained in some classical complex group as a maximal compact subgroup: $U(n)$ in $GL(n, \mathbb{C})$, $SU(n)$ in $SL(n, \mathbb{C})$, $SpU(n)$ in $Sp(n, \mathbb{C})$ and so on. This connection makes it possible to study finite-dimensional representations of complex groups, starting from the information on representations of compact groups obtained in the preceding section. The first main result is as follows:

Theorem V. *Finite-dimensional representations of classical complex Lie groups are semisimple.*

We explain the idea of the proof using $GL(n)$ as an example; we make one simplifying assumption: we suppose that in the representation $\rho: GL(n) \to Aut(L)$, the entries $r_{ij}(g)$ of the transformation matrix $\rho(g)$ are rational functions of the entries of the matrix g (in fact, this is always the case, not just for $GL(n)$, but for all classical groups). Suppose that L has a subspace M invariant under all transformations $\rho(g)$ for $g \in GL(n)$. In view of the fact proved above that representations of the groups $U(n)$ are semisimple, there exists a subspace N invariant under all $\rho(g)$ with $g \in U(n)$ such that $L = M \oplus N$. In a basis obtained by adjoining together bases of M and N, the transformations $\rho(g)$ for $g \in GL(n)$ have matrixes of the form

$$\begin{bmatrix} A(g) & C(g) \\ 0 & B(g) \end{bmatrix}$$

Here the entries $c_{ij}(g)$ of $C(g)$ are rational functions of the entries of g by the above assumption, and are equal to 0 if $g \in U(n)$. Everything therefore reduces to the proof of the following lemma, which is not related to representation theory.

Lemma. *A function $F(Z)$ which is a rational function of the entries of a variable matrix $Z \in GL(n)$ and takes the value 0 for all $Z \in U(n)$ is identically 0.*

As is well known (and can be checked at once), any matrix Z with $\det(E + Z) \neq 0$ can be written as

$$Z = (E - T)(E + T)^{-1},$$

and Z is unitary if and only if $T^* = -T$. Set $F(Z) = G(T)$. Then our assertion reduces to proving that a rational function $G(T)$ of the entries of a matrix T which is 0 on skew-Hermitian matrixes is identically 0. Set $T = X + iY$. The condition $T^* = -T$ takes the form $X^t = -X$, $Y^t = Y$, where t denotes transposition. Our function is a rational function of the elements x_{ij} (for $i < j$) of a skew-symmetric matrix and y_{ij} (for $i \leqslant j$) of a symmetric matrix Y which are independent real variables. Since the rational function is 0 for all real values of the variables, it is identically 0, that is $G(T) = 0$ for all T of the form $X + iY$ with $X^t = -X$, $Y^t = Y$, where X and Y are any complex matrixes. But any matrix T can be represented in this form, setting $X = \frac{1}{2}(T + T^t)$, $Y = \frac{1}{2i}(T - T^t)$.

The proof of Theorem V is just one example of a general method of studying representations of classical complex Lie groups. This method is an analogue of the analytic continuation of real functions into the complex domain, and is called the *unitary trick*. Restricting a representation of such a group G to its maximal compact subgroup K, we get a representation of K. Conversely, from a representation of K we get a representation of G if we let the real parameters on which a matrix $k \in K$ depends take complex values.

We thus get a 1-to-1 correspondence between representations of G and K. For example, the irreducible representations of $\mathrm{SL}(2, \mathbb{C})$ can be written in exactly the same formulas (7) with the only difference that the entries of the matrix $\begin{bmatrix} \alpha & \beta \\ \gamma & \delta \end{bmatrix}$ now take any complex values satisfying $\alpha\delta - \beta\gamma = 1$. (Incidentally, the fact that they are irreducible follows at once from this.) Because of the relation between $\mathrm{SL}(2, \mathbb{C})$ and the Lorentz group, this description is important for physicists.

All we have said so far might give the impression that the theory of representations of the classical complex Lie groups is completely analogous to that of representations of compact groups. In actual fact, this is very far from the case; the theory of representations of any noncompact Lie group is related with certain completely new phenomena.

As in the compact case, there exists an invariant differential n-form on a group G (where $n = \dim G$), and using this we can define the regular representation of G in $L^2(G)$. The regular representation again 'breaks up into irreducibles', but now these words have a different meaning. The irreducible representations are, in general, infinite-dimensional, and depend on continuously varying parameters, so that the situation arising here is of a 'continuous spectrum' type. The regular representation decomposes not as a sum, but as an 'integral' of irreducible representations. For example, the characters of the additive group $\mathbb{R}$ of real numbers are of the form

$$\chi_\lambda(x) = e^{2\pi i\lambda x} \quad \text{with } \lambda \in \mathbb{R},$$

and the 'decomposition' of the regular representation reduces to the representation of functions as Fourier integrals (rather than as Fourier series, as in the case of the compact group $\mathbb{R}/2\pi\mathbb{Z}$). Both the regular representation and the irreducible representations into which it decomposes are unitary. It follows from this that in the majority of cases they cannot be finite-dimensional. For example if G is simple, then a nonidentical representation must be an embedding, and cannot be an embedding into some group $\mathrm{U}(n)$ since G is noncompact and $\mathrm{U}(n)$ is compact. As a rule, not all irreducible unitary representations occur in the decomposition of the regular representation. But even those which arise are not contained in the regular representation as subrepresentations: just as a point of the continuous spectrum of an operators does not correspond to any eigenvector. The exceptional cases when irreducible representations are contained in the regular representation are very interesting; they are an analogue of the discrete part of the spectrum. Of this type, for example are the representations (for $n \geqslant 0$) of the group $\mathrm{SL}(2, \mathbb{R})$ acting on the space of analytic functions $f(z)$ in the upper (or lower) half-plane with the inner product

$$(f_1, f_2) = \int_{\mathbb{C}^+} f_1(z) \cdot \overline{f_2(z)} y^n \, dx \wedge dy \quad \text{for } z = x + iy,$$

by the formulas

$$T_g(f)(z) = (\beta z + \delta)^{-n-2} f\left(\frac{\alpha z + \gamma}{\beta z + \delta}\right), \quad \text{where } g = \begin{bmatrix} \alpha & \beta \\ \gamma & \delta \end{bmatrix}.$$

The very construction suggests that these are related to the theory of automorphic functions.

§ 18. Some Applications of Groups

A. Galois Theory

Galois theory studies the 'symmetries', that is, the automorphisms of finite extensions; see § 6 for the definition of finite extensions and their simplest properties. For simplicity, we will assume that the fields under consideration are of characteristic 0, although in fact all the main results hold in much greater generality, for example, also for finite fields.

Every finite extension L/K is of the form $K(\alpha)$ where α is a root of an irreducible polynomial $P(t) \in K[t]$ (under the assumption that the fields have characteristic 0), and the degree of $P(t)$ is equal to $[L : K]$. Hence Galois theory can be treated in terms of polynomials (as Galois himself did), although a treatment in these terms is not invariant, in the sense that different polynomials $P(t)$ can generate the same extension L/K.

An *automorphism* of an extension L/K is an automorphism σ of the field L which fixes all the elements of K. All the automorphisms of a given extension form a group $\mathrm{Aut}(L/K)$ under composition. An automorphism σ of an extension $K(\alpha)/K$ is uniquely determined by the element $\sigma(x)$ to which α is taken: any element of $K(\alpha)$ can be written in the form $\sum_{i=0}^{n-1} a_i\alpha^i$ with $a_i \in K$, so if $\sigma(\alpha) = \beta$ then $\sigma(\sum a_i\alpha^i) = \sum a_i\beta^i$. On the other hand, if $\sigma(\alpha) = \beta$ and $P(\alpha) = 0$ with $P(t) \in K[t]$ then also $P(\beta) = 0$. Hence

$$|\mathrm{Aut}(L/K)| \leqslant \deg P(t) = [L:K]. \tag{1}$$

The bigger the group $\mathrm{Aut}(L/K)$, the more symmetric the extension L/K. The limiting case is when equality holds in (1), that is, when

$$|\mathrm{Aut}(L/K)| = [L:K].$$

An extension L/K with this property is called a *Galois extension*. By what we said above, for this to happen it is necessary that the irreducible polynomial $P(t)$ of which α is a root (if $L = K(\alpha)$) factorises over L into linear factors; it can be proved that this is also sufficient. In § 12 we gave an example of an extension which is not a Galois extension, and is even 'maximally asymmetric', in the sense that $\mathrm{Aut}(L/K) = \{e\}$. The possibility of applying group theory to the structure of fields is based on the fact that Galois extension nevertheless provide sufficiently complete information.

Theorem I. *Every finite extension is contained in a Galois extension.*

It is not difficult to give a recipe for constructing a Galois extension $\bar{L}/K$ containing a given extension L/K: suppose that $L = K(\alpha)$, where $P(\alpha) = 0$ with $P(t) \in K[t]$; then we must do the following. Over L, write $P(t) = (t - \alpha)P_1(t)$ with $P_1(t) \in L[t]$, then construct an extension $L_1 = L(\alpha_1)$ with $P_1(\alpha_1) = 0$, and proceed in the same way until $P(t)$ factorises into linear factors. Among all Galois extensions $\bar{L}/K$ containing a given extension L/K there exists a minimal one, contained in all others.

For a Galois extension L/K, the group $\mathrm{Aut}(L/K)$ is called the *Galois group* of L/K; it is denoted by $\mathrm{Gal}(L/K)$. By definition

$$|\mathrm{Gal}(L/K)| = [L:K].$$

The Galois group of a finite extension L/K is defined as the Galois group of the smallest Galois extension $\bar{L}/K$ containing L/K; the Galois group of an irreducible polynomial $P(t) \in K[t]$ is the Galois group of the extension $L/K = K(\alpha)$ with $P(\alpha) = 0$.

If $L = K(\alpha)$ with $P(\alpha) = 0$ then the smallest Galois extension $\bar{L}/K$ containing L/K is obtained by successively adjoining to K the roots of the polynomial $P(t)$, as described above. Any automorphism $\sigma \in \mathrm{Gal}(L/K)$ is determined by which elements it maps the roots α_i of $P(t)$ to. On the other hand, it can only take them into roots of the same polynomial. Hence σ performs permutations of the roots of $P(t)$, and the whole group $\mathrm{Gal}(L/K)$ acts on the set of these roots. For example, for the 'asymmetric' extension $L = \mathbb{Q}(\sqrt[3]{2})$ considered in § 12, $\alpha = \sqrt[3]{2}$, and $P(t) = t^3 - 2 = (t - \alpha)(t^2 + \alpha t + \alpha^2)$; but the polynomial $t^2 + \alpha t + \alpha^2$ does not have roots in $\mathbb{Q}(\sqrt[3]{2})$. We set $\bar{L} = L(\alpha_1)$ where $\alpha_1^2 + \alpha\alpha_1 + \alpha^2 = 0$, and hence $\alpha_1 = \left(\frac{-1 + \sqrt{-3}}{2}\right)\alpha$. It follows from this that $\bar{L} = L(\sqrt{-3})$, and any element of $\bar{L}$ can be written as $\xi + \eta\sqrt{-3}$ with $\xi, \eta \in \mathbb{Q}(\sqrt[3]{2})$. Obviously, an automorphism $\sigma \in \mathrm{Aut}(\bar{L}/\mathbb{Q})$ is determined by the values of $\sigma(\sqrt[3]{2})$ and $\sigma(\sqrt{-3})$; at the same time, $(\sigma(\sqrt[3]{2}))^3 = 2$, so that

$$\sigma(\sqrt[3]{2}) = \varepsilon^k \sqrt[3]{2} \quad \text{for} \quad k = 0, 1 \text{ or } 2,$$

where $\varepsilon = \dfrac{-1 + \sqrt{-3}}{2}$ is a 3rd root of 1; and

$$\sigma(\sqrt{-3}) = \pm\sqrt{-3}.$$

It is easy to verify that any combinations of these values for $\sigma(\sqrt[3]{2})$ and $\sigma(\sqrt{-3})$ really define an automorphism of $\bar{L}/\mathbb{Q}$, so that $|\mathrm{Aut}(\bar{L}/\mathbb{Q})| = 6$, and the extension $\bar{L}/\mathbb{Q}$ is Galois since $[\bar{L} : \mathbb{Q}] = 6$. Its Galois group acts on the roots of the polynomial $t^3 - 2$, and obviously gives any permutation of them, so that in this case $\mathrm{Gal}(\bar{L}/\mathbb{Q}) \cong \mathfrak{S}_3$. We write out explicitly the action of $\mathrm{Gal}(\bar{L}/\mathbb{Q})$ on the roots of $x^3 - 2$ as a table (the roots are ordered as $\sqrt[3]{2}$, $\varepsilon\sqrt[3]{2}$ and $\varepsilon^2\sqrt[3]{2}$).

$\sigma(\sqrt[3]{2})$	$\sqrt[3]{2}$	$\sqrt[3]{2}$	$\varepsilon\sqrt[3]{2}$	$\varepsilon^2\sqrt[3]{2}$	$\varepsilon\sqrt[3]{2}$	$\varepsilon^2\sqrt[3]{2}$
$\sigma(\sqrt{-3})$	$\sqrt{-3}$	$-\sqrt{-3}$	$\sqrt{-3}$	$\sqrt{-3}$	$-\sqrt{-3}$	$-\sqrt{-3}$
Permutation of roots	(1)	(23)	(123)	(132)	(12)	(13)

At the heart of Galois theory is a remarkable relation between the subextensions $K \subset L' \subset L$ of a Galois extension L/K and the subgroups of its Galois group $G = \mathrm{Gal}(L/K)$. For any subgroup $H \subset \mathrm{Gal}(L/K)$, we write $L(H)$ for the subfield of L made up of all the elements of L invariant under all automorphisms in H; and for a subfield L' with $K \subset L' \subset L$ we write $G(L')$ for the subgroup $\mathrm{Aut}(L/L')$ of $\mathrm{Gal}(L/K)$.

II. Fundamental Theorem of Galois Theory. *The maps $H \mapsto L(H)$ and $L' \mapsto G(L')$ are mutually inverse; they define a 1-to-1 correspondence between the subgroups $H \subset \mathrm{Gal}(L/K)$ and subfields L' with $K \subset L' \subset L$. This correspondence reverses inclusions: $H \subset H_1$ if and only if $L(H_1) \subset L(H)$. Moreover, $[L(H) : K] = (G : H)$. For $K \subset L' \subset L$, the extension L'/K is Galois if and only if $G(L') \subset G$ is a normal subgroup; in this case, $\mathrm{Gal}(L'/K) \cong G/G(L')$.*

The classic illustration of the methods of Galois theory is their application to the question of solving equations by radicals; the foundation of this is the natural interpretation of a radical $\sqrt[n]{a}$ in Galois theory.

Suppose that the ground field K contains all the nth roots of 1, that is, the polynomial $x^n - 1$ splits into linear factors, and that $x^n - a$ is irreducible, so that $[K(\sqrt[n]{a}):K] = n$. It follows from the above discussion that in this case $K(\sqrt[n]{a})/K$ is a Galois extension, and all of its automorphisms $\sigma \in \mathrm{Gal}(K(\sqrt[n]{a})/K) = G$ are determined by the fact that $\sigma(\sqrt[n]{a}) = \varepsilon\sqrt[n]{a}$, where $\varepsilon^n = 1$. In other words, setting $\sigma(\sqrt[n]{a})/\sqrt[n]{a} = \chi(\sigma)$, we get a character of G, and this character is faithful (that is, its kernal is e), so that the group G is cyclic. The field $K(\sqrt[n]{a})$ as a vector space over K defines a representation of G, which must break up into 1-dimensional representations. In fact

$$K(\sqrt[n]{a}) = K \oplus K\cdot(\sqrt[n]{a}) \oplus K\cdot(\sqrt[n]{a})^2 \oplus \cdots \oplus K\cdot(\sqrt[n]{a})^{n-1},$$

where $\sigma((\sqrt[n]{a})^r) = \chi^r(\sigma)\cdot(\sqrt[n]{a})^r$. Thus the radicals $(\sqrt[n]{a})^r$ correspond to a decomposition of the representation of the cyclic group G on L into 1-dimensional representations. This picture is reversible: suppose that L/K is an extension with cyclic Galois group G of order n; then in the same way as above, we should have

$$L = K\alpha_1 \oplus \cdots \oplus K\alpha_n \quad \text{with} \quad \sigma(\alpha_r) = \chi^r(\sigma)\cdot\alpha_r,$$

where $\chi(\sigma)$ is a character which is a generator of the character group of G. It is not hard to deduce from this that $\alpha_r = \alpha_1^r c_r$ with $c_r \in K$, $\alpha_1 \in L$ and $\alpha_1^n = a \in K$ so that $L = K(\sqrt[n]{a})$.

Thus if all the nth roots of 1 are contained in K, a radical extension $K(\sqrt[n]{a})$ is precisely an extension with a cyclic Galois group. From this, using Theorem II and very simple properties of solvable groups, one can prove the following result.

Theorem III. *An extension L/K is contained in an extension field Λ/K that can be obtained by successively adjoining radicals (that is, such that*

$$\Lambda = \Lambda_1 \supset \Lambda_2 \supset \cdots \supset \Lambda_r = K \text{ where } \Lambda_{i-1} = \Lambda_i(\sqrt[n]{\lambda_i}) \text{ with } \lambda_i \in \Lambda_i)$$

if and only if its Galois group is solvable.

It was this result that led to the notion of solvable groups, and indeed to the term itself.

Consider, for example the field of rational functions $k(t_1,\ldots,t_n) = L$ and the subfield K of symmetric functions. As is well known, $K = k(\sigma_1,\ldots,\sigma_n)$ where σ_i are the elementary symmetric functions, and $\sigma_i \mapsto y_i$ defines an isomorphism of $k(\sigma_1,\ldots,\sigma_n)$ with $k(y_1,\ldots,y_n)$. Obviously, $\mathrm{Gal}(L/K) = \mathfrak{S}_n$, and consists of all permutations of the variables $t_1, \ldots, t_n$. But the t_i are roots of the equation $x^n - \sigma_1 x^{n-1} + \cdots \pm \sigma_n = 0$. Applying the isomorphism $\sigma_i \mapsto y_i$, we can say that the Galois group of the equation

$$x^n - y_1 x^{n-1} + \cdots \pm y_n = 0 \tag{2}$$

over the field $k(y_1, \ldots, y_n)$, where $y_1, \ldots, y_n$ are independent variables, is the symmetric group $\mathfrak{S}_n$.

The equation (2) is called the *generic equation of degree n.*

Putting together the criterion of Theorem III and well-known facts about the structure of the groups $\mathfrak{S}_n$ (§ 13, Theorem I), we get the following result.

Theorem IV. *The generic equation of degree n is solvable in radicals for* $n = 2$, 3, *or* 4, *and not solvable for* $n \geqslant 5$.

The structure of the formulas for solving equations of degrees $n = 2, 3$ and 4 in radicals can also be predicted from properties of the groups $\mathfrak{S}_n$ for $n = 2, 3$ and 4 (§ 13, Theorem I).

B. The Galois Theory of Linear Differential Equations (Picard-Vessiot Theory)

We consider a differential equation

$$y^{(n)} + a_1 y^{(n-1)} + \cdots + a_n y = 0, \tag{3}$$

with coefficients meromorphic functions of one complex variable in some domain. Write K for the field $\mathbb{C}(a_1, \ldots, a_n)$ and L for the field of all rational functions of $a_1, \ldots, a_n$ and of the n linearly independent solutions of (3) and all of their derivatives.

A *differential automorphism* of the extension L/K is an automorphism of L which leaves fixed the elements of K and commutes with differentiation of elements of L. The group of all differential automorphisms of L/K is the *differential Galois group* of L/K or of the equation (3).

Since a differential automorphism commutes with differentiation and leaves fixed the coefficients of (3), it takes solutions of (3) into other solutions. Since solutions of (3) form an n-dimensional vector space, the differential Galois group of (3) is isomorphic to some subgroup of $\mathrm{GL}(n, \mathbb{C})$. It can be proved that this subgroup is an algebraic matrix group (see § 15). In this way there arises a version of Galois theory in which finite extensions are replaced by differential extensions of the type considered above, and finite groups by algebraic groups. This version also has a complete analogue of the fundamental theorem of Galois theory. It turns out that the analogue of solvability by radicals is *solvability by quadratures.* For example, $y = \int a(x)\,dx$ is a solution of the equation $y'' - (a'/a)y' = 0$. In this case the differential automorphisms are of the form $y \mapsto y + c$ with $c \in \mathbb{C}$, and hence the Galois group is isomorphic to $\mathbb{G}_a$ (the group of elements of K under addition, see § 15). The functions $y = \exp(\int a(x)\,dx)$ is a solution of $y' - ay = 0$. Differential automorphisms of this are of the form $y \mapsto cy$ with $c \in \mathbb{C}^*$, so that the differential Galois group is isomorphic to $\mathbb{G}_m$ (the group of elements of K under multiplication, see § 15).

By analogy with classical Galois theory, we have the following result.

Theorem V. *The roots of* (3) *can be expressed in terms of its coefficients by means of rational operations, taking integrals, exponentials of integrals and solving algebraic equations if and only if the differential Galois group has a normal series with quotients* $\mathbb{G}_a$, $\mathbb{G}_m$ *and finite groups.*

For example, it is not hard to find the Galois group of the equations $y'' + xy = 0$; it is just $\mathrm{SL}(2, \mathbb{C})$. Since this group has $\{\pm E\}$ as its only normal subgroup, and the quotient group $\mathrm{PSL}(2, \mathbb{C}) = \mathrm{SL}(2, \mathbb{C})/\{\pm E\}$ is simple, the equation $y'' + xy$ cannot be solved by quadratures.

C. Classification of Unramified Covers

Let X be a connected manifold, $x_0 \in X$ a marked point, and $G = \pi(X, x_0)$ the fundamental group of X; a manifold Y with a marked point $y_0 \in Y$ and a continuous map $p\colon Y \to X$ such that $p(y_0) = x_0$ is a *finite-sheeted cover* if every point $x \in X$ has a neighbourhood U whose inverse image $p^{-1}(U)$ breaks up as a disjoint union of n open subsets U_i such that $p\colon U_i \to U$ is a homeomorphism for each i. The number n is the same for every point x, and equals the number of inverse images of x; it is called the *degree* of the cover.

A map $p\colon Y \to X$ defines a homomorphism $p_*\colon \pi(Y, y_0) \to \pi(X, x_0)$ in a natural way, taking a map $\varphi\colon I \to Y$ that defines a loop into its composite with p. Write $G(Y)$ for the image of p_*. We have the following analogue of the fundamental theorem of Galois theory.

Theorem VI. *The map* $p\colon (Y, y_0) \mapsto G(Y)$ *defines a 1-to-1 correspondence between connected unramified covers of finite degree and subgroups of* G *of finite index. The degree of a cover* $(Y, y_0) \to (X, x_0)$ *equals the index* $(G : G(Y))$. *A chain of covers* $(Z, z_0) \to (Y, y_0) \to (X, x_0)$ *corresponds to an inclusion* $G(Z) \subset G(Y)$ *of subgroups. If* $G(Y)$ *is a normal subgroup of* G *then the quotient group* $F = G/G(Y)$ *acts on* Y *without fixed points, permuting the inverse images of points of* X, *and* $X = F\backslash Y$.

The analogy with the fundamental theorem of Galois theory is so strong that one feels the desire to try to establish some kind of direct connections. In some cases this is in fact possible. Suppose that X is an algebraic variety over the complex number field, and $p\colon Y \to X$ an unramified cover. Using the local homeomorphism $p\colon U_i \to U$ between open sets $U_i \subset Y$ and $U \subset X$, which exists by the definition of unramified cover, we can transfer the complex structure from X to Y. Thus Y has a uniquely defined structure of complex analytic manifold. It can be proved that with this structure Y is isomorphic to an algebraic variety. We arrive at a situation which is analogous to that considered at the end of § 6. If X and Y are irreducible, the map $p\colon Y \to X$ induces a map $p^*\colon \mathbb{C}(X) \to \mathbb{C}(Y)$ on the corresponding rational function fields, and p^* is a field homomorphism, hence an inclusion of fields. Thus $\mathbb{C}(X) \subset \mathbb{C}(Y)$, and it can be proved that this is a finite extension, with $[\mathbb{C}(Y):\mathbb{C}(X)] = (G:G(Y))$. We see that G provides us with a description of certain finite extensions of the field $\mathbb{C}(X)$. However, these are not all extensions of $\mathbb{C}(X)$, but only the 'unramified' ones. A general finite extension $L/\mathbb{C}(X)$ is also of the form $L = \mathbb{C}(Y)$, where Y is an algebraic variety, and there exists a map $p\colon Y \to X$ inducing the inclusion of fields $\mathbb{C}(X) \subset \mathbb{C}(Y)$; but in general the map p is ramified in some subvariety $S \subset X$, that is, for $x \in S$ the number of inverse images $p^{-1}(x)$ is smaller than the degree of the extension $\mathbb{C}(Y)/\mathbb{C}(X)$. A description of such extensions can be found by similar methods if we consider the group $\pi(X \setminus S)$.

In the case that X is a compact complex irreducible algebraic curve, the space X is homeomorphic to an oriented surface. If the genus of this surface is g then $\pi(X)$ has $2g$ generators $x_1, \ldots, x_{2g}$ with the single defining relation

$$x_1 x_2 x_1^{-1} x_2^{-1} x_3 x_4 x_3^{-1} x_4^{-1} \ldots x_{2g-1} x_{2g} x_{2g-1}^{-1} x_{2g}^{-1} = 1 \tag{4}$$

(see § 14, Example 7). Thus the subgroups of finite index of this explicitly defined group describe the unramified extensions $Y \to X$.

We mention a result which is similar in outward appearance. Let K be a finite extension of the p-adic number field $\mathbb{Q}_p$ (§ 7, Example 7), containing the pth roots of 1. Suppose that K contains a primitive p^eth root of 1, but not the p^{e+1}th roots of 1, and that $p^e \neq 2$. Set $n = [K:\mathbb{Q}_p]$.

Theorem VII. *Finite Galois extensions L/K of K for which $[L:K]$ is a power of p are in 1-to-1 correspondence with normal subgroups of index a power of p of the group with $n+2$ generators $\sigma_1, \ldots, \sigma_{n+2}$ and the single defining relation*

$$\sigma_1^{p^e} \sigma_1 \sigma_2 \sigma_1^{-1} \sigma_2^{-1} \ldots \sigma_{n+1} \sigma_{n+2} \sigma_{n+1}^{-1} \sigma_{n+2}^{-1} = 1 \tag{5}$$

(*n is necessarily even*).

Despite the amazing similarity between relations (5) and (4), the reason behind this parallelism is far from clear.

D. Invariant Theory

Let $G = \mathrm{GL}(n, \mathbb{C})$ be the group of linear transformations of an n-dimensional vector space L, and $T(L)$ a space of tensors of some definite kind over L. This set-up defines a representation $\varphi\colon G \to \operatorname{Aut} T(L)$ of G in $T(L)$. Of special interest are the polynomial functions F defined on $T(L)$ and invariant under the action of G: these express intrinsic properties of tensors of $T(L)$, independent of a choice of coordinate system in L (if we interpret elements $g \in \mathrm{GL}(n, \mathbb{C})$ as passing to a different coordinate system). It is convenient to weaken somewhat this requirement: an intrinsic property of a tensor is often expressed by the vanishing of some polynomial which gets multiplied by a constant under the action of an element g:

$$\varphi(g)F = c(g)F. \tag{6}$$

It is easy to see that $c(g)$ is some power of the determinant of g, and (6) is equivalent to the condition that $\varphi(g)F = F$ for $g \in \mathrm{SL}(n, \mathbb{C})$. Polynomials with this property are called *invariants* of $\mathrm{SL}(n, \mathbb{C})$. For example, if $T(L) = S^2(L)$ is the space of quadratic forms then the discriminant of a quadratic form is an invariant.

In what follows we consider the simplest case, when $T(L) = S^m(L)$ is the space of forms of degree m over L. The ring $A = \mathbb{C}[S^m(L)]$ of all polynomials of the coefficients of forms contains the subring B of invariants. We illustrate an application of the main facts of group representation theory by giving the proof of one of the fundamental results of *invariant theory*.

VIII. First Fundamental Theorem. *The ring of invariants is finitely generated over* $\mathbb{C}$.

This is based on the following simple lemma.

Lemma. *If $I \subset B$ is an ideal of the ring of invariants then $IA \cap B = I$.*

The proof is related to the fact that the polynomial ring is graded (compare § 6, Theorem V): $A = A_0 \oplus A_1 \oplus A_2 \oplus \cdots$; and B is also graded as a subring of A. Each of the space A_i defines a representation of $G = \mathrm{SL}(n, \mathbb{C})$. Every element $a \in A$ is contained in some finite-dimensional invariant subspace $\tilde{A} \subset A$ (for example, in $\bigoplus_{i \leqslant n} A_i$, where $n = \deg a$). Since representations of this group are semisimple (§ 17, Theorem V), $\tilde{A}$ splits as a direct sum $\tilde{A} = \bar{A} \oplus \bar{B}$, where $\bar{A}$ is the sum of all the irreducible representations appearing in $\tilde{A}$ distinct from the trivial representation, and $\bar{B} \subset B$; then $\bar{A}$ is a finite-dimensional subspace invariant under G such that $\bar{A} \cap B = 0$. Let $a = \bar{a} + b$, with $\bar{a} \in \bar{A}$ and $b \in B$. For $x \in IA$, let $x = \sum_l i_l a_l$ with $i_l \in I$ and $a_l \in A$. Setting $a_l = \bar{a}_l + b_l$ with $\bar{a}_l \in \bar{A}_l$ and $b_l \in B$, where $\bar{A}_l$ is the subspace corresponding to $\bar{A}$ for the elements a_l, we get $x = \sum i_l b_l + \sum i_l \bar{a}_l$. But $\sum i_l b_l \in I$ and $i_l \bar{A}_l$ is an invariant subspace in which a representation isomorphic to $\bar{A}_l$ is induced. Hence $\sum i_l \bar{a}_l \in \sum i_l \bar{A}_l$. From the basic properties of semisimple modules (see § 10, Theorem I) it follows that the module $\sum i_l \bar{A}_l$ contains only simple submodules isomorphic to one of the $i_l \bar{A}_l$. In particular $(\sum i_l \bar{A}_l) \cap B = 0$, and if $x \in B$ then $\sum i_l \bar{a}_l = 0$ and $x \in I$. This proves the lemma.

The fundamental theorem is now obvious. Sending I to IA is a map of ideals of B into ideals of A, and by the lemma, distinct ideals go into distinct ideals. But then A Noetherian implies that B is Noetherian; and a graded Noetherian ring is finitely generated over $B_0 = \mathbb{C}$ (§ 6, Theorem V).

It was specifically for the proof of this theorem that Hilbert introduced the idea of a Noetherian ring, and proved the Noetherian property of the polynomial ring (although this may sound absurd, since Emmy Noether, in whose honour the term Noetherian was subsequently introduced, was still a baby at the time Hilbert published his work on invariant theory).

E. Group Representations and the Classification of Elementary Particles

In the last two decades, a great deal of enthusiasm on the part of theoretical physicists has gone into attempts to use the representation theory of Lie groups to work out a unified point of view on the enigmatic picture of the elementary particles, which have been discovered up to the present in large numbers. Of the three types of interactions considered in physics, electromagnetic, weak and strong, we will only discuss strong interactions, which relate to nuclear forces. Particles taking part in strong interactions, the *hadrons*, are *mesons* (intermediate particles) and *baryons* (heavy particles).

We start with the remarks given at the end of our treatment of § 17, Example 9. The three spectral lines of Figure 39 (*triplets* in physical terminology) arose as a result of the violations of the original symmetry with respect to SO(3), which reduces to the subgroup $H \cong \mathrm{SO}(2)$. The restriction of the 3-dimensional representation ρ_1 of SO(3) to the subgroup H is no longer irreducible, and decomposes as three 1-dimensional representations, which correspond to the observed lines. This picture, as a model, will be the basis of all the ideas we discuss in what follows; if we have a set of r particles with similar properties, we can attempt to represent the set as a degeneration, related to a lessening of symmetry. Mathematically, this relates to the fact that the states of all the particles under consideration form spaces $L_1, \ldots, L_r$, in which one and the same group H acts by representations $\rho_1, \ldots, \rho_r$. We look for a bigger group $G \supset H$ having an irreducible representation in $L_1 \oplus \cdots \oplus L_r$ in such a way that the restriction of this representation to the subgroup H is equivalent to $\rho_1 \oplus \cdots \oplus \rho_r$.

The first step is to consider the pair consisting of the proton p and neutron n. The proton and neutron have the same spin, and very close (but not identical) masses:

$$\text{mass of proton} = 938.2 \text{ MeV} = 1.6726 \times 10^{-24} \text{ gm}$$

$$\text{mass of neutron} = 939.8 \text{ MeV} = 1.6749 \times 10^{-24} \text{ gm}.$$

They have different charges, but this only manifests itself in considering electromagnetic interactions, which are ignored in the present theory. In this context, Heisenberg proposed already in the 1930s to consider the proton and neutron as two quantum states of one particle, the *nucleon*. Correspondingly, they will be denoted by N^+ and N^0, and the nucleon by N. According to general principles of quantum mechanics, we get as the state space of the nucleon a 2-dimensional complex space L with a Hermitian metric: N^+ and N^0 correspond to a basis of L. The symmetries of this space form the group U(2). From now on we consider its subgroup SU(2), which has very similar properties, and omit the physical arguments which justify this restriction. In a system consisting of many nucleons, the state space will be of the form $L \otimes L \otimes \cdots \otimes L$; the group SU(2) has a representation on this. The (tautological) representation of SU(2) on L was denoted (according to the classification given in § 17, Example 9) by $\rho_{1/2}$. The representation in the state space of the new system will be $T^p(\rho_{1/2})$. We know by § 17, Example 9 that all the irreducible representations of SU(2) are of the form ρ_j, where $j \geqslant 0$ is an integer or half-integer, and $\dim \rho_j = 2j + 1$. Moreover, the Clebsch-Gordan formula § 17, (10) allows us to decompose the representation $\rho_j \otimes \rho_{j'}$ into irreducible representations. Hence we can find the decomposition into irreducibles of our representation $T^p(\rho_{1/2})$. This gives a lot of physical information. The point is that by the quantum-mechanical dictionary of § 1, the probability of passing from a state given by a vector φ to a state given by ψ is equal to $|(\varphi, \psi)|$ (assuming that $|\varphi| = |\psi| = 1$). But the decomposition of a representation into irreducible representations can always be taken to be orthogonal, and this means that states which are represented by vectors transforming according to different irreducible representations, cannot pass into one another: this is the so-called *law of forbidden interactions*. Furthermore, in addition to the indexes j of the irreducible representations ρ_j into which $T^p(\rho_{1/2})$ splits, we can also write down actual bases of the subspaces in which these representations are realised, that is, we can transform the matrix $T^p(\rho_{1/2}(g))$ into a direct sum of matrixes $\rho_j(g)$. This allows us to find the probability of passing between different states of the system.

Now it is natural to apply the same train of thought to the study of other elementary particles. It turns out that these turn up in groups of 2, 3 or 4, and the masses within one group are very close (although 'lone' or singlet particles also occur). Among the baryons we have for example, in addition to the nucleons already considered, a singlet Λ-hyperon (of mass 1115 MeV), a Ξ-doublet (that is, two particles Ξ^+ and Ξ^- of masses 1314 and 1321 MeV) and a Σ-triplet (that is, three particles Σ^+, Σ^0, Σ^- of masses 1189, 1192 and 1197 MeV). The same happens with mesons: there are singlets η-meson, φ-meson and ω-meson, the K and K^*-doublets and the doublets of their antiparticles, and π and ρ-triplets. It is natural to apply the same ideas to these, proposing that particles of one group are quantum states of the same particle, whose quantum states are 1-dimensional for singlets, 2-dimensional for doublets, and 3-dimensional for triplets, and correspond to representations ρ_0, $\rho_{1/2}$ and ρ_1 of SU(2). In the case of several particles the tensor products of representations again arises, which split into irreducibles according to the Clebsch-Gordan formula. All of these arguments, working in the framework of the group SU(2), form the theory of *isotopic spin* or isospin: a state which transforms under the irreducible representation ρ_j of SU(2) is assigned isotopic spin j. This theory has justified itself very well; thus the existence of the triplet of π-mesons was predicted on the basis of this theory, and these were subsequently discovered.

The boldest step of all is the following. In order to be consistent, we should apply the same arguments to all baryons. They form an octet, that is, there are 8 of them: the singlet Λ, the doublets N and Ξ and the triplet Σ. We should propose that their varied nature arises only from violation of some higher symmetries. Physicists put things differently: they propose that there exist an idealised 'superstrong interaction' with respect to which all properties of these particles are identical. For mathematicians, this is the problem of finding some group G containing a subgroup H isomorphic to SU(2), and an 8-dimensional irreducible representation ρ of G such that the restriction of ρ to H splits into ρ_0 (corresponding to Λ), $\rho_{1/2}$ (corresponding to N), another copy of $\rho_{1/2}$ (corresponding to Ξ) and ρ_1 (corresponding to Σ).

Such a group and representation do exist, namely $G = \mathrm{SU}(3)$ with its adjoint representation on the space $M_3^0(\mathbb{C})$ of all 3×3 matrixes of trace 0, where the matrix $g \in \mathrm{SU}(3)$ defines the transformation $x \mapsto g^{-1}xg$ for $x \in M_3^0(\mathbb{C})$; $\dim_{\mathbb{C}} M^3(\mathbb{C}) = 9$, and hence $\dim_{\mathbb{C}} M_3^0(\mathbb{C}) = 8$. We write out this representation in matrix form; write a matrix $x \in M_3(\mathbb{C})$ in the form

$$x = \begin{pmatrix} A & B \\ C & \alpha \end{pmatrix},$$

where A is a 2×2 matrix, B a 2×1 matrix, C a 1×2 matrix and α a number. The condition $\mathrm{Tr}\, x = 0$ means that $\alpha = -\mathrm{Tr}\, A$. In SU(3), consider the subgroup H of matrixes of the form $\begin{pmatrix} U & 0 \\ 0 & 1 \end{pmatrix}$. Then obviously, $U \in \mathrm{SU}(2)$, so that H is isomorphic to SU(2). Since

$$\begin{pmatrix} U & 0 \\ 0 & 1 \end{pmatrix}^{-1} \begin{pmatrix} A & B \\ C & \alpha \end{pmatrix} \begin{pmatrix} U & 0 \\ 0 & 1 \end{pmatrix} = \begin{pmatrix} U^{-1}AU & U^{-1}B \\ CU & \alpha \end{pmatrix},$$

the subdivision of the matrix into blocks A, B, C and α gives a splitting of the adjoint representation into two 2-dimensional representations and a 4-dimensional representation. The 2-dimensional representations obviously coincide with $\rho_{1/2}$. The 4-dimensional representation is reducible, since $M_2(\mathbb{C})$ splits into scalar matrixes plus matrixes of trace 0. As a result of this, the 4-dimensional representation splits into a 1-dimensional representation, consisting of matrixes of the form $\begin{pmatrix} -\alpha & & 0 \\ & -\alpha & \\ 0 & & 2\alpha \end{pmatrix}$, and a 3-dimensional one, consisting of matrixes $\begin{pmatrix} A & 0 \\ 0 & 0 \end{pmatrix}$ with $\mathrm{Tr}\, A = 0$. This is the required decomposition.

To return from the idealised picture, described by representations of SU(3), to the real baryons is a problem of perturbation theory. Writing the perturbed Hamiltonian on the basis of heuristic, but natural, considerations leads to a situation in which the answer depends on two arbitrary constants. A suitable choice of these two constants allows us to get a good approximation for the known masses of the four groups of baryons $(\Lambda, N, \Xi, \Sigma)$. Moreover, the same approach has turned out to be applicable to mesons, in which there are two distinguished octets: the pseudoscalar mesons (η, K, their antiparticles and π) and vector mesons (φ, K^*, their antiparticles and ρ), leading to the same problem of representation theory.

A new situation arises in considering the other group of baryons, which are also classified by isotopic spin. These are the doublet of Ξ^*-hyperons, the triplet of Σ^*-hyperons and the quadruplet of Δ-hyperons. According to the ideology discussed above, the problem is to find a 9-dimensional representation of SU(3) whose restriction to SU(2) splits as $\rho_{1/2} \oplus \rho_1 \oplus \rho_{3/2}$. There is no such representation. However, there exists a 'nearby' representation of SU(3), namely $S^3\rho$, the third symmetric power of the tautological representation ρ of SU(3) in 3-space. If ρ acts on the space of linear forms in variables x, y, z, then $S^3\rho$ acts on the 10-dimensional space L of homogeneous cubic polynomials in x, y, z. Consider the subgroup $H \subset \mathrm{SU}(3)$ isomorphic to SU(2) which fixes z, and acts on x and y by the tautological 2-dimensional representation $\rho_{1/2}$ of SU(2). Then we have the decomposition $L = L_3 \oplus L_2 z \oplus L_1 z^2 \oplus \mathbb{C}z^3$, where L_i is the space of homogeneous polynomials of degree i in x and y (so that $\dim L_i = i + 1$). We get a decomposition of $S^3\rho$ restricted to H:

$$\rho_{3/2} \oplus \rho_1 \oplus \rho_{1/2} \oplus \rho_0.$$

This differs from what we wanted by the summand ρ_0. It is natural to propose that this summand corresponds to yet another particle, which would have to be included in our family of baryons. From group-theoretical considerations we can predict certain properties of this particle, for example its mass. A particle of this kind has indeed been discovered: it is called the Ω^--hyperon.

Finally, we can attempt to make sense of all these ideas starting from general properties of representations of SU(3). We know (§ 17, Theorem IV, (4)) that all representations of this group can be obtained from the irreducible decomposition of arbitrary tensor products of two representations: the tautological representation ρ and the contragredient representation $\hat{\rho}$ (for SU(3), as opposed to SU(2), $\hat{\rho}$ is not equivalent to ρ).

Hence the question arises: don't these elementary representations correspond to certain 'even more elementary' particles? These conjectural particles are called *quarks* and *antiquarks*; their existence is supported by a series of experiments.

Many very important questions remain, however, beyond the reach of a theory based on SU(3). For this reason one considers also symmetries based on introducing other groups, for example SU(6). Similar ideas have been widely developed over the last twenty years, finding applications also outside the domain of strong interactions. But at this point the author's scant information on these matters breaks off.

§ 19. Lie Algebras and Nonassociative Algebra

A. Lie Algebras

Natural and important algebraic systems having all the properties of rings with the exception of the associativity of multiplication appeared very long ago, although the algebraic nature of these objects did not immediately become apparent. In § 5 we gave a description of vector fields on a manifold as first order linear differential operators $\mathscr{D}(F) = \sum p_i \frac{\partial F}{\partial x_i}$, or *derivations* of the rings of functions on manifolds, that is, maps $\mathscr{D}\colon A \to A$ such that

$$\mathscr{D}(a + b) = \mathscr{D}(a) + \mathscr{D}(b),$$

$$\mathscr{D}(ab) = a\mathscr{D}(b) + b\mathscr{D}(a) \tag{1}$$

and

$$\mathscr{D}(\alpha) = 0 \text{ for constants } \alpha.$$

The composite $\mathscr{D}_1\mathscr{D}_2$ of two differential operators is of course again a differential operator, but if $\mathscr{D}_1$ and $\mathscr{D}_2$ are first order operators, then $\mathscr{D}_1\mathscr{D}_2$ is a second order operator, since second derivatives will appear in it (this is especially clear for operators with constant coefficients: because of the isomorphism $\mathbb{R}\left[\frac{\partial}{\partial x_1}, \ldots, \frac{\partial}{\partial x_n}\right] \cong \mathbb{R}[t_1, \ldots, t_n]$, the point here is just that the product of two polynomials of degree 1 is of degree 2). However, there is a very important expression in $\mathscr{D}_1$ and $\mathscr{D}_2$ which is again a first order operator, the so-called *commutator*

$$[\mathscr{D}_1, \mathscr{D}_2] = \mathscr{D}_1\mathscr{D}_2 - \mathscr{D}_2\mathscr{D}_1. \tag{2}$$

The fact that the commutator is again a first order operator is most easily seen by interpreting $\mathscr{D}_1$ and $\mathscr{D}_2$ as derivations of the ring of functions, and checking by substitution that if $\mathscr{D}_1$ and $\mathscr{D}_2$ satisfy (1) then so does $[\mathscr{D}_1, \mathscr{D}_2]$. In coordinates, if $\mathscr{D}_1 = \sum P_i \frac{\partial}{\partial x_i}$ and $\mathscr{D}_2 = \sum Q_i \frac{\partial}{\partial x_i}$ then

$$[\mathscr{D}_1, \mathscr{D}_2] = \sum R_i \frac{\partial}{\partial x_i}, \quad \text{where} \quad R_i = \sum_k \left(P_k \frac{\partial Q_i}{\partial x_k} - Q_k \frac{\partial P_i}{\partial x_k}\right). \tag{3}$$

One sees directly from this that $[\mathscr{D}_1, \mathscr{D}_2]$ is a first order operator, but the definition (2) has the advantage that it is intrinsic, that is, does not depend on the choice of coordinate system $x_1, \ldots, x_n$, while this cannot be seen directly from the expression (3). Via the interpretation as differential operators, the commutator operation can be transferred to vector fields. Here it is called the *Poisson bracket*, and it also denoted by $[\theta_1, \theta_2]$.

The vector space of vector fields together with the bracket operation [,] is very similar to a ring. Indeed, if we interpret [,] as a multiplication, then all the axioms of a ring (or even of an algebra) will be satisfied, with the single exception of the associativity of multiplication; instead of which, the bracket operation satisfies its own specific identities:

$$[\mathscr{D},\mathscr{D}] = 0$$

and

$$[[\mathscr{D}_1,\mathscr{D}_2],\mathscr{D}_3] + [[\mathscr{D}_2,\mathscr{D}_3],\mathscr{D}_1] + [[\mathscr{D}_3,\mathscr{D}_1],\mathscr{D}_2] = 0.$$

These follow easily from the definition (1). The second of these is called the *Jacobi identity*. It is a substitute for associativity, and as we will see, is closely related to associativity.

A set $\mathscr{L}$ having two operations of addition $a + b$ and commutation (or *bracket*) $[a, b]$ is called a *Lie ring* if it satisfies all the axioms of a ring except for the associativity of multiplication, in place of which

$$[a, a] = 0,$$

and

$$[[a,b],c] + [[b,c],a] + [[c,a],b] = 0 \tag{4}$$

for all a, b, $c \in \mathscr{L}$. If $\mathscr{L}$ is a vector space over a field K and $[\gamma a, b] = [a, \gamma b] = \gamma[a, b]$ for all a, $b \in \mathscr{L}$ and $\gamma \in K$ then $\mathscr{L}$ is called a *Lie algebra* over K; the element $[a, b]$ is called the *commutator* of a and b. It follows from the relations $[a, a] = 0$ that $[b, c] = -[c, b]$ for all b, c (you need to set $a = b + c$).

Example 1. Vector fields on a manifold with the Poisson bracket operation form a Lie algebra (over $\mathbb{R}$ or $\mathbb{C}$, depending on whether the manifold is real or complex analytic).

Example 2. All the derivations of a ring A form a Lie ring with respect to the commutation operation (2). If A is an algebra over a field $K \subset A$ then derivations satisfying the condition $\mathscr{D}(\alpha) = 0$ for $\alpha \in K$ form a Lie algebra over K. The verification of this is the same as for differential operators.

Example 3. Let A be an associative, but not necessarily commutative, ring. For a, $b \in A$ we set $[a, b] = ab - ba$. With this bracket, A becomes a Lie ring. If A is an algebra over a field K, then we get a Lie algebra over the same field. The verification of this is again the same as for differential operators.

If $A = M_n(K)$ is the $n \times n$ matrix algebra then the algebra we obtain is the *general linear Lie algebra*, denoted by $\mathfrak{gl}(n, K)$ or $\mathfrak{gl}(n)$.

Note that the case of first order linear differential operators does not quite fit in Example 3, but we can take A to be the ring of all linear differential operators, and choose a subspace $\mathscr{L} \subset A$ of first order operators, which although not a subring (because it is not closed under multiplication ab), is closed under $[a, b] = ab - ba$. Obviously we then get a Lie algebra. The analogous method applied to the algebra $A = M_n(K)$ gives the following new important examples; (the property of being closed under commutation is easily checked).

Example 4. Consider the subspace $\mathscr{L} \subset M_n(K)$ consisting of all matrixes of trace 0; $\mathscr{L}$ is called the *special linear Lie algebra* and denoted by $\mathfrak{sl}(n, K)$ or $\mathfrak{sl}(n)$.

Example 5. $\mathscr{L} \subset M_n(K)$ consists of all skew-symmetric matrixes a, with

$$a^* = -a, \tag{5}$$

where * denotes transposition. $\mathscr{L}$ is called the *orthogonal Lie algebra*, and denoted by $\mathfrak{o}(n, K)$ or $\mathfrak{o}(n)$.

Example 6. Suppose $K = \mathbb{C}$, and that $\mathscr{L} \subset M_n(\mathbb{C})$ is again characterised by (5), but where * denotes Hermitian transposition. Then $\mathscr{L}$ is a Lie algebra over $\mathbb{R}$, called the *unitary Lie algebra*, and denoted by $\mathfrak{u}(n)$. Imposing the additional condition $\operatorname{Tr} a = 0$, we get the *special unitary Lie algebra*, denoted by $\mathfrak{su}(n)$.

Example 7. Suppose $K = \mathbb{H}$ is the algebra of quaternions, and that $\mathscr{L} \subset M_n(\mathbb{H})$ is again characterised by (5), where now * denotes quaternionic Hermitian transposition. Then the Lie algebra $\mathscr{L}$ over $\mathbb{R}$ is called the *unitary symplectic Lie algebra*, and denoted by $\mathfrak{spu}(n)$.

Example 8. Let J be a $2n \times 2n$ nondegenerate skew-symmetric matrix over a field K, and $\mathscr{L} \subset M_{2n}(K)$ the set of $a \in M_{2n}(K)$ for which

$$aJ + Ja^* = 0. \tag{6}$$

$\mathscr{L}$ is called the *symplectic Lie algebra*, and is denoted by $\mathfrak{sp}(2n, K)$ or $\mathfrak{sp}(2n)$.

The origin of the terms introduced in Examples 4–8 will become clear shortly.

We say that a Lie algebra $\mathscr{L}$ is *finite-dimensional* if it is finite-dimensional as a vector space over K; the dimension over K is called the *dimension* of $\mathscr{L}$ and denoted by $\dim \mathscr{L}$ or $\dim_K \mathscr{L}$.

For example, the Lie algebra of vector fields in a region of 3-space is infinite-dimensional over $\mathbb{R}$, since a vector field can be written as $A\frac{\partial}{\partial x} + B\frac{\partial}{\partial y} + C\frac{\partial}{\partial z}$, where A, B, C are any differentiable functions. The algebra $\mathfrak{gl}(n)$ and the algebras of Examples 4–8 are finite-dimensional:

$$\dim \mathfrak{gl}(n) = n^2, \quad \dim \mathfrak{sl}(n) = n^2 - 1, \quad \dim \mathfrak{o}(n) = \frac{n(n-1)}{2},$$

$$\dim_{\mathbb{R}} \mathfrak{u}(n) = n^2, \quad \dim_{\mathbb{R}} \mathfrak{su}(n) = n^2 - 1,$$

$$\dim_{\mathbb{R}} \mathfrak{spu}(n) = 2n^2 + n, \quad \dim_K \mathfrak{sp}(n, K) = 2n^2 + n.$$

An *isomorphism* of Lie rings and algebras is defined exactly as for associative rings. For example, it is well known that all $2n \times 2n$ nondegenerate skew-symmetric matrixes are conjugate (over a field K). It follows easily from this that the algebras defined by condition (6) for different matrixes J are isomorphic (which is why J is not indicated in the notation $\mathfrak{sp}(2n, K)$). Here is a less trivial example of an isomorphism.

Example 9. The vectors of Euclidean 3-space under the vector product operation [,] form a Lie algebra $\mathscr{L}$ over $\mathbb{R}$ (the Jacobi identity (4) is well known in this case). Suppose we assign to each vector a the linear map $\varphi_a(x) = [a, x]$. The scalar triple product formula shows that φ_a is skew-symmetric, that is $\varphi_a^* = -\varphi_a$. On the other hand, for any skew-symmetric linear map φ of $\mathbb{R}^3$, there exists a vector $c \in \mathbb{R}^3$ with $|c| = 1$ such that $\varphi(c) = 0$. Then φ also induces a skew-symmetric transformation in the plane orthogonal to c, that is (if $\varphi \neq 0$) a rotation through 90° and multiplication by a number k. It is easy to check that then $\varphi = \varphi_a$ where $a = kc$. Thus $a \mapsto \varphi_a$ is a 1-to-1 linear map of $\mathscr{L}$ to $\mathfrak{o}(3)$. It follows at once from the Jacobi identity that this map is an isomorphism of Lie algebras. Thus $\mathscr{L}$ is isomorphic to $\mathfrak{o}(3)$.

By analogy with the case of associative algebras, we define the notions of *subalgebra, homomorphism* and *ideal* of Lie algebras (because of the relation $[b, a] = -[a, b]$, there is no distinction between left, right and 2-sided ideals), and of *simple algebra* and the *quotient algebra* by an ideal. The analogue of the homomorphisms theorem holds. We say that a Lie algebra (or ring) $\mathscr{L}$ is *Abelian* (or *commutative*) if $[a, b] = 0$ for all $a, b \in \mathscr{L}$. As in the associative case, for a finite-dimensional algebra with basis $e_1, \ldots, e_n$ the bracket operation is defined by *structure constants* c_{ijk}, with

$$[e_i, e_j] = \sum c_{ijk} e_k.$$

B. Lie Theory

The subject matter here is the study from the infinitesimal viewpoint of Lie groups in a neighbourhood of the identity, that is, differential calculus on the level of groups. The analogue of differentiation is to associate with a Lie group a certain Lie algebra; one also determines to what extent a Lie group can be reconstructed from the corresponding Lie algebra, thus constructing an analogue of the integral calculus.

We start our treatment of this theory (as it arose historically) with the example of a Lie group G acting on a manifold X. Such an action is given by a map

$$\varphi: G \times X \to X \tag{7}$$

(see § 12). Introducing coordinates $u_1, \ldots, u_n$ in a neighbourhood of the identity $e \in G$ and $x_1, \ldots, x_m$ in a neighbourhood of $x_0 \in X$, we specify the map by functions

$$\varphi_1(u_1, \ldots, u_n; x_1, \ldots, x_m), \ldots, \varphi_m(u_1, \ldots, u_n; x_1, \ldots, x_m).$$

For definiteness we assume in the following that these are real analytic, and similarly, we assume that the group law of the Lie group is real analytic. Other versions (n times differentiable, or complex analytic functions) are considered in exactly the same way.

If G is the group $\mathbb{R}$ of real numbers under addition, then the action (7) defines a 1-parameter group of transformations of X. In mechanics, $\varphi(x, t)$ for $t \in \mathbb{R}$ and $x \in X$ can be interpreted as the motion of a point of configuration space X as time t varies, and 'infinitesimally small' motions have been considered for a very long time. By this, one means the velocity field θ of the transformation $\varphi(t, x)$ at time $t = 0$; in coordinates, this is

$$\theta_i = \left.\frac{\partial \varphi_i(t, x_1, \ldots, x_n)}{\partial t}\right|_{(t=0)} \qquad \text{for } i = 1, \ldots, m.$$

The corresponding differential operator is defined by the condition

$$\mathscr{D}(F)(x) = \frac{\partial}{\partial t}(F(\varphi(t, x))|_{(t=0)}.$$

In the case of an arbitrary group G, Hermann Weyl proposes to imagine the manifold X as being filled with a material admitting motions that correspond to the action (7) of G. Here also, we can consider velocity fields of the corresponding motions. Each such field is determined by a vector ξ of the tangent space $T_{e,G}$ to G at the identity e. The action (7) defines a map of the tangent spaces

$$(d\varphi)_{(e,x)}\colon T_{e,G} \oplus T_{x,X} \to T_{x,X},$$

where $T_{x,X}$ is the tangent space to X at x. For any vector $\xi \in T_{e,G}$, the vector $(d\varphi)_{(e,x)}(\xi \oplus 0) \in T_{x,X}$ defines the required vector field θ_ξ on X. It is easy to see that in coordinates this is given by the formulas $\dfrac{\partial \varphi_i}{\partial \xi}$ for $i = 1, \ldots, m$, and is analytic. The corresponding differential operator on X is of the form $\mathscr{D}_\xi(F)(x) = \dfrac{\partial F(\varphi(g, x))}{\partial \xi}$ (differentiating with respect to the argument g). One sees from this that it is intrinsically defined (independently of the choice of the coordinate system). The map $\xi \mapsto \theta_\xi$ is obviously linear in ξ and hence defines a finite-dimensional space $\mathscr{L}$ of vector fields on X. The basic fact is that the finite-dimensional family $\mathscr{L}$ of vector fields constructed in this way is closed under Poisson brackets, and therefore defines a certain Lie algebra.

We explain the reason for this in the only case that will appear in what follows: when φ is the left regular action, that is, when $X = G$ and $\varphi(g_1, g_2) = g_1 g_2$ (for $g_1 \in G$ and $g_2 \in X = G$). The left regular action commutes with the right regular action: the left action of $g_1 \in G$ is of the form $g \mapsto g_1 g$, and the right action for $g_2 \in G$ of the form $g \mapsto g g_2^{-1}$, so the fact that these commutes just expresses the associativity of the group law: $(g_1 g) g_2^{-1} = g_1 (g g_2^{-1})$. From this, by an obvious formal verification, we see easily that for any $\xi \in T_{e,G}$ the vector field $\theta_\xi(x) = (d\varphi)_{(e,x)}(\xi \oplus 0)$ is also invariant under the right regular action. In other words, the tangent vector $\theta_\xi(g g_1^{-1})$ is obtained from the tangent vector $\theta_\xi(g)$ by means of the differential $d(g_1^{-1})$ of the right regular action $g \mapsto g g_1^{-1}$. Vector fields θ with this property are said to be *right-invariant* (see the remark in § 15 following the definition of a Lie group); such a field is uniquely determined by the vector $\theta(e)$, and any tangent vector $\eta \in T_{e,G}$ defines a vector field θ with $\theta(e) = \eta$: the vector $\theta(g)$ is then obtained from η by the right translation taking e to g. Thus the vector space of right-invariant vector fields on G is isomorphic to the space $T_{e,G}$. The space of vector fields $\mathscr{L} = \{\theta_\xi(\cdot) = (d\varphi)_{(e,\cdot)}(\xi \oplus 0) | \xi \in T_{e,G}\}$ constructed above consists as we have just said of right-invariant fields, and hence is isomorphic to a subspace of $T_{e,G}$. But the map $\xi \mapsto \theta_\xi$ has no kernel, as one sees easily, so that

$\dim \mathscr{L} = \dim T_{e,G}$, and therefore $\mathscr{L}$ consists of all the right-invariant vector fields. Finally, that the set of all right-invariant vector fields is closed under commutation follows from the obvious relation: if f is a transformation of a manifold X taking a point x to y, and θ', θ'' are vector fields on X then

$$(df)_x[\theta'_x, \theta''_x] = [(df)_x\theta'_x, (df)_x\theta''_x].$$

Thus the family $\mathscr{L}$ of vector fields we have constructed is a Lie algebra, called the *Lie algebra* of G and denoted by $\mathscr{L}(G)$. We have obtained the result:

Theorem. *The Lie algebra $\mathscr{L}(G)$ of a group G consists of all vector fields*

$$\theta_\xi(\cdot) = (d\varphi)_{(e,\cdot)}(\xi \oplus 0) \qquad \text{for } \xi \in T_{e,G},$$

where $\varphi\colon G \times G \to G$ is the group law of G. It also coincides with the set of differential operators of the form $\mathscr{D}_\xi(F)(g) = \dfrac{\partial}{\partial \xi} F(\varphi(g, \gamma))$, where $\xi \in T_{e,G}$, and differentiation is with respect to the second argument γ. Finally, $\mathscr{L}(G)$ equals the algebra of right-invariant vector fields on G, or of right-invariant first order differential operators.

The structure constants of $\mathscr{L}(G)$ can be expressed very explicitly in terms of the coefficients of the group law $\varphi(x, y)$ of G in a neighbourhood of e. Since $\xi \in T_{e,G}$ is uniquely determined by the values $\mathscr{D}_\xi(x_i)(e)$ for coordinates $x_1, \ldots, x_n$, we only have to find these values for the commutators $[\mathscr{D}_\xi, \mathscr{D}_\eta]$. From the fact that $\varphi(x, e) = x$ and $\varphi(e, y) = y$ it follows that the terms of degree 1 and 2 in the series for $\varphi(x, y)$ are of the form

$$\varphi(x, y) = x + y + B(x, y) + \cdots, \tag{8}$$

where $B(x, y)$ is linear both in x and in y. A simple substitution shows that $\mathscr{D}_\xi\mathscr{D}_\eta(x_i)(e) = B(\xi, \eta)_i$, where $B(\xi, \eta)_i$ is the ith coordinate. Hence

$$[\xi, \eta] = B(\xi, \eta) - B(\eta, \xi). \tag{9}$$

(8) and (9) show that the degree 1 terms in the group law $\varphi(x, y)$ are the same for all Lie groups of the same dimension (they are the same as those of $\mathbb{R}^n$). But the degree 2 terms define the Lie algebra $\mathscr{L}(G)$.

The invariant nature of the definition of the Lie algebra $\mathscr{L}(G)$ make a number of its natural properties almost obvious. If $f\colon G \to H$ is a homomorphism of Lie groups, then df defines a homomorphism $\mathscr{L}(G) \to \mathscr{L}(H)$, whose kernel is the Lie algebra of the kernel of f. If H is a closed subgroup of a Lie group G then $\mathscr{L}(H)$ is a subalgebra of $\mathscr{L}(G)$, and if H is a normal subgroup then $\mathscr{L}(H)$ is an ideal of $\mathscr{L}(G)$ and $\mathscr{L}(G/H) = \mathscr{L}(G)/\mathscr{L}(H)$. If $\varphi\colon G \times X \to X$ is an action of a Lie group on a manifold X then the family $\mathscr{L} = \{\theta_\xi(\cdot) = (d\varphi)_{(e,\cdot)}(\xi \oplus 0) \mid \xi \in T_{e,G}\}$ of vector fields on X defined above is a Lie algebra, and is a homomorphic image of the Lie algebra of G: $\mathscr{L} = \mathscr{L}(G)/\mathscr{L}(N)$ where N is the kernel of the action.

Formula (9) shows that the Lie algebra of an Abelian group is commutative.

Example 10. Let $G = \mathrm{GL}(n)$. In a neighbourhood of the identity matrix we can write $A = E + X$, and if $B = E + Y$ then $AB = E + X + Y + XY$. Thus in (8), we have $B(X, Y) = XY$, and (9) shows that the commutator of elements X and Y in the Lie algebra $\mathscr{L}(\mathrm{GL}(n))$ is of the form $XY - YX$, that is, $\mathscr{L}(\mathrm{GL}(n)) = \mathfrak{gl}(n)$. If X is the matrix $E + (x_{ij})$, where x_{ij} are coordinate functions, then a vector $\xi \in T_{e,G}$ goes to the matrix $\dfrac{\partial X}{\partial \xi} = \left(\dfrac{\partial x_{ij}}{\partial \xi}\right)$.

Example 11. Now let $G = \mathrm{SL}(n)$. Then $G \subset \mathrm{GL}(n)$ is the subgroup defined by $\det(E + X) = 1$. A tangent vector ξ to $\mathrm{GL}(n)$ is tangent to $\mathrm{SL}(n)$ if $\dfrac{\partial}{\partial \xi}(\det(E + X)) = 0$. But, as is well known,

$$\frac{\partial}{\partial \xi}(\det(E + X)) = \mathrm{Tr}\left(\frac{\partial X}{\partial \xi}\right).$$

Hence for ξ tangent to $\mathrm{SL}(n)$ we have $\mathrm{Tr}\dfrac{\partial X}{\partial \xi} = 0$, and therefore $\mathscr{L}(\mathrm{SL}(n)) = \mathfrak{sl}(n)$.

Example 12. Similarly, if $G = \mathrm{O}(n)$ and $E + X \in G$ then $(E + X)(E + X^*) = E$. Hence

$$\left[\left\{\frac{\partial}{\partial \xi}(E + X)\right\}(E + X^*) + (E + X)\left\{\frac{\partial}{\partial \xi}(E + X^*)\right\}\right]\Bigg|_{(X=0)} = 0,$$

or $\left(\dfrac{\partial X}{\partial \xi}\right) + \left(\dfrac{\partial X}{\partial \xi}\right)^* = 0$. Thus $\mathscr{L}(\mathrm{O}(n)) = \mathfrak{o}(n)$.

Example 13. As we indicated at the beginning of § 15, the group $\mathrm{SO}(3)$ is the configuration space for a rigid body moving with a fixed point: the motion of such a body gives a curve $g(t) \in \mathrm{SO}(3)$. A tangent vector $\dfrac{dg}{dt}$ belongs to the tangent space $T_{g(t)}$ to $\mathrm{SO}(3)$ at the point $g(t)$. We can transform it by right translation g^{-1} to the vector $\gamma(t) = \dfrac{dg}{dt}g^{-1} \in T_e$, that is, to an element of the Lie algebra $\mathfrak{o}(3)$. From the fact that $g(t)$ is orthogonal, that is $g(t)g(t)^* = e$, it follows that $\dfrac{dg}{dt}g^* + g\left(\dfrac{dg}{dt}\right)^* = 0$, that is $\gamma(t)^* = -\gamma(t)$, according to Example 12. If some point of the body moves according to the law $x(t) = g(t)(x_0)$ then obviously $\dfrac{dx}{dt} = \dfrac{dg(t)}{dt}(x_0) = \gamma(t)g(t)(x_0) = \gamma(t)(x(t))$. By Example 9, corresponding to the transformation $\gamma(t)$ there is a vector $\omega(t)$ such that γ reduces to vector multiplication by $\omega(t)$. Hence $\dfrac{dx(t)}{dt} = [\omega(t), x(t)]$. This equation shows that at each instant t the velocity of points of the body are the same as under the rotation with constant angular velocity $\omega(t)$; the vector $\omega(t)$ is called the *instantaneous angular velocity*.

We could transform the vector $\frac{dg}{dt}$ *by a left translation into* $\tilde{\gamma}(t) = g^{-1}\frac{dg}{dt} \in T_e$. It is easy to see that $\tilde{\gamma} = g^{-1}\gamma g$, and the corresponding vector $\tilde{\omega} = g^{-1}\omega$ is the instantaneous angular velocity in a coordinate system rigidly attached to the body.

Exactly the same argument as in Examples 11 and 12 allows us to determine the Lie algebras of the other Lie groups we know:

$$\mathscr{L}(\mathrm{U}(n)) = \mathfrak{u}(n), \quad \mathscr{L}(\mathrm{SU}(n)) = \mathfrak{su}(n), \quad \mathscr{L}(\mathrm{Sp}(n) = \mathfrak{sp}(n), \quad \text{and} \quad \mathscr{L}(\mathrm{SpU}(n)) = \mathfrak{spu}(n).$$

We recall again that all the preceding arguments are also applicable to complex Lie groups: the corresponding Lie algebras $\mathscr{L}(G)$ are Lie algebras over $\mathbb{C}$, as one sees easily. In particular,

$$\mathscr{L}(\mathrm{GL}(n, \mathbb{C})) = \mathfrak{gl}(n, \mathbb{C}), \qquad \mathscr{L}(\mathrm{SL}(n, \mathbb{C})) = \mathfrak{sl}(n, \mathbb{C}),$$
$$\mathscr{L}(\mathrm{O}(n, \mathbb{C})) = \mathfrak{o}(n, \mathbb{C}) \quad \text{and} \quad \mathscr{L}(\mathrm{Sp}(2n, \mathbb{C})) = \mathfrak{sp}(2n, \mathbb{C}).$$

We proceed to the second part of Lie theory, to the question of the extent to which a Lie group G can be reconstructed from its Lie algebra $\mathscr{L}(G)$. Here there are two possible way of stating the problem. Firstly, we could study the group law φ only in some neighbourhood of the identity of the group. If we introduce coordinates $x_1, \ldots, x_n$ in this neighbourhood, then the group law is given by n power series

$$\varphi(x, y) = (\varphi_1(x_1, \ldots, x_n; y_1, \ldots, y_n), \ldots, \varphi_n(x_1, \ldots, x_n; y_1, \ldots, y_n)).$$

These must satisfy the associativity relation $\varphi(x, \varphi(y, z)) = \varphi(\varphi(x, y), z))$ and the existence of an identity $\varphi(x, 0) = \varphi(0, x) = x$ (the existence of an inverse, that is, of a power series $\psi(x)$ satisfying $\varphi(x, \psi(x)) = \varphi(\psi(x), x) = x$ follows easily from this by the implicit function theorem). Geometrically, this formulation of the problem corresponds to the study of *local Lie groups*, that is, analytic group laws defined in some neighbourhood V of 0 in $\mathbb{R}^n$ (the product is an element of $\mathbb{R}^n$, but possibly not of the same neighbourhood), and satisfying the associativity axiom and the existence of an identity, which is 0. We say that two local Lie groups defined in neighbourhoods V_1 and V_2 are *isomorphic* if there exists neighbourhoods $V_1' \subset V_1$ and $V_2' \subset V_2$ of 0 and a diffeomorphism $f \colon V_1' \to V_2'$ taking the first group law into the second. A homomorphism of local Lie groups is defined similarly. Under this formulation of the question the answers is very simple.

Lie's Theorem. *Every Lie algebra $\mathscr{L}$ is the Lie algebra of some local Lie group. A local Lie group is determined up to isomorphism by its Lie algebra. Every Lie algebra homomorphism $\varphi \colon \mathscr{L}(G_1) \to \mathscr{L}(G_2)$ between the Lie algebras of two local Lie groups is of the form $\varphi = (df)_e$ where $f \colon G_1 \to G_2$ is a homomorphism of local Lie groups, uniquely determined by this condition.*

The most elementary and striking form of this theorem is obtained if we use formula (9) for the commutation in Lie algebras. The theorem shows that already the degree 2 terms $B(x, y)$ in the group law determines the group law uniquely up to isomorphism (that is, up to analytic coordinate transformations). If we consider $\varphi(x, y)$ as a formal power series, then the theorem takes on a purely algebraic character, without any analysis involved. It holds for 'formal group laws' over an arbitrary field of characteristic 0. We will return later to this algebraic aspect of Lie theory.

When passing to global Lie groups, that is, to the by now usual definition given in § 15, things become a little more complicated. Indeed, already the additive group $\mathbb{R}$ and the circle group $\mathbb{R}/\mathbb{Z}$ are not isomorphic, although they both have the same Lie algebra, the Abelian 1-dimensional algebra. However, the ideal situation is reestablished if we restrict attention to connected and simply connected groups (compare § 14, Example 7).

A theorem proved by Cartan asserts that the above statement of Lie's theorem continues to hold word-for-word if we replace the term 'local Lie group' by 'connected and simply connected Lie group'. (In the assertion on homomorphisms $\mathscr{L}(G_1) \to \mathscr{L}(G_2)$, it is enough for G_1 to be simply connected.)

Lie theory can also be applied to the study of connected but non-simply connected Lie groups, since the universal cover $\tilde{G}$ of a group G can be made into a group (in a unique way) such that $G = \tilde{G}/N$, where N is a discrete normal subgroup contained in the centre of G. This gives a construction of all connected Lie groups having the same Lie algebra. An example of a representation in the form $G = \tilde{G}/N$ is $G = \mathrm{O}(n)$, $\tilde{G} = \mathrm{Spin}(n)$, $N = \{E, -E\}$; or $G = \mathrm{PSL}(n, \mathbb{C})$, $\tilde{G} = \mathrm{SL}(n, \mathbb{C})$ and $N = \{\varepsilon E \mid \varepsilon^n = 1\}$.

C. Applications of Lie Algebras

Most of the applications are based on Lie theory, which reduces many questions of the theory of Lie groups to similar questions on Lie algebras, which are as a rule simpler. Thus the most direct method of deducing the classification of simple Lie groups discussed in § 16 consists of the classification of simple Lie algebras and then application of Lie-Cartan theory. For example, it is proved that over the complex numbers field, there exist the following simple finite-dimensional Lie algebras: $\mathfrak{sl}(n, \mathbb{C})$, $\mathfrak{o}(n, \mathbb{C})$, $\mathfrak{sp}(n, \mathbb{C})$ and a further 5 exceptional algebras of dimensions 78, 133, 248, 14 and 52, denoted respectively by E_6, E_7, E_8, G_2 and F_4. By Lie-Cartan theory this gives the classification of complex connected simple Lie groups. Each Lie algebra corresponds to one simply connected group G, for example $\mathfrak{sl}(n, \mathbb{C})$ corresponds to $\mathrm{SL}(n, \mathbb{C})$; and a quotient group of the form G/N, where N is a discrete normal subgroup contained in the centre of G, has the same Lie algebra. Since for each of these groups the centre Z is itself finite, we obtain together with each simply connected group a finite number of quotient groups, as we mentioned in § 16.

In exactly the same way, the theory of simple real Lie groups reduces to the theory of simple Lie algebras over $\mathbb{R}$. Their study is carried out by methods analogous to those discussed in § 11, where we studied simple algebras and division algebras over non-algebraically closed fields. More precisely, exactly as we did in § 11 for associative algebras, one can also define the operation of extension of the ground field $\mathscr{L}_{K'} = \mathscr{L} \otimes_K K'$ in the case that $\mathscr{L}$ is a Lie algebra over a field K, and K' an extension of K. One proves that if $\mathscr{L}$ is a simple algebra over $\mathbb{R}$, then $L_{\mathbb{C}}$ is either a simple algebra over $\mathbb{C}$, or a direct sum of two isomorphic simple algebras. Thus the problem of studying simple algebras over $\mathbb{R}$ reduces to a similar problem over $\mathbb{C}$. It is precisely in this way that we can give substance to the notion of 'real analogues of a complex Lie group G' as discussed in § 16.

Finally, we indicate the connection of Lie algebras with mechanics.

Example 14. Consider the motion under inertia of a rigid body with a fixed point. Since no external forces act on the body, the law of conservation of angular momentum gives

$$\frac{dJ}{dt} = 0. \tag{10}$$

Suppose that the motion is described by a curve $g(t) \in \mathrm{SO}(3)$, as in Example 13. Introducing the angular momentum $\tilde{J} = g^{-1}J$ in a system of coordinates rigidly attached to the body, we rewrite (10) after obvious transformations in the form

$$\frac{d\tilde{J}}{dt} + [\tilde{\omega}, \tilde{J}] = 0. \tag{11}$$

These equations (there are 3 of them corresponding to the 3 coordinates of the vector $\tilde{J}$) are called the *Euler equations*. They can be viewed as equations for $\tilde{J}$, since the relation between $\tilde{\omega}$ and $\tilde{J}$ is realised by the inertia tensor I,

$$\tilde{J} = I(\tilde{\omega}), \tag{12}$$

where I is a symmetric linear transformation independent of t. The transformation I determines the kinetic energy by the formula

$$T = \tfrac{1}{2}(\tilde{J}, \tilde{\omega}) = \tfrac{1}{2}(I(\tilde{\omega}), \tilde{\omega}), \tag{13}$$

and is therefore positive definite.

In Formula (11), [,] denotes the vector product, but according to Example 9, it can also be interpreted as the commutator bracket of the algebra $\mathfrak{o}(3)$. In this form the equations (11) can be generalised to a very wide class of Lie groups and Lie algebras. Under this, the energy T is interpreted as a Riemannian metric on a Lie group G, invariant under left translations (since the transformation I in (13) is constant); it is defined by a symmetric transformation $\mathscr{L}(G) \to \mathscr{L}(G)^*$. Since (by analogy with the case of a rigid body and the group SO(3)) we take for $\tilde{\omega}$ an element of the Lie algebra $\mathscr{L}(G)$, we have $\tilde{J} \in \mathscr{L}(G)^*$, and for it we can write down the equation (11). It turns out that in a number cases such equations have interesting physical meaning. For example, the case of the group G of all motions of Euclidean 3-space (that is, $G = O(3)\cdot T$ where T is the group of translations) corresponds to the inertial motion of a body in an ideal fluid. The case of the group SO(4) also has physical meaning. But the most interesting is the case of the 'infinite-dimensional Lie group' of all diffeomorphisms of a manifold, which has the Lie algebra of all vector fields as its Lie algebra. This case is related to phenomena such as the motion of an ideal fluid. However, it does not fit into the standard theory of Lie algebras and groups, and the theory would seem to be at present on a heuristic level.

D. Other Nonassociative Algebras

The theory of Lie algebras shows in a very convincing way that deep and important results of the theory of rings are not necessarily connected with the requirement of associativity. Lie algebras are perhaps the most vivid examples of nonassociative rings of importance for the whole of mathematics. But there are others.

Example 15. As discussed in § 8, Example 5, quaternions can be written in the form $z_1 + z_2 j$, where z_1 and z_2 are complex numbers. In this form, the multiplication of quaternions is very simple to describe: if we assume that all the ring axioms are satisfied (and that operations on complex numbers are the same as in $\mathbb{C}$), then we only have to specify that $j^2 = -1$ and $jz = \bar{z}j$. We can attempt to go further in the same direction, defining an 8-dimensional algebra consisting of elements $q_1 + q_2 l$ where q_1, q_2 are quaternions, and l a new element. It turns out that in this way we arrive at an interesting nonassociative division algebra. If we postulate all the ring axioms, except for associativity of multiplication, then we only need to specify the products $q_1 \cdot q_2$, $q_1 \cdot (q_2 l)$, $(q_1 l)\cdot q_2$ and $(q_1 l)\cdot(q_2 l)$. We will assume that quaternions multiply as elements of $\mathbb{H}$, and set in addition

$$q_1(q_2 l) = (q_2 q_1)l, \quad (q_1 l)\cdot q_2 = (q_1 \overline{q_2})l$$

and

$$(q_1 l)\cdot(q_2 l) = -(\overline{q_2} q_1),$$

where $\bar{q}$ denotes the quaternion conjugate of q.

In other words multiplication is defined by the formula

$$(p_1 + p_2 l)(q_1 + q_2 l) = p_1 q_1 - \overline{q_2} p_2 + (q_2 p_1 + p_2 \overline{q_1})l.$$

All the ring axioms with the exception of the associativity of multiplication can easily be checked. The element $1 \in \mathbb{H}$ is the identity of the new ring. It is an 8-dimensional algebra over $\mathbb{R}$: a basis is given for example by $\{1, i, j, k, l, il, jl, kl\}$.

The algebra just constructed is called the *Cayley algebra* or the *algebra of octavions*, or the *Cayley numbers*, and denoted by $\mathbb{O}$.

For $u \in \mathbb{O}$ with $u = q_1 + q_2 l$, set

$$\bar{u} = q_1 - q_2 l, \quad |u| = \sqrt{|q_1|^2 + |q_2|^2} \quad \text{and} \quad \operatorname{Tr} u = \operatorname{Re} q_1.$$

It is easy to see that $u\bar{u} = \bar{u}u = |u|^2$, and that $\operatorname{Tr} uv = \operatorname{Tr} vu$. The element u satisfies the quadratic equation

$$u^2 - (\operatorname{Tr} u)u + |u|^2 = 0. \tag{14}$$

Theorem. *The algebra $\mathbb{O}$ has the following property, a weakening of the requirement of associativity: the product of 3 elements does not depend on the distribution of brackets if 2 of the elements coincide. In other words,*

$$u(uv) = (uu)v, \quad (uv)v = u(vv), \quad u(vu) = (uv)u$$

(the third of these is a consequence of the first two).

Rings satisfying this conditions are called *alternative rings*. It can be proved that in an alternative ring, the subring generated by any two elements is associative.

It follows from the properties given above that for $u \neq 0$ the element $u^{-1} = |u|^{-2}\bar{u}$ is an inverse of u, and $u(u^{-1}v) = v$, $(vu^{-1})u = v$, that is, $\mathbb{O}$ is a *(nonassociative) division algebra.*

It is not hard to prove that $|uv| = |u| \cdot |v|$. In the basis $1, i, j, k, l, il, jl, kl$ this gives a curious identity

$$\left(\sum_{i=1}^{8} x_i^2\right)\left(\sum_{i=1}^{8} y_i^2\right) = \left(\sum_{i=1}^{8} z_i^2\right),$$

where z_i are integral bilinear forms in $x_1, \ldots, x_8$ and $y_1, \ldots, y_8$.

The existence of the algebra $\mathbb{O}$ is the reason underlying a whole series of interesting phenomena of 'low-dimensional' geometry (in dimensions 6, 7 and 8). For example, we observe that for any plane $E = \mathbb{R}u + \mathbb{R}v \subset \mathbb{O}$, the set of all elements $w \in \mathbb{O}$ for which $wE \subset E$ defines a subalgebra $\mathbb{C}(E)$ isomorphic to the complex numbers; this is easy to verify (you need to use the fact that $\mathbb{O}$ is alternative; the subalgebra $\mathbb{C}(E)$ is spanned by 1 and $\alpha = vu^{-1}$). Any 6-dimensional subspace $F \subset \mathbb{O}$ can be given by the equations $\operatorname{Tr}(xu) = 0$ for $u \in E$, where E is some plane. It follows from this that $\alpha F \subset F$ for $\alpha \in \mathbb{C}(E)$ (you need to use the easily verified relation $\operatorname{Tr}(u(vw)) = \operatorname{Tr}((uv)w)$). Thus every 6-dimensional subspace $F \subset \mathbb{O}$ has a natural structure of 3-dimensional vector space over $\mathbb{C}$. In particular, if $X \subset \mathbb{O}$ is a smooth 6-dimensional manifold, then this applies also to its tangent spaces at different points, and the resulting complex structure varies smoothly with the point. We say that a manifold with this property is *almost complex*. The standard example of an almost complex manifold is a complex analytic manifold. We see that *any* orientable 6-*dimensional sub-*

manifold in $\mathbb{R}^8$ *is almost complex*. However, it is very rare to have a complex structure on such a manifold. For example, the almost complex structure on S^6 arising in this way is not defined by any complex structure.

Another application of the Cayley numbers relates to the exceptional simple Lie groups (see § 16), for example, the compact ones. Namely, the compact exceptional simple Lie group G_2 is isomorphic to the connected component of the automorphism group of the algebra $\mathbb{O}$. The groups E_6 and F_4 are also realised as the 2-dimensional 'projective' and 'orthogonal' groups related to $\mathbb{O}$.

Finally, the algebra of Cayley numbers plays a special role from a purely algebraic point of view. A *generalised Cayley algebra* is an 8-dimensional algebra consisting of elements of the form $q_1 + q_2 l$, where q_1 and q_2 belong to some generalised quaternion algebra over a field K (see § 11), and the multiplication is given by

$$(p_1 + p_2 l)(q_1 + q_2 l) = p_1 q_1 + \gamma \overline{q_2} p_2 + (q_2 p_1 + p_2 \overline{q_1}) l,$$

where $\gamma \neq 0$ is some element of K. This algebra is always alternative and simple. It is a nonassociative division algebra if and only if the equation $q_1\overline{q_1} - \gamma q_2 \overline{q_2} = 0$ has no nonzero solutions in the quaternion algebra.

Theorem. *Any alternative division algebra is either associative or isomorphic to some generalised Cayley algebra. Any alternative simple ring is either associative or isomorphic to a generalised Cayley algebra.*

As with associative rings, we can construct projective planes with 'coordinates' in an arbitrary alternative division algebra. This is one of the simplest example of non-Desarguian projective planes (see § 10). They have a simple geometric characterisation: a certain weakened form of Desargues' theorem should hold.

There are certain other types of nonassociative algebras, for which a fairly complete theory can be constructed (at least under the assumptions of finite dimensionality and simplicity), and which have mathematical applications. But no general theory of nonassociative algebras exists at present (at least, not in the sense of a theory to be placed alongside that of associative algebras or Lie algebras). Perhaps such a theory is just not possible? Indeed, an arbitrary finite-dimensional nonassociative algebra is given by a multiplication table c_{ijk} with absolutely no restrictions, so is an arbitrary tensor $C \in \mathscr{L} \otimes \mathscr{L} \otimes \mathscr{L}^*$, defined up to transformations of $\mathrm{Aut}(\mathscr{L})$. But the question then arises: under what types of natural restrictions should such a theory exist? How to understand from a unified point of view the theory of simple associative algebras, Lie algebras, alternative algebras and certain other types? A test problem could be the structure of nonassociative division algebras over $\mathbb{R}$. Here there is a remarkable fact: *such division algebras can have dimensions* 1, 2, 4 *or* 8 *only*. However, an algebraic proof of this fact is not known. The existing proof is topological, based on the study of topological properties of the map $(\mathbb{R}^n \setminus 0) \times (\mathbb{R}^n \setminus 0) \to (\mathbb{R}^n \setminus 0)$ defined by the multiplication of the algebra (which one identifies with $\mathbb{R}^n$).

§ 20. Categories

The notion of categories, together with certain notions related to them, forms a mathematical language having a specific nature as compared with the standard language of set theory, and imparting a somewhat different character to mathematical constructions. Our description starts with an example.

We will use *diagrams*, representing sets by points, and maps from a set X to a set Y by arrows from the point for X to the point for Y. If a diagram has points representing sets $X, A_1, A_2, \ldots, A_n, Y$, and maps $f_1: X \to A_1, f_2: A_1 \to A_2, \ldots, f_{n+1}: A_n \to Y$, then $f_{n+1} \ldots f_2 f_1$ is a certain map from X to Y, the *composite* of $f_1, \ldots, f_{n+1}$. If for all sets X and Y appearing in the diagram, and for any choice of the sets A_i and maps f_i the resulting composite map from X to Y arising is the same, we say that the diagram is *commutative*. Examples of commutative diagrams: the diagrams

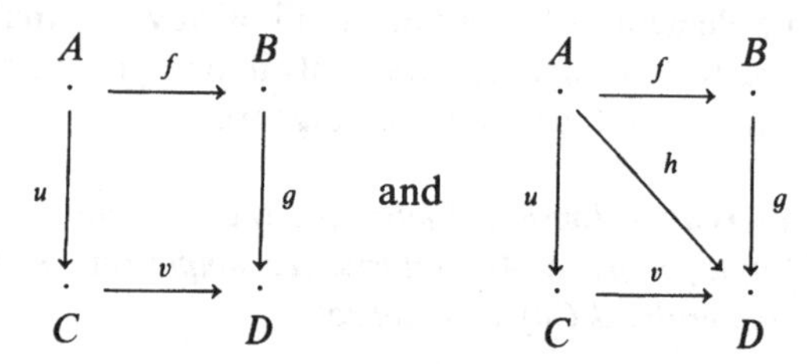

are commutative if $vu = gf$ and $h = vu = gf$ respectively.

We now proceed to the example itself. In set theory, there are two operations defined on arbitrary sets X and Y (these are not considered to be subsets of a fixed set): the *sum* or disjoint union of sets, denoted by $X + Y$, and the *product* denoted by $X \times Y$, consisting of pairs (x, y) with $x \in X$ and $y \in Y$. These operations can be described not by a construction, as we have just done, but by their general properties. For example, for the sum $X + Y$ we have two inclusion maps $f: X \to X + Y$ and $g: Y \to X + Y$, and the following *universal mapping property* holds: for any set Z and maps $u: X \to Z$ and $v: Y \to Z$, there exists a unique map $h: X + Y \to Z$ for which the diagram

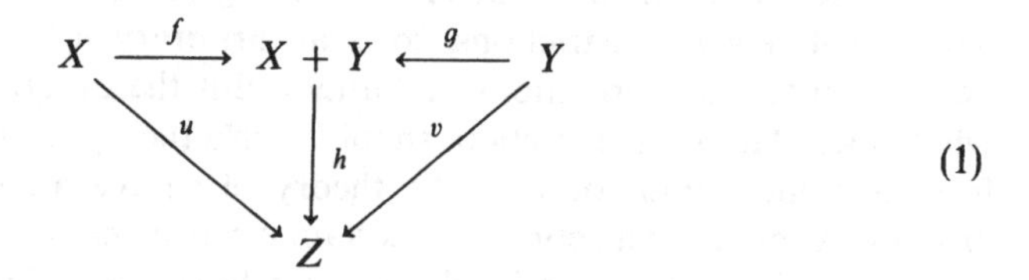

is commutative. In exactly the same way, the product $X \times Y$ has projection maps $f: X \times Y \to X$ and $g: X \times Y \to Y$, and for any set Z and maps $u: Z \to X$ and $v: Z \to Y$, there exists a unique map $h: Z \to X \times Y$ for which the diagram

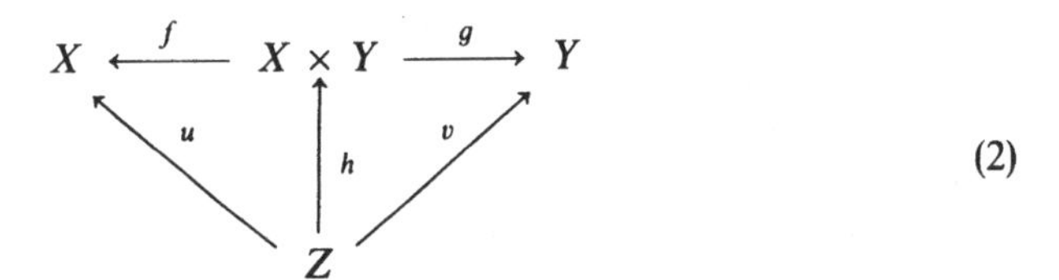

is commutative (that is, $fh = u$ and $gh = v$). Obviously $h(z) = (u(z), v(z))$.

As a next step we can consider sets in which certain special types of maps are distinguished, and consider what constructions are defined by requiring that universal mapping properties (1) or (2) hold. For topological spaces and continuous maps between them, we obtain, of course, the notions of disjoint union and product of topological spaces. The case of groups, where only group homomorphisms are considered as maps, is more interesting. For Abelian groups, and more generally for modules over a given ring, the direct sum $A \oplus B$ has both embeddings $A \to A \oplus B$ and $B \to A \oplus B$ and canonical projections $A \oplus B \to A$ and $A \oplus B \to B$, and both of the universal mapping properties of (1) and (2) are satisfied (of course, u, v and h are now homomorphisms, rather than arbitrary maps). Thus here the analogues of the two operations of set theory, disjoint union and product, come together. But this is not the case for non-Abelian groups: on the direct product $G \times H$, there are canonical homomorphisms $G \times H \to G$ and $G \times H \to H$ which satisfy the universal mapping property (2); but although the inclusion maps $f: G \to G \times H$ and $g: H \to G \times H$ are defined, the universal mapping property (1) does not hold. To see this, it is enough to take a group K having two subgroups isomorphic to G and H, but whose elements do not commute with one another, and let u and v be isomorphisms of G and H with these subgroups. Since elements $g \in G$ and $h \in H$ commute in $G \times H$, the diagram

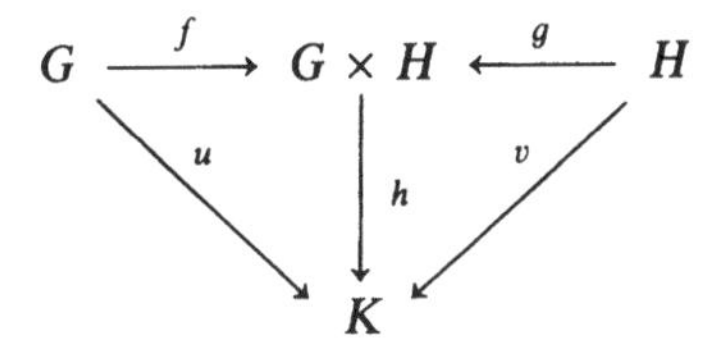

will not be commutative for any homomorphism h. Nevertheless, there does exist a construction of a group satisfying the universal mapping property (1). It is called the *free product* of G and H. This is the group generated by two subgroups G' and H' isomorphic to G and H, with no relations between the elements of G' and H' other than the relations holding in G' and H' individually. A precise definition can be given by analogy with the definition of a free group (§ 14, Example 6). In particular, the free group S_2 on two generators is the free product of two infinite cyclic groups.

Let us treat another variation on the same theme: we consider as sets commutative rings which are algebras over a given ring K, and as maps K-algebra homomorphisms between them. The direct sum $A \oplus B$ with its canonical projections $f: A \oplus B \to A$ and $g: A \oplus B \to B$ satisfies the universal mapping property (2). But although there are natural inclusion maps $f: A \to A \oplus B$ and $g: B \to A \oplus B$, the analogue of the universal mapping property (1) does not hold: the point is that for $a \in A$ and $b \in B$, we always have $ab = 0$ in $A \oplus B$; however, there might exist a ring C containing subrings A', B' isomorphic to A, B, but for which the relation $ab = 0$ does not always hold; then a homomorphism $h: C \to A \oplus B$ with the required property does not exist. It is easy to see that in this case the tensor product $A \otimes_K B$ (see § 12, Example 3) is a ring with the required property.

In conclusion, let us check that in all the cases we have considered, the construction satisfying the universal mapping property (1) or (2) is unique. Suppose for example that we are considering the diagram (1), and that for given sets X and Y we have two such sets R and S. Then the diagram

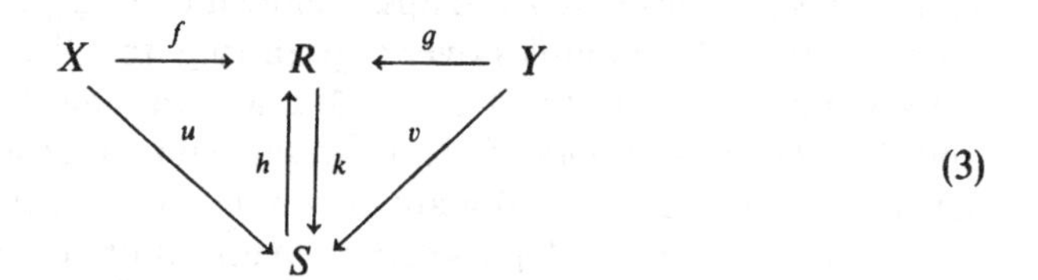

(3)

must be commutative. Hence $f = hu$ and $u = kf$, that is $f = (hk)f$, and similarly $g = (hk)g$. The requirement that the map h in (1) is unique, applied to the trivial case $S = R$, implies that hk is the identity map of R to itself; in the same way, we prove that kh is the identity map of S to itself. Hence R and S are isomorphic.

Now we draw the moral from the example we have considered. All the preceding arguments involved sets and certain maps between them. But we never needed to consider what kind of elements our sets were made up of, or how these elements transformed under our maps; the only thing we needed was that maps can be composed, and that different maps can be compared with one another—as we saw particularly vividly when using the commutativity of diagrams in the final argument concerned with diagram (3). This is the approach axiomatised in the notion of a category.

A *category* $\mathscr{C}$ consists of the following data:

(a) A set $Ob(\mathscr{C})$, whose elements are called the *objects* of $\mathscr{C}$;

(b) for any $A, B \in Ob(\mathscr{C})$, a set $H(A, B)$ whose elements are the *morphisms* in $\mathscr{C}$ from A to B;

(c) for any $A, B, C \in Ob(\mathscr{C})$, and any $f \in H(A, B)$ and $g \in H(B, C)$, a morphism $h \in H(A, C)$ called the *composite* of f and g, and written gf;

(d) for any $A \in Ob(\mathscr{C})$ a morphism $1_A \in H(A, A)$, called the *identity morphism*.

The above data must satisfy the conditions:

$$h(gf) = (hg)f \quad \text{for} \quad f \in H(A, B),\ g \in H(B, C) \text{ and } h \in H(C, D);$$

and

$$f 1_A = 1_B f = f \quad \text{for} \quad f \in H(A, B).$$

Thus in category theory, an object $A \in Ob(\mathscr{C})$ is characterised not as a set, in terms of the elements it is made up of, but in terms of its relations with other objects $B \in Ob(\mathscr{C})$. Thus we are interested primarily in its 'relations', rather than its 'construction'.

In the following examples of categories we take the point of view of 'naive set theory', ignoring logical contradictions that arise when we operate with such notions as 'all sets', 'all groups', and the like. Methods (involving the notion of a *class*) have been established by specialists to allow us to get around these contradictions (at least in the opinion of the majority of specialists).

Example 1. The category $\mathscr{S}et$ whose objects are arbitrary sets and whose morphisms are arbitrary maps between sets.

Example 2. The category whose objects are arbitrary subsets of a given set X, and whose morphisms are the inclusion maps between them (so that $H(A, B)$ is either empty or consists of a single element). Variation: X is a topological space, the objects are the open subsets, and the morphisms are inclusions between them.

Example 3. The category $\mathscr{T}op$ whose objects are topological spaces, and whose morphisms are continuous maps between them. Variations: the objects are differentiable (or analytic) manifolds, and the morphisms are differentiable (or analytic) maps between them. Another important variation: the objects are topological spaces (X, x_0) with a marked point x_0, and the morphisms are continuous maps $f\colon X \to Y$ taking the marked point of X into the marked point of Y, that is, $f(x_0) = y_0$. This category is denoted by $\mathscr{T}op_0$.

Example 4. The category $\mathscr{H}ot$ of topological spaces up to homotopy equivalence. Two continuous maps $f\colon X \to Y$ and $g\colon X \to Y$ between topological spaces are *homotopic* if one can be continuously deformed into the other, that is, if there exists a continuous map $h\colon X \times I \to Y$ (where $I = [0, 1]$ is the unit interval) such that $h(x, 0) = f(x)$ and $h(x, 1) = g(x)$. Two spaces X and Y are *homotopy equivalent* or have the same *homotopy type* if there exist continuous maps $f\colon X \to Y$ and $g\colon Y \to X$ such that gf is homotopy equivalent to the identity map 1_X of X and fg to 1_Y. The objects of $\mathscr{H}ot$ are topological spaces, and morphisms between them are continuous maps up to homotopy equivalence; spaces with the same homotopy type become isomorphic in $\mathscr{H}ot$. The category $\mathscr{H}ot_0$ is defined by analogy with Example 3.

Example 5. The category $\mathscr{M}od_R$ whose objects are modules over a given ring R, and whose morphisms are homomorphisms between them, that is $H(M, N) = \mathrm{Hom}_R(M, N)$. The category $\mathscr{M}od_{\mathbb{Z}}$ of Abelian groups is denoted by $\mathscr{A}b$.

Example 6. The category of groups: the objects are arbitrary groups, and the morphisms homomorphisms between them.

Example 7. The category of rings: the objects are arbitrary rings, and the morphisms homomorphisms between them. Variations: we consider only commutative rings, or only algebras over a given ring A (in the latter case, we take the morphisms to be A-algebra homomorphisms).

Example 8. We now give an example of a category whose morphisms are not defined as maps between sets. This example is related to the formal group laws, mentioned in the previous section in connection with Lie theory. The more standard terminology is a *formal group*: a formal group is an n-tuple $\varphi = (\varphi_1, \ldots, \varphi_n)$ of formal power series

$$\varphi_i(X, Y) = \varphi_i(x_1, \ldots, x_n; y_1, \ldots, y_n)$$

in two sets of variables, $X = (x_1, \ldots, x_n)$ and $Y = (y_1, \ldots, y_n)$, with coefficients in an arbitrary field K, and satisfying the conditions

$$\varphi(X, \varphi(Y, Z)) = \varphi(\varphi(X, Y), Z),$$

$$\varphi(0,0) = 0, \quad \text{and} \quad \varphi(X, 0) = \varphi(0, X) = X.$$

The number n is called the *dimension* of the formal group φ. A *homomorphism* of an n-dimensional group φ into an m-dimensional group ψ is an m-tuple $F = (f_1, \ldots, f_m)$ of formal power series in n variables such that

$$F(0) = 0, \quad \text{and} \quad \psi(F(X), F(Y)) = F(\varphi(X, Y)).$$

The objects of our category are formal groups defined over a given field K, and the morphisms are homomorphisms between them. If K is a field of characteristic 0, the study of our category reduces completely by Lie theory to the study of the category of finite-dimensional Lie algebras over K and homomorphisms between them. But if char $K = p > 0$, a new domain with quite specific properties arises. The theory of 1-dimensional formal groups is already far from trivial, and has important applications in algebraic geometry, number theory and topology.

Example 9. A category with a single object $\mathcal{O}$. In this case, the category is determined by the set $H(\mathcal{O}, \mathcal{O})$, which is an arbitrary set with an associative operation (composition) and an identity element. This algebraic notion is called a *semigroup with unit*.

Example 10. *The dual category* $\mathscr{C}^*$. For every category $\mathscr{C}$, the dual category $\mathscr{C}^*$ has the same objects as $\mathscr{C}$, but the morphisms $H(A, B)$ in $\mathscr{C}^*$ are given by $H(B, A)$ in $\mathscr{C}$, and the composite of two morphisms f and g in $\mathscr{C}^*$ is defined as the composite of g and f in $\mathscr{C}$. If we imagine a category as being a single diagram, in which the objects are represented as points and morphisms as arrows between them, then $\mathscr{C}^*$ is obtained from $\mathscr{C}$ by reversing the direction of the arrows.

The notion of the dual category leads to a certain duality in category theory. Namely, any notion or assertion of category theory can be applied to $\mathscr{C}^*$ to give a dual notion or assertion in $\mathscr{C}$, obtained from the first by 'reversing the arrows'.

Returning to the example treated at the beginning of this section, we can now define operations on objects in any category analogous to the sum and product of sets. For this, we need to use diagrams (1) and (2), which are meaningful in any category; the corresponding objects, if they exist, are called the *category-theoretical sum* and *product* of objects. As we have seen, these do not always exist. For example, the sum does not exist in the category of finite groups, but does exist in the category of all groups. However, if the sum or product exists, then they are unique up to isomorphism (objects A and B are *isomorphic* if there exist morphisms $f \in H(A, B)$ and $g \in H(B, A)$ such that $gf = 1_A$ and $fg = 1_B$): the proof given above is meaningful in any category. We can say that in the category of modules, sum and product both coincide with the direct sum of modules; in the category of groups, sum coincides with free product, and product with direct product of groups; in the category of commutative rings, sum coincides with tensor product and product with direct sum. In the category of topological spaces, sum and product coincide with the same operations on sets. In the category of topological spaces with a marked point, the product of spaces (X, x_0) and (Y, y_0) is their ordinary product $X \times Y$ with marked point (x_0, y_0); but the sum is the so-called *bouquet* $X \vee Y$, consisting of X and Y glued together at the points x_0 and y_0. For example, the bouquet of two circles is a 'figure-eight' (Figure 40).

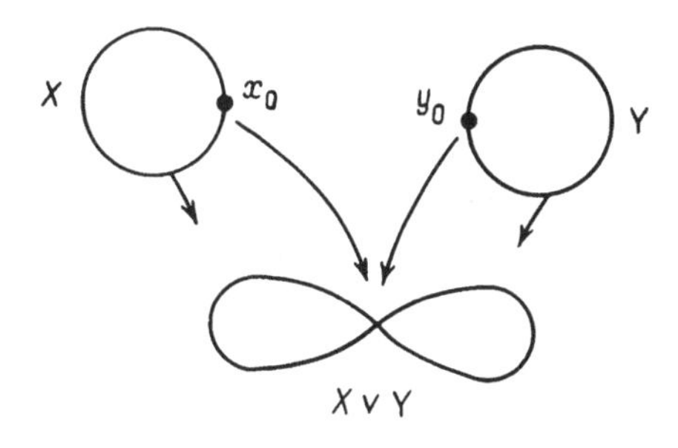

Fig. 40

Since diagrams (1) and (2) which serve as the definition of sum and product are obtained from one another by reversing the arrows, these notions are dual to one another, that is, they go into one another on passing from the category $\mathscr{C}$ to the dual category $\mathscr{C}^*$.

The idea of an 'invariant' or 'natural' construction in the language of categories is expressed through the notion of a functor. A *covariant functor* from a category $\mathscr{C}$ to a category $\mathscr{D}$ consists of two maps (denoted by the same letter), a map $F\colon Ob(\mathscr{C}) \to Ob(\mathscr{D})$, and for any A, $B \in \mathscr{C}$ a map $F\colon H(A, B) \to H(F(A), F(B))$, satisfying the conditions:

$$F(1_A) = 1_{F(A)} \quad \text{for} \quad A \in Ob(\mathscr{C})$$

$$\text{and} \quad F(fg) = F(f)F(g) \quad \text{whenever } fg \text{ is defined in } \mathscr{C}.$$

For example, in the category of vector spaces $\mathscr{C}$, the map $E \mapsto T^r E$ taking a vector space to its r-fold tensor product is obviously compatible with linear maps:

if $f: E \to E'$ is a linear map then $f(x_1 \otimes \cdots \otimes x_r) = f(x_1) \otimes \cdots \otimes f(x_r)$ defines a linear map $T^rE \to T^rE'$. It is easy to see that together these two maps define a covariant functor $F = T^r$, the r-fold tensor product; this is a functor from $\mathscr{C}$ to itself.

A *contravariant functor* is also given by a map $F: Ob(\mathscr{C}) \to Ob(\mathscr{D})$, but this time, for $A, B \in Ob(\mathscr{C})$ it defines a map

$$F: H(A, B) \to H(F(B), F(A))$$

(in the reverse order), with the conditions

$$F(1_A) = 1_{F(A)} \quad \text{and} \quad F(fg) = F(g)F(f)$$

(also in the reverse order). A typical example of a contravariant functor is the operation of taking a vector space into its dual vector space.

Example 11. Let A be a commutative ring and M, N two A-modules. Fixing the module N, we set $F_N(M) = M \otimes_A N$, and take a homomorphism $f: M \to M'$ into the homomorphism $F(f): M \otimes_A N \to M' \otimes_A N$ given by $F(f)(m \otimes n) = f(m) \otimes n$; then F_N is a covariant functor from $\mathscr{M}od_A$ into itself. We set $G_N(M) = \mathrm{Hom}_A(M, N)$, and for $f: M \to M'$ and $\varphi \in \mathrm{Hom}_A(M', N)$, write $G(f)(\varphi)$ for the composite φf; then G_N is a contravariant functor from $\mathscr{M}od_A$ into itself. If R is a noncommutative ring then $G_N(M) = \mathrm{Hom}_R(M, N)$ is a contravariant functor from $\mathscr{M}od_R$ into $\mathscr{A}b$.

Here are some more examples, in which we only indicate the effect of the functor on $Ob(\mathscr{C})$: the reader will easily guess its action on the sets $H(A, B)$.

Example 12. The standard constructions of topology are functors. Consider the *path space* $H(I, X)$ of a topological space X: this is the set of continuous maps $\varphi: I \to X$ of the interval $I = [0, 1]$ into X. The topology of $H(I, X)$ is determined by the requirement that for each open set $U \subset X$, the set $\{\varphi | \varphi(I) \subset U\}$ should be open. Since for any map $f \in H(X, Y)$ composing with f takes a path $\varphi \in H(I, X)$ into a path $f\varphi \in H(I, Y)$, it follows that $H(I, X)$ is a covariant functor from the category $\mathscr{T}op$ into itself. Most frequently it is considered on the category $\mathscr{T}op_0$ of topological spaces (X, x_0) with a marked point. Then by definition $H(I, X)$ consists only of those maps $\varphi: I \to X$ for which $\varphi(0) = x_0$. Finally, if in $H(I, X)$ we consider only those maps φ for which $\varphi(0) = \varphi(1) = x_0$ (that is $H(S^1, X)$, where S^1 is the circle with a marked point), then we get ΩX, the *loop space* of X. All of these covariant functors carry over naturally to functors on the homotopy category $\mathscr{H}ot$ of Example 4.

Example 13. The majority of topological invariants are groups, and are functors from the category $\mathscr{T}op$ or $\mathscr{H}ot$ into the category of groups or of Abelian groups. Thus the fundamental group $\pi(X)$ (§ 14, Example 7) is a covariant functor from the category of topological spaces with a marked point into the category of groups; the homotopy groups $\pi_n(X)$ for $n \geqslant 2$, the homology groups $H_n(X, A)$ and the cohomology groups $H^n(X, A)$ (the definition of which will be discussed in § 21 below) are functors from the same category into the category of Abelian groups. The functors π_n and H_n are covariant and H^n are contravariant. All of these objects are invariant with respect to homotopy equivalence, and pass to the category $\mathcal{H}ot$.

Example 14. An important functor is the space of functions $\mathscr{F}(X, \mathbb{R})$ on a set X (say real-valued). Here a number of variations are possible: if X is discrete (for example, finite) then we consider all maps, if X is a topological space, continuous maps, if X has a measure, square-integrable functions, and so on. Since a map $f: X \to Y$ takes a function $\varphi \in \mathscr{F}(Y, \mathbb{R})$ into $\varphi f \in \mathscr{F}(X, \mathbb{R})$, it follows that $F(X) = \mathscr{F}(X, \mathbb{R})$ is a contravariant functor from the category of sets (or topological spaces, or spaces with measure, ...) into the category of vector spaces. It is precisely due to the fact that $\mathscr{F}(X, \mathbb{R})$ is a functor that any transformation of X corresponds to an invertible linear transformation of $\mathscr{F}(X, \mathbb{R})$, and if G is a transformation group of X then $\mathscr{F}(X, \mathbb{R})$ has a representation of G defined on it. In particular, if $X = G$ and we consider the action of G on itself by left translations, then $\mathscr{F}(X, \mathbb{R})$ is the regular representation of G.

Example 15. In an arbitrary category $\mathscr{C}$, any object $A \in Ob(\mathscr{C})$ defines a covariant functor h_A from $\mathscr{C}$ to the category of sets $\mathscr{S}et$; we set $h_A(X) = H(A, X)$, and for any $f \in H(X, Y)$, we define the map $h_A(f): H(A, X) \to H(A, Y)$ as composition with f, that is $h_A(f)(g) = fg$ for $g \in H(A, X)$. In the same way $h^A(X) = H(X, A)$ defines a contravariant functor. We have already met the functors $h^N(M) = \mathrm{Hom}_R(M, N)$ on the category $\mathscr{M}od_R$ (Example 11) and $h_{S^1}(X) = \Omega X$ on $\mathscr{T}op_0$ (Example 12).

The functors h_A and h^A are useful in a very general situation when we want to carry over some construction defined for sets to any categories. If Φ denotes a set-theoretical construction and Ψ is the construction in a category $\mathscr{C}$ we are looking for, then we require that the functors $h_{\Psi(A)}$ and $\Phi(H_A)$ are equivalent (applying the functor h^A in place of h_A gives a different dual construction). Here we say that two functors F_1 and F_2 from $\mathscr{C}$ to $\mathscr{S}et$ are *equivalent* if for any $X \in Ob(\mathscr{C})$ we can define an invertible map $\varphi_X: F_1(X) \to F_2(X)$ such that for any $Y \in Ob(\mathscr{C})$ and any $f \in H(X, Y)$ the diagram

$$\begin{array}{ccc} F_1(X) & \xrightarrow{\varphi_X} & F_2(X) \\ \downarrow{\scriptstyle F_1(f)} & & \downarrow{\scriptstyle F_2(f)} \\ F_1(Y) & \xrightarrow{\varphi_Y} & F_2(Y). \end{array}$$

is commutative. For example, it is easy to see that the definition of sum $A + B$ in a category reduces to the requirement that the functors $h_{A+B}(X)$ and $h_A(X) \times h_B(X)$ are equivalent. The product is related in a similar way to the functor h^A.

As an application, we discuss the very important notion of a *group* (or *group object*) in a category $\mathscr{C}$. We will suppose that products exist in $\mathscr{C}$. A group law on an object $G \in Ob(\mathscr{C})$ is defined as a morphism $\mu \in H(G \times G, G)$; the operation of passing to the inverse is determined by specifying a morphism $\iota \in H(G, G)$. Finally, the existence of a unit element is simplest to state if we assume that in $\mathscr{C}$ there exists a *final object* $\mathcal{O} \in Ob(\mathscr{C})$, that is, an object such that $H(A, \mathcal{O})$ consists of a single element for every $A \in Ob(\mathscr{C})$ (a single point in $\mathscr{Set}$ or $\mathscr{Top}$, the zero group in $\mathscr{Ab}$, etc.). Then the identity element is defined as a morphism $\varepsilon \in H(\mathcal{O}, G)$. These three morphisms, μ, ι and ε should be subject to a number of conditions which express the associativity of multiplication and the other group axioms; these can be stated as requirements that certain diagrams be commutative. But we will not list these here, since a simpler requirement which is equivalent to all of these conditions is that for any $A \in Ob(\mathscr{C})$ they define a group law on the set $H(A, G)$ (maps into a group themselves form a group!), and for any $f \in H(A, B)$ the composition map $H(B, G) \to H(A, G)$ is a homomorphism. In other words, the functor h^G should be a functor from $\mathscr{C}$ into the category of groups.

For example, a Lie group is a group object in the category of (differentiable or complex analytic) manifolds.

Example 16. Consider the loop space ΩX over a topological space X with a marked point x_0 (Example 12). Composition of loops, as defined in § 14, Example 7, defines a continuous map $\mu: \Omega X \times \Omega X \to \Omega X$, reversing a loop defines a map $\iota: \Omega X \to \Omega X$, and the loop reduced to x_0 defines a unit element. This data does not define a group: for example, the product of a loop with its inverse is not equal to the unit loop, but only homotopic to it. But in the category $\mathscr{Hot}_0$ (Example 4) we do obtain a group, as one checks easily.

Example 17. We are going to use the operation of contracting to a point a closed subset A of a topological space X. By this we mean the topological space, denoted by X/A, which set-theoretically consists of $X \setminus A$ plus a single extra point a; the contraction map $p: X \to X/A$ is then defined as being the identity on $X \setminus A$ and taking A to a. An open subset of X/A is defined to be a set whose inverse image under p is open in X.

The *suspension* of a topological space X is the space ΣX obtained from the cylinder $X \times I$ (where $I = [0, 1]$) by contracting its top and bottom faces $X \times 0$ and $X \times 1$ to two points (see Figure 41).

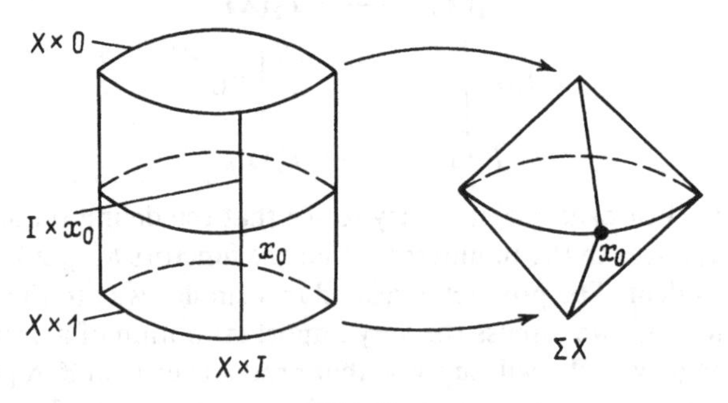

Fig. 41

In the category of spaces with a marked point, we consider the *reduced suspension* SX obtained from ΣX by contracting also $x_0 \times I$ to a point, which is taken as the marked point of SX. The properties of SX are simplest to state using the *smash product* operation $X \wedge Y$ in $\mathscr{T}\!op_0$ and $\mathscr{H}\!ot_0$. By definition, for spaces (X, x_0) and (Y, y_0) we set

$$X \wedge Y = X \times Y/(X \times y_0 \cup x_0 \times Y).$$

It is easy to check that this operation is distributive with respect to the sum operation $X \vee Y$ (see Figure 40), that is,

$$X \wedge (Y \vee Z) = (X \wedge Y) \vee (X \wedge Z)$$

and (4)

$$(X \vee Y) \wedge Z = (X \wedge Z) \vee (Y \wedge Z).$$

In particular, the reduced suspension is given by $SX = S^1 \wedge X$ where S^1 is the circle, obtained from $[0, 1]$ by glueing the points 0 and 1. It is easy to see that $SS^1 = S^1 \wedge S^1 = S^2$, and more generally $SS^n = S^{n+1}$ (where S^n is the n-dimensional sphere).

Identifying the points 0 and 1/2 in the circle $S^1 = I/\{0, 1\}$ gives a map $S^1 \to S^1 \vee S^1$ (Figure 42),

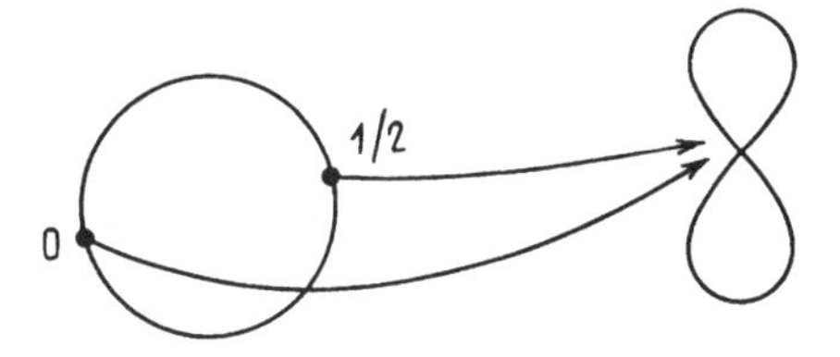

Fig. 42

and hence in $\mathscr{T}\!op_0$ and $\mathscr{H}\!ot_0$, a map

$$SX \to SX + SX$$

(since $\vee$ is the sum in this category, $SX = S^1 \wedge X$ and $\wedge$ is distributive). By duality we have a morphism $\mu\colon (SX) \times (SX) \to SX$ in the dual category. Reversing the circle, corresponding to the symmetry of S^1 with respect to the point 1/2 defines a map $S^1 \to S^1$ and hence $\iota\colon SX \to SX$. Mapping the whole space to a point defines a unit element. It is easy to see that in this way SX defines a group object in the category $\mathscr{H}\!ot_0^*$.

Example 18. The groups in $\mathscr{H}\!ot_0$ and $\mathscr{H}\!ot_0^*$ constructed in Examples 16–17 define important invariants which are now ordinary groups; this is for the simple reason that by definition, if G is a group object in a category $\mathscr{C}$ then $H(A, G)$ is a group for any $A \in Ob(\mathscr{C})$. Similarly, if $G \in Ob(\mathscr{C})$ is a group in the dual category $\mathscr{C}^*$ then $H(G, A)$ is a group for any $A \in Ob(\mathscr{C})$. Hence if we make any choice of a topological space R (with marked point), then for any topological space X both $H(X, \Omega R)$ and $H(SR, X)$ will be groups, and functors from the category $\mathscr{H}ot_0$ into the category of groups. We have already met one of these: if R consists of two points, then $SR = S^1$ (Figure 43) and $H(SR, X)$ is the fundamental group $\pi(X)$. But since $S^n = SS^{n-1}$, also $H(S^n, X)$ is a group for any $n \geqslant 1$. This is denoted by $\pi_n(X)$ and is called the nth *homotopy group* of X. In particular, $\pi(X) = \pi_1(X)$.

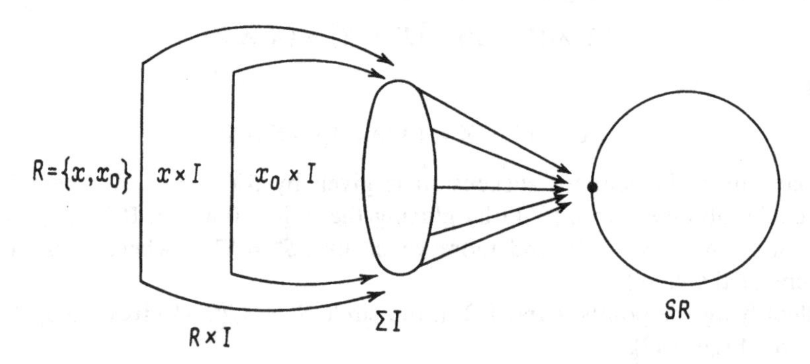

Fig. 43

For $n \geqslant 2$ the group $\pi_n(X)$ is Abelian, and the reason for this is also categorical: it consists of the fact that for $n \geqslant 2$, $S^n = S(SS^{n-2})$, and for any spaces R and X the group $H(S(SR), X)$ is Abelian. In fact, the representation $SSR = S^1 \wedge S^1 \wedge R$ allows us to define two maps

$$SSR \to SSR \vee SSR$$

using the map $S^1 \to S^1 \vee S^1$ of Figure 42 for either the first factor S^1 in $SSR = S^1 \wedge S^1 \wedge R$ or the second. From this we get two group laws on $H(S(SR), X)$, which we denote by $\cdot$ and $*$. They have a distributivity property: $(f \cdot g) * (u \cdot v) = (f * u) \cdot (g * v)$; this can be checked very easily from the definitions using the distributivity property (4) of $\wedge$. Moreover, both operations have the same unit element e. But it follows formally from this that the two operations coincide and are commutative:

$$f \cdot g = (f * e) \cdot (e * g) = (f \cdot e) * (e \cdot g) = f * g,$$

and

$$g \cdot f = (e * g) \cdot (f * e) = (e \cdot f) * (g \cdot e) = f * g = f \cdot g.$$

Similarly, $H(X, \Omega\Omega R)$ is an Abelian group for any spaces R and X.

We can also give a definition of cohomology groups in the same spirit. For this one proves that for any choice of an Abelian group A there exists a sequence of spaces K_n for $n = 1, 2, 3, \ldots$ with marked points, such that

$$\pi_m(K_n) = 0 \quad \text{for} \quad m \neq 0, n$$

$$\pi_n(K_n) = A, \tag{5}$$

and

$$K_{n-1} = \Omega K_n \text{ (in } \mathscr{H}ot_0).$$

Then just as before, the set $H(X, K_n)$ is an Abelian group; this is called the nth *cohomology group* of X with coefficients in A, and denoted by $H^n(X, A)$. (This is not the most natural method of defining cohomology groups, nor the original one: this will be discussed in the next section.) For the n-sphere S^n, we have

$$H^m(S^n, A) = 0 \quad \text{for} \quad m \neq 0, n,$$

$$H^n(S^n, A) = A,$$

and

$$S^n = SS^{n-1} \quad \text{(in } \mathscr{H}ot),$$

so that the spheres S^n are in this sense analogous to the spaces K_n (with the groups π_i instead of H^i).

Many of the most splendid achievements of topology (for example, those connected with the study of the groups $\pi_m(S^n)$) are based on the ideology which we have tried to hint at in the preceding constructions: the category $\mathscr{H}ot_0$ can to a significant extent be treated as an algebraic notion, in many respects analogous, for example, to the category of modules, and the intuition of algebra can successfully be applied to it.

§21. Homological Algebra

A. Topological Origins of the Notions of Homological Algebra

The algebraic aspect of homology theory is not complicated. A *chain complex* is a sequence $\{C_n\}_{n \in \mathbb{Z}}$ of Abelian groups (most often $C_n = 0$ for $n < 0$) and connecting homomorphisms $\partial_n\colon C_n \to C_{n-1}$, called *boundary maps*; a *cochain complex* is a sequence $\{C^n\}_{n \in \mathbb{Z}}$ of Abelian groups and homomorphisms $d_n\colon C^n \to C^{n+1}$, called *coboundary maps* or *differentials*. The boundary homomorphisms of a chain complex must satisfy the condition $\partial_n \partial_{n+1} = 0$ for all $n \in \mathbb{Z}$, and the coboundary of a cochain complex the condition $d_{n+1} d_n = 0$. Thus a complex is defined not just by the system of groups, but also by the homomorphisms, and we will for example denote a chain complex by $K = \{C_n, \partial_n\}$.

The condition $\partial_n\partial_{n+1} = 0$ in the definition of a chain complex shows that $\partial_{n+1}(C_{n+1})$, the image of ∂_{n+1}, is contained in the kernel of ∂_n, that is, $\operatorname{Im}\partial_{n+1} \subset \operatorname{Ker}\partial_n$. The quotient group

$$H_n(K) = \operatorname{Ker}\partial_n/\operatorname{Im}\partial_{n+1}$$

is called the *nth homology group* of the chain complex $K = \{C_n, \partial_n\}$, and denoted by $H_n(K)$. Similarly, for a cochain complex $K = \{C^n, d_n\}$, we have $\operatorname{Im} d_{n-1} \subset \operatorname{Ker} d_n$; the group $\operatorname{Ker} d_n/\operatorname{Im} d_{n-1}$ is called the *nth cohomology group* of K, and denoted by $H^n(K)$.

Here are the two basic situations in which these notions arise.

Example 1. An *n-dimensional simplex* or *n-simplex* is the convex hull of $n + 1$ points in Euclidean space not lying in a $(n - 1)$-dimensional subspace. A *complex* is a set made up of simplexes meeting along whole faces, such that the complex contains together with a simplex also all of its faces, and such that every point belongs to only a finite number of simplexes. As a topological space, a complex is determined by its set of vertexes, together with the data of which vertexes form simplexes. Thus this is a finite method of specifying a topological space, analogous to defining a group by generators and relations. A topological space X homeomorphic to a complex is called a *polyhedron*, and a homeomorphism of X with a complex is called a *triangulation* of X. Thus a triangulation is a partition of a space into pieces which are homeomorphic to simplexes, and which 'fit together nicely'. Figure 44 shows a triangulation of the sphere. Spaces which arise in practice usually admit triangulations; for example, this is the case for differentiable manifolds. But there are many such triangulations, just as there are many representations of a group by generators and relations.

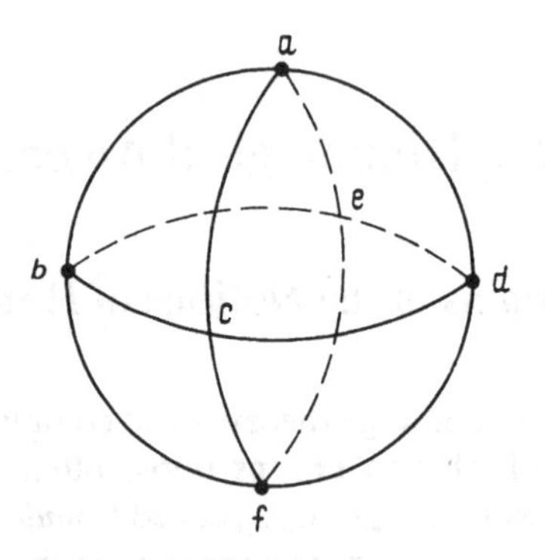

Fig. 44

We can associate with each complex X a chain complex $K = \{C_n, \partial_n\}_{n\geqslant 0}$. Here $C_n = \bigoplus \mathbb{Z}\sigma_i$ is the free $\mathbb{Z}$-module with generators corresponding to n-simplexes σ_i of X. To define the homomorphisms ∂_n, each simplex σ_i is oriented, that is, a definite ordering $\sigma_i = \{x_0, \ldots, x_n\}$ of its vertexes is chosen. We then set

$$\partial_n \sigma_i = \sum_{k=0}^{n} (-1)^k \varepsilon_k \sigma_i^k,$$

where σ_i^k is the simplex $\{x_0, \ldots, x_{k-1}, x_{k+1}, \ldots, x_n\}$, and $\varepsilon_k = 1$ or -1, depending on whether $x_0, \ldots, x_{k-1}, x_{k+1}, \ldots, x_n$ is an even or odd permutation of the sequence of vertexes of σ_i^k in the chosen orientation; ∂_n is then extended to the whole group C_n by additivity. The property $\partial_n \partial_{n+1} = 0$ is easy to check. Elements $x_n \in \operatorname{Ker} \partial_n$ are called *cycles*, and $y_n \in \operatorname{Im} \partial_{n+1}$ *boundaries*; the groups $H_n(K)$ are called the *homology groups* of X, and denoted by $H_n(X)$. The geometric meaning of an element $x \in H_n(X)$ is that it is a closed n-dimensional piece of the space X, and two pieces are identified if together they bound an $(n+1)$-dimensional piece. For example, in Figure 45, (a) c and c' are closed curves on the torus defining the same element of $H_1(X)$, and in Figure 45, (b) the curve d on the 'double torus' is the zero element.

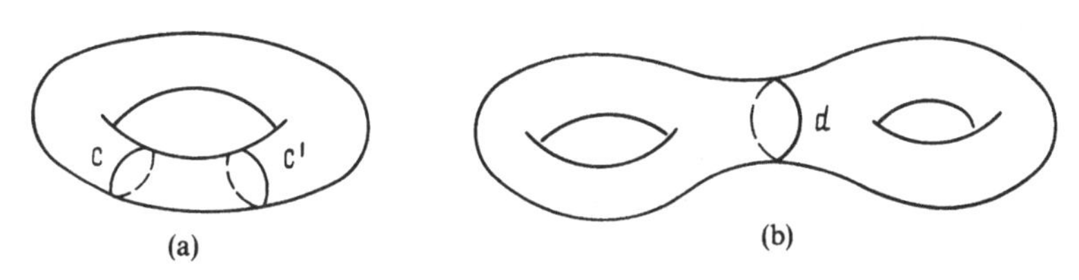

Fig. 45

The basic property of the groups $H_n(X)$, which is completely nonobvious from the definition we have given, is that they do not depend on the triangulation of the polyhedron X, but only on X as a topological space. Moreover, they define covariant functors from the category $\mathscr{H}ot$ to the category $\mathscr{A}b$ of Abelian groups. In other words, a continuous map $f: X \to Y$ induces homomorphisms $f_{*n}: H_n(X) \to H_n(Y)$, depending only on the homotopy class of f, and satisfying the conditions in the definition of a functor.

It is precisely the 'functorial' character of the groups $H_n(X)$ which make them so useful in topology: they determine a 'projection' of the topology into algebra. We give a very simple example. It is easy to prove that the n-dimensional sphere S^n satisfies

$$H_0(S^n) \cong \mathbb{Z} \cong H_n(S^n)$$

and

$$H_k(S^n) = 0 \quad \text{for } k \neq 0 \text{ or } n;$$

on the other hand, the n-dimensional ball B^n has the homotopy type of a point, since it can be contracted radially to the centre, and so $H_k(B^n) = 0$ for $k \neq 0$. We now give a proof of the famous Brouwer fixed point theorem: every continuous map $\Phi: B^n \to B^n$ has a fixed point. For otherwise, for any point $x \in B^n$, draw the ray from $\Phi(x)$ to x and write $f(x)$ for its intersection with the boundary S^{n-1} of B^n. Then f is a continuous map, and $f(x) = x$ for $x \in S^{n-1}$; that is, $f \circ i = 1$, where i is the inclusion $i: S^{n-1} \hookrightarrow B^n$ of S^{n-1} as the boundary sphere, and 1 is the identity map of S^{n-1}. Now functoriality gives a map $f_{*(n-1)}: H_{n-1}(B) \to H_{n-1}(S)$, which must be the zero map (since $H_{n-1}(B) = 0$). But on the other hand, $f_{*(n-1)} \circ i_{*(n-1)} = 1$, which gives a contradiction.

Alongside the chain complex $K = \{C_n, \partial_n\}$ constructed above for an arbitrary polyhedron X, we can take any Abelian group A and construct the chain complex $K \otimes_{\mathbb{Z}} A = \{C_n \otimes_{\mathbb{Z}} A, \partial_n\}$ and the cochain complex $\mathrm{Hom}(K, A) = \{\mathrm{Hom}(C_n, A), d_n\}$; recall that the functor $F(C) = C \otimes_{\mathbb{Z}} A$ is covariant and $G(C) = \mathrm{Hom}(C, A)$ is contravariant (see §20, Example 11). The group $H_n(K \otimes A)$ is denoted by $H_n(X, A)$, and $H^n(\mathrm{Hom}(K, A)) = H^n(X, A)$. These are called the *homology and cohomology groups* of X *with coefficients in* A. Homology is covariant and cohomology is a contravariant functor from $\mathscr{Hot}$ to the category $\mathscr{Ab}$. The groups $H^n(X, A)$ have already been mentioned at the end of §20.

Example 2. Let X be a differentiable manifold and Ω^r the space of differential r-forms on X. If $\varphi \in \Omega^r$ is a form, written in local coordinates as $\varphi = \sum f_{i_1 \dots i_r} dx_{i_1} \wedge \cdots \wedge dx_{i_r}$, then the *differential* of φ is the form

$$d\varphi = \sum df_{i_1 \dots i_r} \wedge dx_{i_1} \wedge \cdots \wedge dx_{i_r};$$

this expression does not depend on the choice of the coordinate system and defines a homomorphism $d_r: \Omega^r \to \Omega^{r+1}$. The relation $d_{r+1} d_r = 0$ is not hard to verify. Therefore $K = \{\Omega^r, d_r\}$ is a cochain complex; its cohomology $H^r(K)$ is called the *de Rham cohomology* of X and denoted by $H^r_{DR}(X)$. By analogy with the exterior algebra $\bigwedge(E) = \bigoplus \bigwedge^k(E)$ of a vector space E (see §5, Example 12), we can consider the graded ring $\Omega(X) = \bigoplus \Omega^r$ of all differential forms on X. The operation of taking exterior products extends from $\Omega(X)$ to the group $H^*_{DR}(X) = \bigoplus H^r_{DR}(X)$, which becomes a graded ring (and a superalgebra).

The connection between Examples 1 and 2 is based on the operation of *integrating a differential forms along a chain.* More precisely, we can find a triangulation of a manifold X sufficiently fine that every simplex σ_i is contained in some coordinate neighbourhood, and sufficiently smooth that the homeomorphism $f_i: \bar{\sigma}_i \to \sigma_i$ between σ_i and the standard simplex $\bar{\sigma}_i$ in Euclidean space is differentiable as often as we like. Then setting $\int_\sigma \varphi = \sum n_i \int_{\sigma_i} \varphi$ for $\sigma = \sum n_i \sigma_i \in C_r$ and $\varphi \in \Omega^r$ reduces the definition of the integral over a chain σ to the definition of the integral $\int_{\sigma_i} \varphi$ over a single simplex. Now using the diffeomorphism $f_i: \bar{\sigma}_i \to \sigma_i$ reduces it to integrating the form $f_i^* \varphi$ over the simplex $\bar{\sigma}_i$ in Euclidean space, that is, to the computation of an ordinary multiple integral.

Stokes' Theorem (Generalised Form). *For a form* $\varphi^{r-1} \in \Omega^{r-1}$ *and a chain* $c_r \in C_r$,

$$\int_{\partial c_r} \varphi^{r-1} = \int_{c_r} d\varphi^{r-1}$$

Since an integral over a chain depends in an additive way on the chain, the proof of this theorem reduces to the case that C_r is a simplex in Euclidean space. In this situation, it reduces to the well-known theorems of Green and Stokes for $r = 1$, 2 or 3, and is proved in the general case in exactly the same way. Introducing the notation $(c, \varphi) = \int_c \varphi$ for $c \in C_r$ and $\varphi \in \Omega^r$, and extending the pairing (c, φ) to $c \in C_r \otimes \mathbb{R}$ turns Stokes' theorem into the assertion that the two operators ∂ on $C_r \otimes \mathbb{R}$ and d on Ω^r are dual maps. Considering the pairing (c, φ) for $\partial c = d\varphi = 0$, it follows from this that (c, φ) vanishes if either $c = \partial c'$ or $\varphi = d\varphi'$, and hence it induces a pairing between $H_r(X, \mathbb{R})$ and $H^r_{DR}(X)$.

de Rham's Theorem. *The pairing thus constructed is a duality between these two spaces, so that de Rham cohomology provides an analytic method of computing the homology of a manifold. An equivalent formulation is that $H^r_{DR}(X)$ is isomorphic to $H^r(X, \mathbb{R})$.*

This isomorphism allows us to transfer the multiplication in the ring $H^*_{DR}(X)$ (see Example 2) to the group $H^*(X, \mathbb{R}) = \bigoplus H^r(X, \mathbb{R})$, which thus becomes a ring, the *cohomology ring* of X. Of course, there is also a method of defining a multiplication in $H^*(X, \mathbb{R})$ which does not use the relation with differential forms, and is not restricted to the case that X is a manifold.

Now we return to the algebraic theory of complexes, restricting ourselves to cochain complexes; the theory of chain complexes is obtained simply by reversing the arrows. A *subcomplex* of $K = \{C^n, d_n\}$ is a complex $K_1 = \{C^n_1, d_n\}$ for which the groups C^n_1 are subgroups of C^n, and the differentials between them are obtained by restricting the differential d_n defined on C^n so that, in particular, $d_n(C^n_1) \subset C^{n+1}_1$. In this situation, the d_n induce differentials on the groups $C^n_2 = C^n/C^n_1$, and we get a complex K_2, called the *quotient* of K by K_1, and denoted by K/K_1.

The cohomology groups of the complexes K, K_1 and $K_2 = K/K_1$ are connected by important relations. If we write $\operatorname{Ker} d_n$ for the kernel of d_n in C^n, then by definition we get

$$H^n(K) = \operatorname{Ker} d_n/d_{n-1}(C^{n-1}) \quad \text{and} \quad H^n(K_1) = (\operatorname{Ker} d_n \cap C^n_1)/d_{n-1}(C^{n-1}_1).$$

Since $C^{n-1}_1 \subset C^{n-1}$ and $d_{n-1}(C^{n-1}_1) \subset d_{n-1}(C^{n-1})$, sending an element of $(\operatorname{Ker} d_n \cap C^{n-1}_1)/d_{n-1}(C^{n-1}_1)$ into its coset modulo the bigger subgroup $d_{n-1}(C^{n-1})$, we get a homomorphism $i_n: H^n(K_1) \to H^n(K)$. Similarly, using the homomorphism $C^n \to C^n_2 = C^n/C^n_1$, we get an equally obvious homomorphism $j_n: H^n(K) \to H^n(K_2)$.

There is another homomorphism, which is rather less obvious. Suppose that $x \in H^n(K_2)$; then x corresponds to an element y of $\operatorname{Ker} d_n$ in C^n/C^n_1. Consider an inverse image $\bar{y}$ of y in C^n. Since $dy = 0$ in C^{n+1}/C^{n+1}_1, we have $d\bar{y} \in C^{n+1}_1$, and from the definition of a complex, it follows that $d\bar{y} \in \operatorname{Ker} d_{n+1}$. It is easy to prove that the coset $d\bar{y} + d_n C^n_1$ defines an element of $H^{n+1}(K_1)$ depending only on the original element x, and not on the choice of the auxilliary elements y and $\bar{y}$, and that this gives a homomorphism $\delta_n: H^n(K_2) \to H^{n+1}(K_1)$.

All the homomorphisms we have constructed can be joined together into an infinite sequence

$$\begin{aligned} \cdots \xrightarrow{j_{n-1}} H^{n-1}(K_2) \xrightarrow{\delta_{n-1}} \\ H^n(K_1) \xrightarrow{i_n} H^n(K) \xrightarrow{j_n} H^n(K_2) \xrightarrow{\delta_n} \\ H^{n+1}(K_1) \xrightarrow{i_{n+1}} \cdots \end{aligned} \tag{1}$$

This sequence has a very important property, which is most simply stated using the following extremely convenient algebraic notion. A sequence of groups and homomorphisms $\cdots \to A_{n-1} \xrightarrow{f_{n-1}} A_n \xrightarrow{f_n} A_{n+1} \xrightarrow{f_{n+1}} \cdots$ is *exact* if for each n the image of f_{n-1} coincides with the kernel of f_n. This condition is equivalent to saying that the $\{A_n, f_n\}$ form a cochain complex whose cohomology is equal to 0. Conversely, the cohomology of any complex measures its failure to be exact. The exactness of a sequence $0 \to A \xrightarrow{f} B$ means simply that f is an embedding of A into B, whereas that of $B \xrightarrow{g} C \to 0$ means that g maps onto the whole of C. Finally, the exactness of $0 \to A \to B \to C \to 0$ is just another way of saying that $A \subset B$ and $C = B/A$; an exact sequence of this form is a *short exact sequence.*

We can now state the basic property of the sequence (1) in the following compact form:

Theorem (Long Exact Cohomology Sequence). *If*

$$0 \to K_1 \to K \to K_2 \to 0$$

is a short exact sequence of cochain complexes (that is, $K_1 \subset K$ and $K_2 = K/K_1$), the sequence (1) *is exact.*

The proof is an almost tautological verification; (1) is called the *long exact cohomology sequence.*

If X is a triangulated topological space, and Y a closed subspace of X made up entirely of some of the simplexes of the triangulation of X, then with $Y \subset X$ we can associate chain complexes K_X and K_Y, such that $K_Y \subset K_X$. Hence we have a short exact sequence of chain complexes

$$0 \to K_Y \to K_X \to K_X/K_Y \to 0$$

and, as one sees very easily, for any Abelian group A, an exact sequence of cochain complexes

$$0 \to \mathrm{Hom}(K_X/K_Y, A) \to \mathrm{Hom}(K_X, A) \to \mathrm{Hom}(K_Y, A) \to 0.$$

The cohomology of the complex $\mathrm{Hom}(K_X/K_Y, A)$ has an interpretation which is already a fact of geometry rather than algebra: for $n \neq 0$,

$$H^n(K_X/K_Y, A) \cong H^n(X/Y, A)$$

where X/Y is the space obtained from X by contracting Y to a point. This assertion will also be true for H^0 if we modify slightly the definition of the complexes K_X in dimension $n = 0$ (for each of X, Y, and X/Y), restricting C_0 to be the set of those 0-dimensional chains $\sum n_i \sigma_i^0$ for which $\sum n_i = 0$.

The resulting cohomology groups are denoted $\tilde{H}^0(X, A)$. The algebraic theorem on the long exact cohomology sequence then gives us the assertion that the sequence

$$\begin{gathered} 0 \to \tilde{H}^0(X/Y, A) \to \tilde{H}^0(X, A) \to \tilde{H}^0(Y, A) \to \cdots \\ \cdots \to \tilde{H}^{n-1}(Y, A) \to \\ \tilde{H}^n(X/Y, A) \to \tilde{H}^n(X, A) \to \tilde{H}^n(Y, A) \to \cdots \end{gathered} \tag{2}$$

is exact, where $\tilde{H}^n(X, A) = H^n(X, A)$ for $n > 0$. From this it follows, for example, that if all $\tilde{H}^n(X, A) = 0$ then the groups $\tilde{H}^n(Y, A)$ and $\tilde{H}^{n+1}(X/Y, A)$ are isomorphic.

This result can be viewed in the following way: $\tilde{H}^0(X, A)$ defines a functor from the category $\mathscr{H}ot$ to the category of Abelian groups. For spaces X, Y and X/Y this functor is related by an exact sequence

$$0 \to \tilde{H}^0(X/Y, A) \to \tilde{H}^0(X, A) \to \tilde{H}^0(Y, A),$$

which however is not exact if we add $\to 0$ onto the right-hand end. But it can be extended to an exact sequence (2) by introducing an infinite number of new functors $\tilde{H}^n(X, A)$. This situation occurs frequently, and the particular case we have considered suggests the general principle: if for some important functor $F(A)$ with values in the category of Abelian groups short exact sequences arise in natural situations, for example of the form $0 \to F(B) \to F(C) \to F(A)$, then we should wonder whether it is not possible to define a family of functors $F^n(A)$ such that $F^0 = F$, and which are related by a long exact sequence of type (1) or (2). This is a completely new method of constructing functors. In the remainder of this section we will give two illustrations of this rather flexible principle. A third realisation of it is the subject matter of the following §22.

B. Cohomology of Modules and Groups

We have already seen in §20, Example 11 that for a fixed module A over a ring R and a 'variable' module M, $F(M) = \operatorname{Hom}_R(M, A)$ is a contravariant functor from the category of R-modules to that of Abelian groups. Therefore an exact sequence of modules

$$0 \to L \xrightarrow{f} M \xrightarrow{g} N \to 0 \tag{3}$$

defines homomorphisms $F(g)\colon \operatorname{Hom}_R(N, A) \to \operatorname{Hom}_R(M, A)$ and $F(f)\colon \operatorname{Hom}_R(M, A) \to \operatorname{Hom}_R(L, A)$. It is easy to check that the sequence

$$0 \to \operatorname{Hom}_R(N, A) \xrightarrow{F(g)} \operatorname{Hom}_R(M, A) \xrightarrow{F(f)} \operatorname{Hom}_R(L, A) \tag{4}$$

is exact, but will not remain exact if we add $\to 0$ onto the right-hand end. This means that the homomorphism $F(f)$ does not have to be onto: this can be seen in the example of the exact sequence $0 \to p\mathbb{Z}/p^2\mathbb{Z} \to \mathbb{Z}/p^2\mathbb{Z} \to \mathbb{Z}/p\mathbb{Z} \to 0$ of $\mathbb{Z}$-modules when $A = \mathbb{Z}/p\mathbb{Z}$.

However, the sequence (4) can be extended preserving exactness. This relates to the groups $\mathrm{Ext}_R(A, B)$ introduced in § 12, Example 2. One can show that, in a similar way to $\mathrm{Hom}_R(A, B)$, the group $\mathrm{Ext}_R(A, B)$ defines for fixed A a covariant functor $G(B) = \mathrm{Ext}_R(A, B)$ from the category $\mathcal{M}od_R$ into $\mathcal{A}b$, and for fixed B a contravariant functor $E(A) = \mathrm{Ext}_R(A, B)$. In view of the exact sequence (3), the module M can be thought of as an element of $\mathrm{Ext}_R(N, L)$, and any homomorphism $\varphi \in \mathrm{Hom}_R(L, A)$ defines a homomorphism $G(\varphi)$: $\mathrm{Ext}_R(N, L) \to \mathrm{Ext}_R(N, A)$. In particular, $G(\varphi)(M) \in \mathrm{Ext}_R(N, A)$, and as a function of φ it defines a homomorphism ∂: $\mathrm{Hom}_R(L, A) \to \mathrm{Ext}_R(N, A)$. One can prove that the sequence

$$0 \to \mathrm{Hom}_R(N, A) \xrightarrow{F(g)} \mathrm{Hom}_R(M, A) \xrightarrow{F(f)} \mathrm{Hom}_R(L, A) \xrightarrow{\partial} \mathrm{Ext}_R(N, A)$$

is exact. But we can include this into the even longer sequence

$$\begin{aligned} 0 \to \mathrm{Hom}_R(N, A) &\xrightarrow{F(g)} \mathrm{Hom}_R(M, A) \xrightarrow{F(f)} \mathrm{Hom}_R(L, A) \\ &\xrightarrow{\partial} \mathrm{Ext}_R(N, A) \xrightarrow{E(g)} \mathrm{Ext}_R(M, A) \xrightarrow{E(f)} \mathrm{Ext}_R(L, A), \end{aligned} \tag{5}$$

(where $E(g)$ and $E(f)$ are the homomorphisms defined by the functor $E(M) = \mathrm{Ext}_R(M, A)$), and the given sequence will also be exact! This of course sustains our hope of extending this to an infinite exact sequence.

We will in fact construct a system of Abelian groups, denoted by $\mathrm{Ext}_R^n(A, B)$; for fixed n and fixed argument B these are contravariant functors of the first argument, and for any exact sequence $0 \to L \to M \to N \to 0$ of modules they are connected by the exact sequence (for all $n \geqslant 0$)

$$\begin{aligned} &\ldots \to \mathrm{Ext}_R^{n-1}(L, A) \to \\ &\mathrm{Ext}_R^n(N, A) \to \mathrm{Ext}_R^n(M, A) \to \mathrm{Ext}_R^n(L, A) \to \\ &\mathrm{Ext}_R^{n+1}(N, A) \to \cdots \end{aligned} \tag{6}$$

Here Ext_R^0 is just Hom_R and Ext_R^1 is Ext_R.

The idea of the construction of such a system of functors is very simple. Suppose that our problem is already solved, and that in addition we know some type of modules (which we will denote by P) for which these functors vanish, that is,

$$\mathrm{Ext}_R^n(P, A) = 0 \quad \text{for all modules } A \text{ and all } n \geqslant 1.$$

Suppose, in addition, that we have managed to fit a module N into an exact sequence $0 \to L \to P \to N \to 0$ with P a module of this type, representing N as a homomorphic image of P. Then from the exact sequence (6) it will follow that the group $\mathrm{Ext}_R^n(N, A)$ is isomorphic to $\mathrm{Ext}_R^{n-1}(L, A)$, and we obtain an inductive definition of our functors.

The problem then is to find the modules P which are to annihilate the as yet unknown functors Ext_R^n. But a part of these functors is known, namely $\mathrm{Ext}_R^1 = \mathrm{Ext}_R$, and we must start by considering modules which annihilate this. We say that a module P such that $\mathrm{Ext}_R(P, A) = 0$ for every module A is *projective*. Stated more simply, this says that if P is represented as a homomorphic image of a module A, then there exists a submodule $B \subset A$ such that $A \cong P \oplus B$ (and the first projection $A \to P$ is just the given homomorphism $A \to P$). Thus when we are dealing with projective modules, we have as it were got back to a semisimple situation. The simplest example of a projective module is a free module. In fact, let F be a free module over a ring R and $x_1, \ldots, x_n$ a system of free generators of F (which we assume finite only to simplify notation). If $0 \to L \xrightarrow{f} M \xrightarrow{g} F \to 0$ is an exact sequence then g maps onto F, so that there exist elements $y_i \in M$ such that $g(y_i) = x_i$. Let $M' = Ry_1 + \cdots + Ry_n$ be the submodule of M which they generate. From the fact that $\{x_i = g(y_i)\}$ is a free system of generators, it follows that the same holds for $\{y_i\}$. From this it follows easily that g maps M' isomorphically to F, and $M = L \oplus M'$, where $L = \operatorname{Ker} g$.

It turns out that the class of projective modules is already sufficient to carry out our program. Any module is a homomorphic image of a free module, and hence a fortiori of a projective module. Let

$$0 \to L \to P \to N \to 0 \tag{7}$$

be a representation of N as a homomorphic image of a projective module P. Suppose that the functors $\mathrm{Ext}_R^r(L, A)$ are already defined for $r \leqslant n - 1$, and set $\mathrm{Ext}_R^n(N, A) = \mathrm{Ext}_R^{n-1}(L, A)$. We omit the definition of the homomorphism $\mathrm{Ext}_R^n(\varphi)\colon \mathrm{Ext}_R^n(L', A) \to \mathrm{Ext}_R^n(L, A)$ corresponding to a homomorphism $\varphi\colon L \to L'$, which is not difficult. It can be proved that both the groups $\mathrm{Ext}_R^n(L, A)$ and the homomorphisms $\mathrm{Ext}_R^n(\varphi)$ do not depend on the choice of the sequence (7) for N, that is, they are well defined. They form a contravariant functor (for fixed n and A), and are related by the exact sequence (6).

Putting together the n steps, we can obtain a definition of the functors Ext_R^n that is not inductive. For this, we represent the module L in (7) in a similar form, that is, we fit it into an exact sequence $0 \to L' \to P' \to L \to 0$ with P' projective, and then do the same with L', and so on. We obtain an infinite exact sequence

$$\cdots \to P_n \xrightarrow{\varphi_n} \cdots \to P_2 \xrightarrow{\varphi_2} P_1 \xrightarrow{\varphi_1} P_0 \to N \to 0 \tag{8}$$

with the P_n projective modules, called a *projective resolution* of N. Applying the functor $\mathrm{Hom}_R(P, A)$ to this, and leaving off the first term, we get a sequence

$$\mathrm{Hom}_R(P_0, A) \xrightarrow{\psi_0} \mathrm{Hom}_R(P_1, A) \xrightarrow{\psi_1} \cdots \to \mathrm{Hom}_R(P_n, A) \xrightarrow{\psi_n} \cdots, \tag{9}$$

which will not necessarily be exact, but which will be a cochain complex (from the fact that $\varphi_n \circ \varphi_{n+1} = 0$ in (8), and it follows from the definition of a functor that $\psi_{n+1} \circ \psi_n = 0$ in (9)). The cohomology of the complex (9) coincides with the groups $\mathrm{Ext}^n_R(N, A)$.

Example 3. Group Cohomology. The most important case in which the groups $\mathrm{Ext}^n_R(N, A)$ have many algebraic and general mathematical applications is when $R = \mathbb{Z}[G]$ is the integral group ring of a group G, so that the category of R-modules is just the category of G-modules. For a G-module A, the group $\mathrm{Ext}^n_{\mathbb{Z}}(\mathbb{Z}, A)$ (where $\mathbb{Z}$ is considered as a module with a trivial G-action) is called the *nth cohomology group* of G with coefficients in A, and denoted by $H^n(G, A)$.

To construct a projective resolution of $\mathbb{Z}$ (or even one just consisting of free modules) is a technical problem that is not particularly difficult. As a result we obtain a completely explicit form of the complexes (8) and (9). Let us write out the second of these. It has groups $C^i(=\mathrm{Hom}_G(P_i, A))$ consisting of arbitrary functions $f(g_1,\ldots,g_n)$ of n elements of G with values in A. The differential $d_n\colon C^n \to C^{n+1}$ is defined as follows:

$$(df)(g_1,\ldots,g_{n+1}) = g_1 f(g_2,\ldots,g_{n+1}) + \sum_{i=1}^{n} (-1)^i f(g_1,\ldots,g_i g_{i+1},\ldots,g_n)$$
$$+ (-1)^{n+1} f(g_1,\ldots,g_n);$$

(to understand the first term, you need to recall that $f(g_2,\ldots,g_{n+1}) \in A$ and that A is a G-module, which explains the meaning of $g_1 f(g_2,\ldots,g_{n+1})$).

We write out the first few cases:

$n = 0\colon f = a \in A$ and $(df)(g) = ga - a$;

$n = 1\colon f(g) \in A$ and $(df)(g_1, g_2) = g_1 f(g_2) - f(g_1 g_2) + f(g_1)$;

$n = 2\colon f(g_1, g_2) \in A$ and $(df)(g_1, g_2, g_3) = g_1 f(g_2, g_3) + f(g_1, g_2 g_3) - f(g_1 g_2, g_3) - f(g_1, g_2)$.

Thus $H^0(G, A)$ is the set of all elements $a \in A$ such that $ga - a = 0$ for all $g \in G$, that is, the set of G-invariant elements of A.

$H^1(G, A)$ is the group of function $f(g)$ of $g \in G$ with values in A satisfying the condition

$$f(g_1 g_2) = f(g_1) + g_1 f(g_2), \tag{10}$$

modulo the group of functions of the form

$$f(g) = ga - a \quad \text{with } a \in A. \tag{11}$$

If the action of G on A is trivial then $H^1(G, A) = \mathrm{Hom}(G, A)$.

$H^2(G, A)$ is the group of functions $f(g_1, g_2)$ of g_1, $g_2 \in G$ with values in A satisfying the condition

$$f(g_1, g_2 g_3) + g_1 f(g_2, g_3) = f(g_1 g_2, g_3) + f(g_1, g_2), \tag{12}$$

modulo the group of functions of the form

$$f(g_1, g_2) = h(g_1 g_2) - h(g_1) - g_1 h(g_2), \tag{13}$$

with $h(g)$ any function of $g \in G$ with values in A.

We give a number of examples of situations in which these groups arise.

Example 4. Group Extensions. Suppose that a group Γ has a normal subgroup N with quotient $\Gamma/N = G$. How can we reconstruct Γ from G and N? We have already met this problem in § 16; recall that we say that Γ is an extension of G by N. We now treat this in more detail in the case that N is an Abelian group, and in this case we denote it by A rather than N.

For any $\gamma \in \Gamma$ and $a \in A$, the element $\gamma a \gamma^{-1}$ is also contained in A. Furthermore, the map $a \mapsto \gamma a \gamma^{-1}$ is an automorphism of A. Since A is commutative, $\gamma a \gamma^{-1}$ is not affected by changing γ in its coset mod A; hence $\gamma a \gamma^{-1}$ depends only on the coset $g \in G = \Gamma/A$ containing γ. We therefore denote it by $g(a)$. Thus G acts on A, and A becomes a G-module (but with the group operation in A written multiplicatively, as in Γ, rather than additively).

We choose any-old-how a representative of each coset $g \in G = \Gamma/A$, and denote it by $s(g) \in \Gamma$. In general $s(g_1)s(g_2) \neq s(g_1 g_2)$, but these two elements belong to the same coset of A; hence there exist elements $f(g_1, g_2) \in A$ such that

$$s(g_1)s(g_2) = f(g_1, g_2)s(g_1 g_2). \tag{14}$$

The elements $f(g_1, g_2)$ cannot be chosen at all arbitrarily. Writing out the associative law $(s(g_1)s(g_2))s(g_3) = s(g_1)(s(g_2)s(g_3))$ in terms of (14) we see that they satisfy the condition

$$f(g_1, g_2 g_3) g_1(f(g_2, g_3)) = f(g_1 g_2, g_3) f(g_1, g_2), \tag{15}$$

which is the relation (12) written multiplicatively. It is easy to check that the structure of A as a G-module and the collection of elements $f(g_1, g_2) \in A$ for g_1, $g_2 \in G$ satisfying (15) already define an extension Γ of G by A. However, there was an ambiguity in our construction, namely the arbitrary choice of the coset representatives $s(g)$. Any other choice is of the form $s'(g) = h(g)s(g)$ with $h(g) \in A$. It is easy to check that using these representatives, we get a new system of elements $f'(g_1, g_2)$, related to the old ones by

$$f'(g_1, g_2) = f(g_1, g_2) h(g_1) g_1(h(g_2)) h(g_1 g_2)^{-1}.$$

Thus taking account of (13), we can say that an extension of G by A is uniquely determined by the structure of A as a G-module and an element of $H^2(G, A)$.

Let us stay for a moment with the case that the element of $H^2(G, A)$ corresponding to an extension is zero. By (14) this means that we can choose coset representatives $s(g)$ for Γ/A in such a way that $s(g_1 g_2) = s(g_1)s(g_2)$. In other words, these coset representatives themselves form a group G' isomorphic to G, and any element $\gamma \in \Gamma$ can be uniquely written in the form $\gamma = ag'$ with $a \in A$ and $g' \in G'$. In this case we say that the extension is *split*, that Γ is a *semidirect product* of A and G, and that G' is a *complement* of A in Γ. For example, the group of motions of the plane is a semidirect product of the group of translations and a group of rotations, and as a complement of the group of translations we can choose the group of rotations around some fixed point. Now for a split extension, how uniquely is a complement determined in general? The point is that we can change the coset representatives by $s'(g) = f(g)s(g)$ with $f(g) \in A$; it is easy to see a necessary condition for these to form a subgroup is the following relation:

$$f(g_1 g_2) = f(g_1) g_1 f(g_2)),$$

which is (10) written multiplicatively. But there exist a 'trivial' way of getting from one complement to another, namely conjugacy by an element $a \in A$, which takes G' into $G'' = aG'a^{-1}$. It is easy to see that $f(g)$ goes under this into $f(g)ag(a)^{-1}$. In view of (11) we deduce that in a semidirect product of A and G, complements of A, up to conjugacy by elements of A, are described by $H^1(G, A)$.

Example 5. The Cohomology of Discrete Groups. Suppose that X is a topological space having the homotopy type of a point, and that a group Γ acts on X discretely and freely (see §14). In this case we can construct a triangulation of X such that Γ acts freely on it. Then the group C_n of n-chains of X will be a free Γ-module, and the chain complex $\{C_n, \partial_n\}$ will be a projective resolution of $\mathbb{Z}$ (in fact made up of free Γ-modules). Now simply putting together definitions shows that for any Abelian group A considered as a Γ-module with trivial action, the cohomology groups $H^n(\Gamma, A)$ have a geometric realisation; they are isomorphic to the cohomology groups $H^n(\Gamma \backslash X, A)$ of the quotient space $\Gamma \backslash X$ (see Example 1). Any group Γ can be realised as a transformation group satisfying the above conditions. In this way we get a geometric interpretation of the groups $H^n(\Gamma, A)$.

This situation is realised in particular if $X = G/K$ where G is a connected Lie group, and K a maximal compact subgroup of G, since then X is homeomorphic to Euclidean space. Let $\Gamma \subset G$ be a discrete group without elements of finite order. Then Γ acts freely on the coset space G/K by left translations, and we have seen that $H^n(\Gamma, A) \cong H^n(\Gamma \backslash G/K, A)$. In particular, since $\Gamma \backslash G/K$ is a finite-dimensional space, $H^n(\Gamma, A) = 0$ for all n from some point on. For the groups $H^n(\Gamma, \mathbb{Z})$ we can introduce the notion of the *Euler characteristic*

$$\chi(\Gamma, \mathbb{Z}) = \sum (-1)^n \operatorname{rank} H^n(\Gamma, \mathbb{Z}).$$

We see that $\chi(\Gamma, \mathbb{Z}) = \chi(\Gamma \backslash G/K)$, where the right-hand side is the *topological Euler characteristic* of the space X, defined by $\chi(X) = \sum (-1)^q \dim_{\mathbb{R}} H^q(X, \mathbb{R})$.

This is applicable in particular to the case when G is an algebraic group over $\mathbb{Q}$, and Γ is an arithmetic subgroup $\Gamma \subset G(\mathbb{Z})$ of finite index in $G(\mathbb{Z})$ (see §15.C). In this case $\chi(\Gamma, \mathbb{Z})$ often has a delicate arithmetic meaning; for example, it may be expressed in terms of values of the Riemann ζ-function at integers. Thus for any subgroup $\Gamma \subset \mathrm{SL}(2, \mathbb{Z})$ of finite index and without any elements of finite order,

$$\chi(\Gamma, \mathbb{Z}) = (\mathrm{SL}(2, \mathbb{Z}) : \Gamma) \cdot \zeta(-1) = -\frac{(\mathrm{SL}(2, \mathbb{Z}) : \Gamma)}{12}.$$

Finally, we mention yet another very important application of group cohomology. Let K be a field and L/K a Galois extension of K with group G (see § 18.A). The group L^* of nonzero elements of L under multiplication is a G-module, and the cohomology groups $H^n(G, L^*)$ have very many applications both in algebraic questions and in arithmetic (when K is an algebraic number field).

C. Sheaf Cohomology

Let X be a topological space and $\mathscr{C}$ the category whose objects are the open subsets of X, and morphisms inclusions between them (§ 20, Example 2). A contravariant functor from $\mathscr{C}$ to the category of Abelian groups is called a *presheaf* of Abelian groups on X. Thus a presheaf $\mathscr{F}$ assigns to each open subset $U \subset X$ an Abelian group $\mathscr{F}(U)$, and to any two open sets $V \subset U$ a homomorphism $\rho_V^U: \mathscr{F}(U) \to \mathscr{F}(V)$ such that $\rho_U^U = 1_{\mathscr{F}(U)}$ and if $W \subset V \subset U$ then $\rho_W^U = \rho_W^V \rho_V^U$.

Example 6. The Presheaf $\mathcal{O}_{\mathscr{C}}$ of Continuous Functions on X. By definition, $\mathcal{O}_{\mathscr{C}}(U)$ is the set of all continuous functions on U, and ρ_V^U is the restriction of functions from U to V. In view of this example, the ρ_V^U are also called *restriction homomorphisms* in the general case.

A presheaf $\mathscr{F}$ is called a *sheaf* if for any open set $U \subset X$ and for any representation $U = \bigcup U_\alpha$ of it as a union of open sets the following two conditions hold:

1. For $s \in \mathscr{F}(U)$, if $\rho_{U_\alpha}^U s = 0$ for each U_α then $s = 0$.
2. If $s_\alpha \in \mathscr{F}(U_\alpha)$ are such that $\rho_{U_\alpha \cap U_\beta}^{U_\alpha} s_\alpha = \rho_{U_\alpha \cap U_\beta}^{U_\beta} s_\beta$ for all α and β then there exists $s \in \mathscr{F}(U)$ such that $s_\alpha = \rho_{U_\alpha}^U s$ for each α.

Sheaves over a given space themselves form a category. A *homomorphism* of a sheaf $\mathscr{F}$ into a sheaf $\mathscr{G}$ is a system of homomorphisms $f_U: \mathscr{F}(U) \to \mathscr{G}(U)$ for all open sets $U \subset X$, such that for all inclusions $V \subset U$ the diagram

$$\begin{array}{ccc} \mathscr{F}(U) & \xrightarrow{f_U} & \mathscr{G}(U) \\ \downarrow{\scriptstyle \rho_V^U} & & \downarrow{\scriptstyle \tilde{\rho}_V^U} \\ \mathscr{F}(V) & \xrightarrow{f_V} & \mathscr{G}(V) \end{array}$$

is commutative, where ρ_V^U and $\tilde{\rho}_V^U$ are the restriction maps for $\mathscr{F}$ and $\mathscr{G}$. We say that $\mathscr{F}$ is a *subsheaf* of $\mathscr{G}$ if $\mathscr{F}(U)$ is a subgroup of $\mathscr{G}(U)$ for every open set $U \subset V$.

The definition of a sheaf indicates that the groups $\mathscr{F}(U)$ are specified by local conditions. For example, the presheaf of continuous functions $\mathcal{O}_{\mathscr{C}}$ is a sheaf, and if X is a differentiable manifold then the sheaf $\mathcal{O}_{\text{diff}} \subset \mathcal{O}_{\mathscr{C}}$ of differentiable functions is the presheaf for which $\mathcal{O}_{\text{diff}}(U)$ consists of all differentiable functions on U (that is, having derivatives up to a given order $n \leqslant \infty$). Similarly, if X is a complex analytic manifold, then the sheaf $\mathcal{O}_{\text{an}}$ of analytic functions is the presheaf for which $\mathcal{O}_{\text{an}}(U)$ consists of all analytic functions on U. All of these examples motivate a general point of view, which unifies the definitions of various types of spaces. The definition should consist of a topological space X, a subsheaf $\mathcal{O}$ of the sheaf of continuous functions on X, and some supply $\mathcal{M}$ of models (such as the cube in $\mathbb{R}^n$ with the sheaf of differentiable functions on it, or the polydisc in $\mathbb{C}^n$ with the sheaf of analytic functions), with the requirement that each point $x \in X$ should have a neighbourhood U such that U together with the restriction to it of the sheaf $\mathcal{O}$ should be isomorphic to one of the models. This conception allows us to find natural definitions of objects which would otherwise not be easy to formulate; for example, complex analytic varieties with singularities (complex spaces). With a suitable modification this leads also to the natural definition of algebraic varieties over an arbitrary field, and their far-reaching generalisations, schemes.

Other examples of sheaves: the sheaf Ω^r of differential forms on a differentiable manifold X (here $\Omega^r(U)$ is the set of all differential forms on U), or the sheaf of vector fields.

Example 7. Let X be a Riemann surface (a 1-dimensional complex analytic manifold), $x_1, \dots, x_k \in X$ any set of points of X and $n_1, \dots, n_k$ any set of positive integers. The formal combination $D = n_1 x_1 + \cdots + n_k x_k$ is called a *divisor* (the condition $n_i \geqslant 0$ is not usually assumed). The *sheaf* $\mathcal{F}_D$ *corresponding to* D is defined as follows: $\mathcal{F}_D(U)$ is the set of meromorphic functions on U having poles only at points $x_1, \dots, x_k$ in U, such that the order of the pole at x_j is at most n_j.

If $f: \mathcal{F} \to \mathcal{G}$ is a sheaf homomorphism then $\mathcal{H}(U) = \operatorname{Ker} f_U$ defines a subsheaf of $\mathcal{F}$, called the *kernel* of f. Defining the image in the same way would not be good: the presheaf $\mathcal{J}'(U) = \operatorname{Im} f_U$ is not a sheaf in general. But it can be embedded into a minimal subsheaf $\mathcal{J}$ of $\mathcal{G}$: $\mathcal{J}(U)$ consists of the elements $s \in \mathcal{G}$ such that for each point $x \in U$ there exists a neighbourhood U_x in which $\rho^U_{U_x} s \in \operatorname{Im} f_{U_x}$. This sheaf $\mathcal{J}$ is called the *image* of f. Now that we have notions of kernel and image, we can define exact sequences of sheaves, repeating word-for-word the definition given for modules. For each sheaf $\mathcal{F}$ and its subsheaf $\mathcal{G}$ we can construct a sheaf $\mathcal{H}$ for which the sequence $0 \to \mathcal{G} \to \mathcal{F} \to \mathcal{H} \to 0$ is exact; $\mathcal{H}$ is called the *quotient sheaf*, $\mathcal{H} = \mathcal{F}/\mathcal{G}$.

The most important invariant of a sheaf $\mathcal{F}$ on a space X is the group $\mathcal{F}(X)$. Usually, this is a set of all the global objects defined by local conditions. For example, for the sheaf of differential forms, it is the group of differential forms on the whole manifold, for the sheaf of vector fields, the group of global vector fields. In Example 7, it is the group of all functions which are meromorphic everywhere on a Riemann surface X, with poles only at $x_1, \dots, x_n$, and of orders at most $n_1, \dots, n_k$. From the definition of homomorphisms of sheaves it follows that taking a sheaf $\mathcal{F}$ into $\mathcal{F}(X)$ is a covariant functor from the category of sheaves to the category of Abelian groups. Let

$$0 \to \mathcal{G} \to \mathcal{F} \to \mathcal{H} \to 0$$

be an exact sequence of sheaves. It is easy to check that the sequence of groups

$$0 \to \mathscr{G}(X) \to \mathscr{F}(X) \to \mathscr{H}(X)$$

is exact, but does not remain exact if we add $\to 0$ onto the right-hand end. Here is an example of this phenomenon. Let X be the Riemann sphere, $\mathcal{O}_{\mathrm{an}}$ the sheaf of analytic functions on X. We define a sheaf $\mathscr{H}$ by taking a finite set of points $\Phi = \{x_1, \ldots, x_k\} \subset X$, assigning to each point x_i one copy $\mathbb{C}_i$ of the group of complex numbers, and setting $\mathscr{H}(U) = \bigoplus \mathbb{C}_j$, where the sum is taken over $x_j \in \Phi \cap U$. Taking a function $f \in \mathcal{O}_{\mathrm{an}}(U)$ into its set of values $\{f(x_j)\} \in \bigoplus \mathbb{C}_j$ at points $x_j \in \Phi \cap U$ defines a homomorphism $\mathcal{O}_{\mathrm{an}} \to \mathscr{H}$ with image $\mathscr{H}$. If $\mathscr{G}$ is the kernel then we get an exact sequence $0 \to \mathscr{G} \to \mathcal{O}_{\mathrm{an}} \to \mathscr{H} \to 0$. In it, $\mathcal{O}_{\mathrm{an}}(X) = \mathbb{C}$ by Liouville's theorem, and $\mathscr{H}(X) = \mathbb{C}^k$ by definition, and hence for $k > 1$ the sequence $\mathcal{O}_{an}(X) \to \mathscr{H}(X) \to 0$ is not exact.

We are thus in the situation which we have already discussed, and our problem is to construct functors F^n from the category of sheaves on X to the category of groups in such a way that $F^0(\mathscr{F}) = \mathscr{F}(X)$, and for a short exact sequence $0 \to \mathscr{G} \to \mathscr{F} \to \mathscr{H} \to 0$ we have a long exact sequence

$$\begin{aligned} &\cdots \to F^{n-1}(\mathscr{H}) \to \\ F^n(\mathscr{G}) \to F^n(\mathscr{F}) &\to F^n(\mathscr{H}) \to \\ F^{n+1}(\mathscr{G}) &\to \cdots . \end{aligned} \tag{16}$$

We will argue in the same way as in constructing the functors Ext^n. Suppose that functors F^n with the required properties have been constructed, that we know a class of sheaves (which we will denote by $\mathscr{Q}$) for which $F^n(\mathscr{Q}) = 0$ for $n \geqslant 1$, and that we have represented a sheaf $\mathscr{F}$ as a subsheaf of such a $\mathscr{Q}$. Then we have an exact sequence $0 \to \mathscr{F} \to \mathscr{Q} \to \mathscr{H} \to 0$ and a corresponding long exact sequence (16). From the fact that $F^n(\mathscr{Q}) = 0$, we get that $F^n(\mathscr{F}) = F^{n-1}(\mathscr{H})$, and this gives an inductive definition of the functors F^n.

Now we proceed to construct sheaves $\mathscr{Q}$ with the required property. Since in particular such a sheaf $\mathscr{Q}$ will satisfy $F^1(\mathscr{Q}) = 0$, if $\mathscr{Q}$ fits into an exact sequence $0 \to \mathscr{Q} \xrightarrow{\varphi} \mathscr{F} \xrightarrow{\psi} \mathscr{G} \to 0$ then $0 \to \mathscr{Q}(X) \xrightarrow{u} \mathscr{F}(X) \xrightarrow{v} \mathscr{G}(X) \to 0$ must be exact (this follows by considering the first 4 terms of the sequence (16)). One class of sheaves with this property is known, the so-called flabby sheaves: a sheaf $\mathscr{Q}$ is *flabby* (or *flasque*) if the restriction homomorphisms $\rho_U^X\colon \mathscr{Q}(X) \to \mathscr{Q}(U)$ maps onto $\mathscr{Q}(U)$ for all $U \subset X$. An elementary argument shows that if $\mathscr{Q}$ is a flabby sheaf and $0 \to \mathscr{Q} \xrightarrow{\varphi} \mathscr{F} \xrightarrow{\psi} \mathscr{G} \to 0$ an exact sequence of sheaves, then $0 \to \mathscr{Q}(X) \xrightarrow{u} \mathscr{F}(X) \xrightarrow{v} \mathscr{G}(X) \to 0$ is exact.

A typical example of a flabby sheaf is the sheaf $\mathscr{Q}$ for which $\mathscr{Q}(U)$ is the set of all (real or complex valued) functions on U; the sheaves $\mathcal{O}_{\mathscr{C}}$, $\mathcal{O}_{\mathrm{diff}}$ and $\mathcal{O}_{\mathrm{an}}$ are all subsheaves of this. In a similar way, any sheaf is a subsheaf of a flabby sheaf. We are now in a situation which is completely analogous to that which we considered in §21.B, and we can give the definition of the new functors F^n first of all by induction: if a sheaf $\mathscr{F}$ fits in an exact sequence $0 \to \mathscr{F} \to \mathscr{Q} \to \mathscr{G} \to 0$ with $\mathscr{Q}$

flabby, then $F^n(\mathscr{F}) = F^{n-1}(\mathscr{G})$; $F^0(\mathscr{F}) = \mathscr{F}(X)$. It can be proved that $F^n(\mathscr{F})$ does not depend on the choice of the exact sequence $0 \to \mathscr{F} \to \mathscr{Q} \to \mathscr{G} \to 0$. They are called the *cohomology groups* of $\mathscr{F}$, and denoted by $H^n(X, \mathscr{F})$.

Putting together the n steps involved in the definition of $H^n(X, \mathscr{F})$, we can obtain a noninductive definition. An exact sequence

$$0 \to \mathscr{F} \to \mathscr{Q}_0 \to \mathscr{Q}_1 \to \cdots \to \mathscr{Q}_n \to \cdots$$

of sheaves in which the $\mathscr{Q}_i$ are flabby is called a *flabby resolution* of $\mathscr{F}$. Applying the functor $F(\mathscr{F}) = \mathscr{F}(X)$ to it and leaving out the first term, we obtain a cochain complex

$$\mathscr{Q}_0(X) \to \mathscr{Q}_1(X) \to \cdots \to \mathscr{Q}_n(X) \to \cdots;$$

the cohomology of this provide the groups $H^n(X, \mathscr{F})$.

Applications of sheaf cohomology are related to certain finiteness theorems concerning them. The first of these shows that in many situations a sheaf has only a finite number of nonzero cohomology groups.

Theorem. *If X is an n-dimensional manifold and $\mathscr{F}$ is any sheaf on X then $H^q(X, \mathscr{F}) = 0$ for any $q > n$.*

The second finiteness theorem relates to the case when all the groups $\mathscr{F}(U)$ are vector spaces (over $\mathbb{R}$ or $\mathbb{C}$), and the restriction homomorphisms ρ_V^U are linear maps. Then the $H^q(X, \mathscr{F})$ are also vector spaces, and one can ask about their dimensions. Of special interest is the dimension of $H^0(X, \mathscr{F}) = \mathscr{F}(X)$, usually the most important invariant. Generally speaking, this dimension is infinite even in the simplest cases; for example, for the sheaves $\mathcal{O}_{\mathscr{C}}$ or $\mathcal{O}_{\text{diff}}$, when $H^0(X, \mathcal{O}_{\mathscr{C}})$ is the space of all continuous, and $H^0(X, \mathcal{O}_{\text{diff}})$ of all differentiable functions on X. However, there are important cases when the corresponding spaces are finite-dimensional. Suppose for example that X is the Riemann sphere. Then $H^0(X, \mathcal{O}_{\text{an}})$ is the space of functions that are holomorphic at every point (including the point at infinity). By Liouville's theorem such a function is constant, so that $H^0(X, \mathcal{O}_{\text{an}}) = \mathbb{C}$ is 1-dimensional. The same thing holds for any compact connected complex analytic manifold X: on it $H^0(X, \mathcal{O}_{\text{an}}) = \mathbb{C}$. It can be proved that in this case all the cohomology groups $H^q(X, \mathcal{O}_{\text{an}})$ are also finite-dimensional over $\mathbb{C}$. The same holds for the sheaf of holomorphic differential forms or holomorphic vector fields on a compact complex analytic manifold, and for the sheaf $\mathscr{F}_D$ of Example 7 if the Riemann surface X is compact. We restrict ourselves here to the above examples, and do not state the general theorem on this subject.

In all cases when X is a finite-dimensional manifold and the spaces $H^q(X, \mathscr{F})$ are finite-dimensional, we can define the *Euler characteristic* of a sheaf $\mathscr{F}$:

$$\chi(X, \mathscr{F}) = \sum (-1)^q \dim H^q(X, \mathscr{F}) \tag{17}$$

(the sum consists of a finite number of terms, in view of the first finiteness theorem stated above).

This new invariant plays a two-fold role. First of all, the Euler characteristic of certain standard sheaves, which are related to the manifold in an intrinsic way, give invariants of the manifold itself. For example for a compact Riemann surface X,

$$\chi(X, \mathcal{O}_{\mathrm{an}}) = 1 - p,$$

where p is the genus of X; note that $1 - p$ is one half of the topological Euler characteristic $\chi(X, \mathbb{R})$ of X. In a similar way, for any compact complex analytic manifold X, the Euler characteristic $\chi(X, \mathcal{O}_{\mathrm{an}})$ gives an important invariant, the *arithmetic genus* of X.

On the other hand, the Euler characteristic turns out to be a 'coarse' invariant, and easy to calculate. In the majority of cases we are mainly interested in the dimension of the first term $\dim H^0(X, \mathcal{F})$ in the sum (17), but this is already a more delicate problem, which can be solved, for example, if we manage to prove that all the remaining terms are zero. Thus, in the case of the sheaf $\mathcal{F}_D$ of Example 7 associated with a divisor $D = \sum n_i x_i$ on a compact Riemann surface X, the Euler characteristic $\chi(X, \mathcal{F}_D)$ does not depend on the individual choice of the points x_i, but only on their number $d = \sum n_i$:

$$\chi(X, \mathcal{F}_D) = \chi(X, \mathcal{O}_{\mathrm{an}}) + d = 1 - p + d. \tag{18}$$

On the other hand, it can be proved that $H^q(X, \mathcal{F}_D) = 0$ for $q \geqslant 2$, and that if $d > 2p - 2$ then also $H^1(X, \mathcal{F}_D) = 0$, so that

$$\dim \mathcal{F}_D(X) = 1 - p + d \quad \text{for} \quad d > 2p - 2. \tag{19}$$

Recall that $\mathcal{F}_D(X)$ is the space of functions which are meromorphic on X and have poles at x_i of order at worst n_i. The equality (19) is first and foremost an existence theorem for such functions. If we have such functions at our disposal, then we can construct maps of Riemann surfaces to one another, and study the question of their isomorphism, and so on. As the simplest example, suppose that $p = 0$ and $D = x$ is one point. From (19) we get that $\mathcal{F}_D(X)$ is 2-dimensional; since the constant functions form a 1-dimensional subspace, we see that there exists a meromorphic function f on X having a pole of order 1 at x. It is not hard to prove that the map defined by this function is an isomorphism of the Riemann surface X with the Riemann sphere; that is, a Riemann surface of genus 0 is analytically isomorphic to the Riemann sphere (or in other terms, conformally equivalent to it).

Example 8. The analogue of the sheaf $\mathcal{F}_D$ of Example 7 can be constructed for an arbitrary n-dimensional complex analytic manifold X. For this we replace the points x_i by $(n - 1)$-dimensional complex analytic submanifolds C_i, set $D = \sum n_i C_i$, and take $\mathcal{F}_D(U)$ to be all functions which are meromorphic on U and have poles only along the submanifolds $U \cap C_i \subset U$ of order $\leqslant n_i$. Even in this much more general set-up, the same principle holds: we can consider the submanifolds C_i as $(2n - 2)$-dimensional cycles, so that D defines a homology class

$\sum n_i C_i$ in $H_{2n-2}(X, \mathbb{Z})$, and the Euler characteristic $\chi(X, \mathscr{F}_D)$ depends only on this homology class (for $n = 1$, the homology class of the 0-dimensional cycle $\sum n_i x_i$ is the number $d = \sum n_i$). The 'coarse' nature of the Euler characteristic is expressed by the fact that it is a topological invariant, depending only on an element of the discrete group $H_{2n-2}(X, \mathbb{Z})$. There is a formula analogous to (18) in this case too, but of course it is much more complicated. The relation (18) is called the *Riemann-Roch theorem*, and the same name is used for its generalisation which we have just mentioned.

§ 22. *K*-theory

A. Topological *K*-theory

We now make further use of the notion of a family of vector spaces $f: E \to X$ over a topological space X introduced at the end of § 5. A *homomorphism* $\varphi: E \to E'$ of a family $f: E \to X$ into a family $f': E' \to X$ is a continuous map φ which takes the fibre $f^{-1}(x)$ into $(f')^{-1}(x)$ and is linear on the fibre for each point $x \in X$. If φ defines an isomorphism of the fibres, then φ is called an *isomorphism.*

For each open subset $U \subset X$ and family $f: E \to X$ of vector spaces, the restriction $f^{-1}(U)$ defines a family of vector spaces over U.

The simplest example is the family $X \times \mathbb{C}^n$, where f is the projection to X, and is called the *trivial family*. The main class of families which we will consider is the class of (complex) *vector bundles*. By this we mean a family which is *locally trivial*, that is, such that every point $x \in X$ has a neighbourhood U for which the family $f^{-1}(U)$ is isomorphic to the trivial family $U \times \mathbb{C}^n$. For an arbitrary continuous map $\varphi: Y \to X$ and a vector bundle E on X, the *inverse image* $\varphi^*(E)$ of E is defined; the fibre of $\varphi^*(E)$ over $y \in Y$ is identified with the fibre of E over $\varphi(y)$. The precise definition is as follows: $\varphi^*(E)$ consists of points $(y, e) \in Y \times E$ for which $\varphi(y) = f(e)$. The set of isomorphism classes of vector bundles over a given space X is denoted by $\mathscr{V}ec(X)$. By means of φ^*, this is a contravariant functor $\mathscr{V}ec$ from the category of topological spaces to the category of sets.

In $\mathscr{V}ec(X)$ we can define operations $E \oplus F$ and $E \otimes F$, which reduce to taking direct sum and tensor products of fibres over a point of X. The operation $\oplus$ is commutative and associative, but it does not define a group, since there is obviously no negative of an element. In other words, $\mathscr{V}ec(X)$ is a commutative semigroup (written additively) with a 0, the fibre bundle $X \times \{0\}$; see § 20, Example 9. We can try to make it into a group in the same way that we construct all the integers from the nonnegative integers, or the rationals from the integers. Here we should note one 'pathological' property of addition in $\mathscr{V}ec(X)$: from $a \oplus c = b \oplus c$ it does not follow that $a = b$. In view of this, the required group consists of all pairs (a, b) with $a, b \in \mathscr{V}ec(X)$, with pairs (a, b) and (a', b') identified if there exists c such that $a \oplus b' \oplus c = a' \oplus b \oplus c$. Addition of pairs is defined component-by-component. It is easy to see that the set of classes of pairs defines a group, in which the class (a, b) equals the difference $(a, 0) - (b, 0)$. The group we obtain is denoted by $K(X)$. Taking a fibre bundle $a \in \mathscr{V}ec(X)$ into the class of the pair $(a, 0)$ we get a homomorphism

$$\mathscr{V}ec(X) \to K(X),$$

and $a, b \in \mathscr{V}ec(X)$ map to the same element of $K(X)$ only if there exists $c \in \mathscr{V}ec(X)$ such that $a \oplus c = b \oplus c$. It is easy to see that $K(X)$ defines a contravariant functor from the category $\mathscr{T}op$ into the category of Abelian groups. For example, if X is a point, then $\mathscr{V}ec(X)$ just consists of finite-dimensional vector spaces; hence an element of $\mathscr{V}ec(X)$ is determined by its dimension $n \geqslant 0$, and $K(X) \cong \mathbb{Z}$. In the general case, the study of the group $K(X)$ is 'linear algebra over a topological space X'.

From now on we restrict attention to the category $\mathscr{C}_0$ of compact topological spaces with a marked point. For these, it can be proved that if $\varphi\colon Y \to X$ is a homotopy equivalence then φ^* defines an isomorphism of the groups $K(X)$ and $K(Y)$. Thus the functor $K(X)$ extends to the category $\mathscr{H}\mathscr{C}_0$ of compact spaces up to homotopy equivalence (with a marked point). If $x_0 \in X$ is a marked point and $f\colon E \to X$ a fibre bundle, then taking E to the dimension of the fibre $f^{-1}(x_0)$ defines a homomorphism

$$\Psi\colon K(X) \to \mathbb{Z}.$$

(we can say that $\Psi = \varphi^*$ where $\varphi\colon x_0 \hookrightarrow X$ is the inclusion). The kernel of Ψ is denoted by $\tilde{K}(X)$. This construction is analogous to the introduction of the group $\tilde{H}^0(X, A)$ in connection with the exact sequence (2) of § 21. It is easy to prove that

$$K(X) = \mathbb{Z} \oplus \tilde{K}(X).$$

If $Y \subset X$ is a closed subset with $x_0 \in Y$ then the inclusion map $f\colon (Y, x_0) \hookrightarrow (X, x_0)$ and the contraction of Y to a point $g\colon X \to X/Y$ (where the image of Y is the marked point of X/Y) define the homomorphisms in a sequence

$$\tilde{K}(X/Y) \to \tilde{K}(X) \to \tilde{K}(Y),$$

and it is not hard to prove that this is exact. We arrive at the question already discussed in § 21 of extending this sequence to an infinite exact sequence.

In the present case, the question can be solved as follows: write SX for the reduced suspension of the space X (for the definition, see § 20, Example 7). By induction, we define $\tilde{K}^0(X) = \tilde{K}(X)$ and $\tilde{K}^{-n}(X) = \tilde{K}^{-n+1}(SX)$. Then the sequence

$$\begin{aligned} &\ldots \to \tilde{K}^{-n-1}(Y) \to \\ \tilde{K}^{-n}(X/Y) \to \tilde{K}^{-n}(X) &\to \tilde{K}^{-n}(Y) \to \\ \tilde{K}^{-n+1}(X/Y) &\to \cdots \end{aligned}$$

is exact (under suitable definitions of the homomorphisms $\tilde{K}^{-n}(Y) \to \tilde{K}^{-n+1}(X/Y)$, which we omit). This definition can be explained as follows. Replace the reduced suspension of X by the suspension ΣX (once more, see § 20, Example 17 for the definition), which has the same homotopy type. Write CX for the cone over X, that is $(X \times I)/(X \times 1)$. Viewing X as the base of the cone $X \times 0 \subset CX$, we can say that $\Sigma X = CX/X$. The cone CX has the homotopy type of a point, since it can be contracted to its vertex; hence the inclusion $X \hookrightarrow CX$ is an analogue of representing a module M as the homomorphic image of a projective module by $f: P \to M$; and CX/X is an analogue of $\operatorname{Ker} f$. Thus our definition is similar to the inductive definition of the functors Ext^n given in § 21.B; or more precisely, dual to it (the inclusion of X and the surjection onto M have changed places), which is indicated by the negative indexes in $\tilde{K}^{-n}(X)$.

A remarkable property of this 'cohomology theory' is that it is periodic

$$\tilde{K}^{-n}(X) \cong \tilde{K}^{-n+2}(X). \tag{1}$$

We cannot stop here to discuss the proof of this *periodicity theorem*. Periodicity allows us to extend our sequence of functors $\tilde{K}^n(X)$ in a natural way to positive values of n, preserving condition (1). The 'cohomology theory' arising in this way is called K-theory. Of course, there are essentially only two functors $\tilde{K}^0(X)$ and $\tilde{K}^1(X)$ in it. By definition $\tilde{K}^0(X) = \tilde{K}(X)$ and $\tilde{K}^1(X) = \tilde{K}(SX)$.

We give another interpretation of the functor $K^1(X)$. For this, we once more replace the reduced suspension SX by the usual suspension ΣX, and recall that it can be obtained by glueing together the two cones

$$C_1X = (X \times [0, 1/2])/(X \times 0) \quad \text{and} \quad C_2X = (X \times [1/2, 1])/(X \times 1)$$

along their bases $X \times 1/2$. On each cone, a fibre bundle E is isomorphic to the trivial bundle (since the cone is contractible). Hence the fibre bundle E over ΣX is obtained by glueing together the bundles $\mathbb{C}^n \times C_1X$ and $\mathbb{C}^n \times C_2X$ along $\mathbb{C}^n \times X$. This glueing is realised by an isomorphism φ_x of the fibres $\mathbb{C}^n$ over the corresponding points of the base of the cone, that is, by a family of maps $\varphi_x \in \mathrm{GL}(n, \mathbb{C})$ for $x \in X$, or a continuous map $X \to \mathrm{GL}(n, \mathbb{C})$ given by $x \mapsto \varphi_x$. From these considerations we get the interpretation we need:

$$\tilde{K}^1(X) = \tilde{K}(SX) = H(X, \mathrm{GL}). \tag{2}$$

Here $H(\ ,\ )$ denotes the set of morphisms in the category $\mathscr{H}\!ot$, and the somewhat indeterminate symbol GL denotes that we must take maps into $\mathrm{GL}(n, \mathbb{C})$ for arbitrarily large n. We can embed the groups $\mathrm{GL}(n, \mathbb{C})$ into one another, $\mathrm{GL}(n, \mathbb{C}) \to \mathrm{GL}(n+1, \mathbb{C})$ by $A \mapsto \begin{pmatrix} A & 0 \\ 0 & 1 \end{pmatrix}$, and take their union; this will be our space GL.

K-theory, like any other cohomology theory, gives a certain projection of the homotopic topology into algebra. In the given case, the projection reproduces the original in a very faithful way, since the K-functors have a whole series of operators arising from the operations of external and symmetric powers of vector

(J-6682/4888/K668/LSK)

spaces, and these operations are functorial, that is, compatible with the maps $\tilde{K}^n(X) \to \tilde{K}^n(Y)$ corresponding to a continuous map $f: Y \to X$. It often happens that, putting all this information together, one gets a contradition; these are theorems on the non-existence of some kind of maps. For example, K-theory provides the simplest proof of the theorem stated at the end of § 19 that a division algebra of finite rank over the real number field has rank either 1, 2, 4 or 8. Another famous application of K-theory is a solution of the old question of how many linearly independent vector fields there exist on the n-dimensional sphere (that is, how many vector fields $\theta_1, \ldots, \theta_k$ such that at any point x the corresponding vectors $\theta_1(x), \ldots, \theta_k(x)$ are linearly independent). If the exact power of 2 dividing $n+1$ is 2^r then there are $2r$ of them if $4|r$, $2r-1$ if r is of the form $4k+1$ or $4k+2$, and $2r+1$ if r is of the form $4k-1$.

But the most beautiful application of K-theory relates to the question of the index of an elliptic operator. Linear differential operators of any finite order on a differentiable manifold X were defined in § 7, Example 3. In a local coordinate system they are of the form

$$\mathscr{D} = \sum_{i_1+\cdots+i_n \le k} a_{i_1\ldots i_n}(x) \frac{\partial^{i_1+\cdots+i_n}}{\partial x_1^{i_1} \ldots \partial x_n^{i_n}}, \tag{3}$$

where $a_{i_1\ldots i_n}(x)$ are differentiable complex-valued functions. A $m_1 \times m_2$ matrix $(\mathscr{D}_{ij})$ of differential operators defines a differential operator on the trivial vector bundles

$$\mathscr{D}: X \times \mathbb{C}^{m_1} \to X \times \mathbb{C}^{m_2}. \tag{4}$$

It is not hard (using local triviality) to extend this definition to operators $\mathscr{D}: E \to F$ acting on any differentiable vector bundles, but for brevity will restrict ourselves to (4).

Let us determine what the operator (4) gives us at one point of a manifold X. For this we need to fix a point x in (3), so that the coefficients $a_{i_1\ldots i_n}(x)$ become constants. The operators $\dfrac{\partial}{\partial x_i} = \xi_i$ are elements of the tangent space T_x of X (see § 5, Example 13), and $\mathscr{D}$ gives us a polynomial $P(\xi, x) = \sum a_{i_1\ldots i_n} \xi_1^{i_1} \ldots \xi_n^{i_n}$ on the cotangent space T_X^*. The operator (4) defines an $m_1 \times m_2$ matrix of such polynomials $(P_{ij}(\xi, x))$. Let k be the maximum of the degrees of all of the polynomials $P_{ij}(\xi, x)$, and $\tilde{P}_{ij}(\xi, x)$ their homogeneous parts of degree k. If $m_1 = m_2 = m$ and $\det(\tilde{P}_{ij}(\xi, x)) \neq 0$ for $\xi \neq 0$ then the operator (4) is *elliptic* at x, and if this holds for all $x \in X$ then it is an *elliptic operator* on X.

Thus an elliptic operator $\mathscr{D}$ defines at each point $x \in X$ and for each $\xi \in T_x^*$ with $\xi \neq 0$ a linear transformation $(\tilde{P}_{ij}(\xi, x)) \in \mathrm{GL}(m)$, which we denote by $\sigma_{\mathscr{D}}(\xi, x)$. The map $(\xi, x) \mapsto \sigma_{\mathscr{D}}(\xi, x)$ is defined for $\xi \in T_x^*$ with $\xi \neq 0$. In other words, it is defined on the manifold $T_X^* \setminus s$, where T_X^* is the cotangent bundle, and s is the zero section of T_X^* consisting of the zero point in each fibre. The vector space $\mathbb{R}^n \setminus \{0\}$ is homeomorphic to $\mathbb{R}_+ \times S^{n-1}$, and hence has the homotopy type of a sphere S^{n-1}. Thus we can say that $\sigma_{\mathscr{D}}(\xi, x)$ is defined on some fibre bundle

S_X over X (not a vector bundle) whose fibres are spheres S^{n-1}, and gives a map

$$\sigma_{\mathscr{D}}: S_X \to \mathrm{GL}(m, \mathbb{C}). \tag{5}$$

This map is called the *symbol* of the elliptic operator $\mathscr{D}$, and its homotopy class is the most important topological invariant of $\mathscr{D}$. By (2), $\sigma_{\mathscr{D}} \in \tilde{K}^1(S_X)$, and this already establishes the connection with K-theory.

We now move on to state the index problem. The operator (4) gives a map $A(X)^{m_1} \to A(X)^{m_2}$, where $A(X)$ is the ring of differentiable functions on X. Its kernel is denoted by $\operatorname{Ker} \mathscr{D} \subset A(X)^{m_1}$, and its image by $\operatorname{Im} \mathscr{D} \subset A(X)^{m_2}$; the quotient $A(X)^{m_2}/\operatorname{Im} \mathscr{D}$ is called the *cokernel* and denoted by $\operatorname{Coker} \mathscr{D}$. It is proved in the theory of elliptic operators that for an elliptic operator $\mathscr{D}$, the spaces $\operatorname{Ker} \mathscr{D}$ and $\operatorname{Coker} \mathscr{D}$ are finite-dimensional. In other words, the space of solutions $\mathscr{D}f = 0$ for $f \in A(X)^m$ is finite-dimensional, and the number of conditions on g for the equation $\mathscr{D}f = g$ to be solvable with $f \in A(X)^m$ is finite. The difference between these dimensions,

$$\operatorname{Ind} \mathscr{D} = \dim \operatorname{Ker} \mathscr{D} - \dim \operatorname{Coker} \mathscr{D} \tag{6}$$

is called the *index* of the elliptic operator $\mathscr{D}$.

The *index theorem* asserts that the index of an elliptic operator $\mathscr{D}$ depends only on its symbol $\sigma_{\mathscr{D}}$, and it gives an explicit formula expressing $\operatorname{Ind} \mathscr{D}$ in terms of $\sigma_{\mathscr{D}}$. A little more precisely, we can say that the space GL has certain special cohomology classes, that do not depend on anything else. The map $\sigma_{\mathscr{D}}$ allows us to transfer these to S_X. On the other hand, S_X also has certain special cohomology classes which are entirely independent of the operator $\mathscr{D}$. Finally, on the cohomology ring $H^*(S_X)$ there is a standard polynomial in all the cohomology classes so far mentioned, which gives a class $\alpha_{\mathscr{D}} \in H^{2n-1}(S_X, \mathbb{Z})$ of maximal dimension $2n - 1 = \dim S_X$. It is well known from topology that $H^{2n-1}(S_X, \mathbb{Z}) = \mathbb{Z}$, and hence the class $\alpha_{\mathscr{D}}$ is given by an integer, which turns out to be equal to $\operatorname{Ind} \mathscr{D}$. Although we have only spoken of operators on trivial bundles, the index theorem holds for elliptic operators on arbitrary fibre bundles.

Already the qualitative fact that the index depends only on the symbol of the operator $\mathscr{D}$ (and not on its more delicate analytic properties) shows that the difference (6) is 'coarse', in the same way that the Euler characteristic §21, (17) is coarse. This is not a chance analogy. The index theorem, applied to complex analytic manifolds and certain very simple operators over them, implies the Riemann-Roch theorem mentioned in §21.C.

B. Algebraic K-theory

We mentioned in §5 the analogy between families of vector spaces $f: E \to X$ and modules over a ring A. In particular, as we have seen, a family $E \to X$ defines a module over the ring $\mathscr{C}(X)$ of continuous functions on X. In this analogy, which are the modules corresponding to vector bundles? There are many arguments indicating that these are the projective modules of finite rank (compare §21.B). First of all, this is shown by the following result.

Theorem I. *Let X be a compact topological space and $E \to X$ a family of vector spaces. The module M over $\mathscr{C}(X)$ corresponding to this family is projective if and only if E is a vector bundle; $E \leftrightarrow M$ defines a 1-to-1 correspondence between vector bundles over X and finitely generated projective modules over $\mathscr{C}(X)$.*

We can also indicate certain algebraic properties of finitely generated projective modules which are analogues of local triviality.

This justifies the introduction (by analogy with §22.A) of the semigroup $\Pi(A)$ whose elements are classes of finitely generated projective modules over a ring A and whose sum is the direct sum of modules. Repeating word-for-word the arguments of §22.A, we can construct a group $K(A)$ and a map $\varphi: \Pi(A) \to K(A)$ such that the set $\varphi(\Pi(A))$ generates $K(A)$, and such that for two projective modules $P, Q \in \Pi(A)$, their images $\varphi(P)$ and $\varphi(Q)$ are equal if and only if there exists a third module $R \in \Pi(A)$ for which $P \oplus R \cong Q \oplus R$.

For any prime ideal $I \subset A$, the ring A/I can be embedded in a field k, and hence there exists a homomorphism $A \to k$ with kernel I. Thus k is an A-module and the module $M \otimes_A k$ is defined. If M is finitely generated, then $M \otimes_A k$ is a finite-dimensional vector space over k. It can be proved that for an integral domain A and a projective module M, the dimension of this space is independent of the choice of ideal I, and for $I = 0$ is equal to rank M (compare §5). The function rank M extends to $K(A)$ and defines a homomorphism $K(A) \to \mathbb{Z}$ whose kernel is denoted by $\tilde{K}(A)$. It is easy to see that $K(A) = \tilde{K}(A) \oplus \mathbb{Z}$.

We consider the groups $K(A)$ and $\tilde{K}(A)$ for some very simple rings.

Theorem II. *If $A = k$ is a field, then $\Pi(k)$ consists of finite-dimensional vector spaces over k, the homomorphism $K(k) \to \mathbb{Z}$ is defined by dimension, and is obviously an isomorphism, so that $\tilde{K}(k) = 0$.*

Theorem III. *If A is an principal ideal domain then for any module M of finite type we have $M \cong M_0 \oplus A^r$ where M_0 is a torsion module (see §6, Theorem II). If M is projective, then it is a direct summand of a free module, and hence is torsion-free. Hence $M_0 = 0$ and $M \cong A^r$, and this again means that $\tilde{K}(A) = 0$.*

Consider the ring A of numbers of the form $a + b\sqrt{-5}$ for $a, b \in \mathbb{Z}$ (§4, Theorem VIII). We saw in §4 that the ideal $P = (3, 2 + \sqrt{-5})$ is not principal. It is not hard to show that $\varphi(P) - \varphi(A) \in \tilde{K}(A)$, and $\varphi(P) \neq \varphi(A)$, so that $\tilde{K}(A) \neq 0$.

Theorem IV. *If A is the ring of integers of any algebraic number field (see the end of §7), the group $\tilde{K}(A)$ is finite and isomorphic to the group of ideal classes of this field (§12, Example 1). In particular, for the ring A of numbers of the form $a + b\sqrt{-5}$ with $a, b \in \mathbb{Z}$ we have $\tilde{K}(A) \cong \mathbb{Z}/2$.*

Theorem V. *If X is a compact topological space then the group $K(\mathscr{C}(X))$ defined in this section is isomorphic to $K(X)$ defined in* §22.A.

The definition of higher K-functors $K_n(A)$ proceeds along already familiar lines. First of all a group $K(I)$ is also defined for an ideal $I \subset A$ (the definition given above is not applicable, since we always considered rings with 1). Then we construct an exact sequence

$$K(I) \to K(A) \to K(A/I) \tag{7}$$

and finally, we define groups $K_n(A)$ such that $K_0 = K$, and which extend (7) to an infinite exact sequence

$$\cdots \to K_{n+1}(A/I) \to K_n(I) \to K_n(A) \to K_n(A/I) \to K_{n-1}(I) \to \cdots$$

(in algebraic K-theory the functors K_n are covariant, so that the index is written as a subscript). We will not state all of these definitions, but merely give interpretations of some of the groups which arise in this way.

The interpretation of the groups $K_1(A)$ is analogous to that which gives relation (2) in the topological case. A continuous map $\varphi: X \to \mathrm{GL}(n)$ is an invertible matrix whose entries are continuous functions of a point of X, that is an element of $\mathrm{GL}(n, \mathscr{C}(X))$, where $\mathscr{C}(X)$ is the ring of continuous functions of X. Thus the natural starting point should be the group $\mathrm{GL}(n, A)$, and as in §22.A, the infinite limit $\mathrm{GL}(A)$ of these as $n \to \infty$. But now we must also interpret the letter H in formula (2), that is, recall that maps $\varphi: X \to \mathrm{GL}$ are considered up to homotopy. This means that we consider the quotient group $\mathrm{GL}(A)/\mathrm{GL}(A)_0$, where $\mathrm{GL}(A)_0$ is the connected component of the identity in $\mathrm{GL}(A)$. What is the analogue of this subgroup in the algebraic case? In many questions, matrixes playing the role of transformations that are 'trivially deformable to the identity' are given by

$$E + aE_{ij}, \tag{8}$$

where E_{ij} is the matrix with 1 in the (i,j)th place and 0 elsewhere; matrixes of the form (8) are called *elementary matrixes*. For example, the proof that the group $\mathrm{SL}(n, \mathbb{R})$ is connected is based on the fact that any element of it is a product of elementary matrixes. The subgroup generated by all elementary matrixes in $\mathrm{GL}(A)$ is written $\mathrm{E}(A)$. A fact which is unexpected, although quite elementary to prove, is that $\mathrm{E}(A)$ is the commutator subgroup of $\mathrm{GL}(A)$. The proof of this uses in an essential way the fact that we are considering the union of all the groups $\mathrm{GL}(n, A)$ for $n = 1, 2, \ldots$; for each individual group this is not true in general. In particular, the group $\mathrm{GL}(A)/\mathrm{E}(A)$ is commutative. This gives what we need:

$$K_1(A) \cong \mathrm{GL}(A)/\mathrm{E}(A).$$

Passing to the determinant gives a homomorphism $\mathrm{GL}(A)/\mathrm{E}(A) \to A^*$ and even a representation

$$K_1(A) = A^* \oplus SK_1(A) \quad \text{where} \quad SK_1(A) = \mathrm{SL}(A)/\mathrm{E}(A)$$

(where admittedly the additive and multiplicative notation are hopelessly confused).

From a course of elementary linear algebra, we know that $\mathrm{SL}(n,k) = \mathrm{E}(n,k)$ for a field k; this follows from Gauss' method of solving linear equations by row and column operations. Hence $SK_1(k) = 0$. If A is an integral domain with a Euclidean algorithm then the main lemma of § 6, which was the basis for the proof of the structure theorem of modules of finite type, gives the same result: $SK_1(A) = 0$. The group K_1 occurs (and first occurred) in topology in the case of the group ring $A = \mathbb{Z}[G]$ of a finite Abelian group G (with G the fundamental group of some manifold). In this case it is often nontrivial. Generally speaking, the homomorphism $\mathrm{GL}(A) \to K_1(A)$ is a kind of 'universal determinant'. In this form it can also be generalised to the case of a noncommutative ring A.

We only describe the group K_2 when $A = k$ is a field. It can then be given by generators $\{a, b\}$ corresponding to any elements a, $b \in k$ with a, $b \neq 0$. The defining relations are of the form:

$$\{a_1 a_2, b\} = \{a_1, b\}\{a_2, b\}, \tag{9_1}$$

$$\{a, b_1 b_2\} = \{a, b_1\}\{a, b_2\}, \tag{9_2}$$

$$\{a, 1 - a\} = 1 \quad \text{for } a \neq 0 \text{ or } 1. \tag{9_3}$$

The group $K_2(k)$ has an especially vivid application to the description of division algebras of finite rank over k (compare § 11 and § 12, Example 3 for the definition of the notions occuring here).

It can be shown that for an arbitrary field k all the elements of the Brauer group $\mathrm{Br}(k)$ have finite order: if $\dim_k D = n^2$ (see § 11, Theorem IV), then the element corresponding to D in the Brauer group has nth power equal to 1. This proves in particular that the generalised quaternion algebras (a, b) introduced in § 11 define elements of $\mathrm{Br}(k)$ of order 2 or 1.

We only give a precise description of the relation between $K_2(k)$ and the Brauer group $\mathrm{Br}(k)$ for elements of order 2 of these groups. In any Abelian group C, the elements satisfying $c^2 = 1$ obviously form a subgroup: we denote it by ${}_2C$. The elements of the form c^2 for $c \in C$ also form a subgroup; we denote it by C^2. We now send any generator $\{a, b\}$ of the group $K_2(k)$ with presentation (9) with a, $b \in k$ with a, $b \neq 0$ into the generalised quaternion algebra (a, b). It is not hard to check that relations (9) are satisfied in the Brauer group: checking (9_1) and (9_2) is a simple exercise in tensor multiplication, and the proof of (9_3) follows from the fact that the algebra (a, b) defines the identity element of $\mathrm{Br}(k)$ if and only if $ax^2 + by^2$ is solvable in k (see § 11, (4)); but $a \cdot 1^2 + (1 - a) \cdot 1^2 = 1$! Thus we get a homomorphism $\varphi_2\colon K_2(k) \to \mathrm{Br}(k)$. As we have seen, $\varphi_2(K_2(k)) \subset {}_2\mathrm{Br}(k)$, and hence $\varphi_2(K_2(k)^2) = 1$. In consequence we get the homomorphism

$$\varphi\colon K_2(k)/K_2(k)^2 \to {}_2\mathrm{Br}(k). \tag{10}$$

The main result is that (10) is an isomorphism. This is a very strong assertion: in view of the description (9) of $K_2(k)$ we get a presentation of ${}_2\mathrm{Br}(k)$ by generators and relations, and indeed for a completely arbitrary field k. There is a similar description for the group ${}_n\mathrm{Br}(k)$, consisting of elements $c \in \mathrm{Br}(k)$ for which $c^n = 1$. Since all elements of $\mathrm{Br}(k)$ are of finite order, $\mathrm{Br}(k) = \bigcup_n \mathrm{Br}(k)$, so that as a result we get an explicit description of the whole of this group.

An increasing role is being played by algebraic K-theory in arithmetic questions. We give some examples which do not more than hint at one direction of such applications: the relation between K-theory and the value of zeta-functions. The classical Riemann zeta-function is defined by the series

$$\zeta(s) = \sum_{n=1}^{\infty} \frac{1}{n^s} \quad \text{for} \quad \operatorname{Re} s > 1,$$

and has an analytical continuation over the whole plane of the complex variable s. It satisfies the Euler identity:

$$\zeta(s) = \prod_p \frac{1}{1 - p^{-s}}, \tag{11}$$

in which the product takes place over all primes p. For a finite field $\mathbb{F}_q$ with q elements we define its zeta-function to be

$$\zeta_{\mathbb{F}_q}(s) = \frac{1}{1 - q^{-s}}.$$

Then Euler's identity (11) can be rewritten in the form

$$\zeta(s) = \prod_p \zeta_{\mathbb{F}_q}(s), \tag{12}$$

This accords very well in the 'functional view of a ring' discussed in §4, according to which we should view $\mathbb{Z}$ as a ring of functions on the set of primes numbers p, with values in $\mathbb{F}_p$. This suggests the definition of a zeta-function analogous to (12) for a wide class of rings.

We now return to K-theory. In the case of finite fields $\mathbb{F}_q$, it can be proved that all the groups $K_n(\mathbb{F}_q)$ are finite for $n \geq 1$. The information about their orders can be written out in the following beautiful form:

$$\frac{|K_{2m}(\mathbb{F}_q)|}{|K_{2m+1}(\mathbb{F}_q)|} = |\zeta_{\mathbb{F}_q}(-m)|, \quad \text{for} \quad m \geq 1.$$

For the case of the ring $\mathbb{Z}$, the facts known at present allow us to suppose some kind of connection between the values of the Riemann zeta-function $\zeta(-m)$ for $m > 0$ and the ratios $|K_{2m}(\mathbb{Z})|/|K_{2m+1}(\mathbb{Z})|$. It is known that $\zeta(-m)$ are rational numbers for odd integers $m > 0$, and 0 for even integers >0; and that the group $K_{2m}(\mathbb{Z})$ and $K_{2m+1}(\mathbb{Z})$ are finite. The relation

$$\frac{|K_{2m}(\mathbb{Z})|}{|K_{2m+1}(\mathbb{Z})|} = |\zeta(-m)|, \quad \text{for odd } m > 0$$

is already false in the simplest case $m = 1$, since $|K_2(\mathbb{Z})| = 2$, $|K_3(\mathbb{Z})| = 48$, but

$\zeta(-1) = -\frac{1}{12}$. However, it is not excluded that it holds up to powers of 2. In any case, it is proved that the denominator of $\zeta(-m)$ divides $|K_{2m+1}(\mathbb{Z})|$. On the other hand, if p is a prime number and $p > m$ then the power of p dividing $|K_{2m}(\mathbb{Z})|$ is not less than that dividing the numerator of $\zeta(-m)$ (under an addition condition on p which conjecturally is always satisfied, and which has been checked for $p < 125{,}000$). Recall that we have already met the values of zeta-functions at negative integers in connections with the cohomology of arithmetic groups (§ 21, Example 5). This is not a coincidence; here a relation with the groups $K_n(\mathbb{Z})$ does in fact exist.

The relations of K-theory with number theory are many and various, but we cannot describe them here in more detail, since this would require us to introduce complicated technical tools.

Comments on the Literature

The whole of this book is based on the interweaving of two themes: the systematic treatment of algebraic notions and theories, and the working out of key examples. References for these two themes are described separately. It is my impression that, as a rule, the first edition of a book may sometimes be fresher and more interesting than subsequent ones, even when these are technically more finished in some respects. For this reason references are to the first editions known to me.

The basic notions of algebra, groups, rings, modules, fields, and the main theories pertaining to these notions, including the theory of semisimple modules and rings and Galois theory, are treated in the classical two-volume textbook of van der Waerden [104 (1930, 1931)]. Although more than half a century has elapsed since the appearance of this remarkable book, it is in no way dated, and for the majority of the questions it treats no better source can be found even today.

The process of isolating out the basic algebraic notions, and recasting algebra in the spirit of an axiomatised approach occupied more than a century, and involved the participation of Gauss, Galois, Jordan, Klein, Kronecker, Dedekind and Hilbert. But fixing the results of this century-old process in the form of the standard language of algebra took scarcely more than a decade, from 1920 to 1930; an especially prominent role in this was played by E. Noether. It was at this time that van der Waerden's book appeared. To get a feeling for the change in the whole spirit of algebra and the manner of its exposition, it is useful to compare van der Waerden's book with Weber's course [105 (1898, 1899)], from which algebra had been studied by previous generations.

Of the more recent literature we must note the books of the Bourbaki series Éléments de mathématiques devoted to algebra [16 (1942–1948)], [17 (1959)]. These books might give the impression that they could serve as textbooks for beginners, since their treatment is almost entirely self-contained, and starts from the simplest definitions. This impression is however entirely illusory in view of their basic principle, to consider the subject in the maximal possible generality, and in view of the complete absence of any material motivating the introduction of notions and the direction in which the theory is developed. However, the specialist may find a wealth of valuable details in them.

From the more specialised texts, we note the course of commutative algebra of Atiyah and Macdonald [5 (1969)], which is written bearing in mind also the interests of nonalgebraists.

The specialised results on the structure of division algebras given in § 11 is treated systematically in the survey of Deuring [33 (1935)] or the book of A. Weil [106 (1935)].

The references we have quoted cover in the main the material of §§ 2–11 of this book. From § 12 we go over to group theory; for the foundations we can here again recommend the book of van der

Waerden and (for a slight extension of the point of view) the chapters on group theory of H. Weyl's classical monograph [107 (1928)]. Although intuitive objects such as transformation groups appear as examples of groups, the example that was most stimulating for the development of the general notion of a group and for transforming group theory into an independent subject was the permutation group of a finite set, more specifically, the set of roots of a polynomial; ideas going back to Lagrange and Abel took concrete form in the works of Galois [45 (1951)]. Very clearly visible in this is the growth of understanding that questions of field theory relate to the Galois group specifically as an abstract group, despite the fact that the group is realised as a concrete permutation group. (Lagrange expressed this idea by saying that 'permutations are the metaphysics of equations'.) A further stimulus was Gauss' systematic use of congruence classes and classes of quadratic forms, and his definition of operations on these [Gauss 46 (1870)], creating the feeling that some general notion lay concealed beneath the surface.

The first known book on the subject of group theory was that of Jordan [70 (1870)], which contains a wealth of examples and ideas, and has not lost its value to the present day; this book treats finite transformation groups only. Starting with the work of Klein and Lie, considerations of infinite discrete and continuous groups come to the fore. Here we first have occasion to refer to Klein's wonderful book 'Lectures on the development of mathematics in the 19th century' [73 (1926)]; the period in the development of group theory under discussion is described here from the point of view of one of its most influential participants. But the book also contains much else of interest on the development of other branches of algebra (and of mathematics as a whole).

The first book on abstract group theory was Burnside's book [21 (1897)], which considered only finite groups. For a long time subsequent treatments only improved it; the most finished treatment was achieved by Speiser [98 (1937)]. A modern course in the theory of finite groups is the 3-volumes text by Huppert and Blackburn [67 (1967), 68 (1982)]. The point of view of infinite groups is given most prominence in Kurosh [77 (1955, 1956)]. An interesting historical survey of the theory of defining groups by generators and relations is contained in [Chandler and Magnus 24 (1981)]. Logical problems arising in this can be found for example in Manin's course [81 (1977)].

The first book on the theory of Lie groups is the 3-volume book of Lie and Engel [79 (1883–1893)]; this book is interesting as a witness of the birth of a new branch of science. A more modern treatment of the main notions can be found in Pontryagin [90 (1938)], and an even more modern one (that is, wherever possible without the use of coordinate systems) in Chevalley [25 (1946)]. A beautiful treatment is also given in [Hochschild 64 (1965)].

For algebraic groups we note the survey of Chevalley [28 (1958)] and the books [A. Borel 12 (1969)], [Humphreys 66 (1975)] and [Springer 99 (1981)]. The classification of simple Lie groups can be found: for compact groups in [Zhelobenko 111 (1970)] and [Pontryagin 90 (1938)], for complex groups in [*Séminaire Sophus Lie* 94 (1955)], for real Lie groups in [Goto and Grosshans 50 (1978)]. The classification of simple algebraic groups is contained in Chevalley's seminar [27 (1958)] and the books [Humphreys 66 (1975)] and [Springer 99 (1981)]. For finite simple groups we note the survey [Tits 103 (1963)] and Gorenstein's book [49 (1982)] (although this book does not contain a proof of the classification—a unified exposition of this does not at present exist).

The foundations of the theory of representations of finite groups were laid by Frobenius; one can consult his collected works [44 (1968)] for this. A more recent treatment can be found in the appropriate sections of H. Weyl's book [107 (1928)] and the early part (§§ 1–8) of [Serre 95 (1967)]. For the representations of compact groups see also the books of [Weyl 107 (1928)], [Pontryagin 90 (1939)], [Chevalley 25 (1946)] and [Zhelobenko 111 (1970)]. A wide survey on the theory of representations of Lie groups is given in [Kirillov 72 (1972)]. An introduction to more recent questions is provided by the conference proceedings [6 (1979)] edited by Atiyah.

H. Weyl's book [109 (1939)] is a classical study in representation theory. It had a strong influence on the subsequent development of this subject. It contains in particular the concept of 'coordinatisation' and the idea of the relation between symmetries and representations which we have used in this book.

Lie theory is included in practically all the textbooks on Lie groups we have quoted. Various stages of its development can be seen in [Lie and Engel 79 (1883, 1888, 1893], [Pontryagin 90 (1939)], [Chevalley 25 (1946)], [Hochschild 64 (1965)].

For the Cayley numbers or octavions, we note the survey of Freudenthal [43 (1951)]. For geometrical applications see the survey [Lawson 78 (1985)].

The proof of the main theorem on nonassociative division algebras over the real number field can be found in [Atiyah 4 (1967)].

The notions of categories and functors were formulated in books of Eilenberg and MacLane [38 (1942)], [39 (1945)], where their significance as a new language for the axiomatisation of mathematics was argued in detail. A systematic treatment of the basic notions of category theory can be found in Chapter II of the book of Hilton and Stammbach [61 (1971)]. Certain aspects of it are considered in the first sections of Grothendieck's article [51 (1957)]. A detailed discussion of the notion of a group object in a category is contained in his book [53 (1961)], Chapter 0, § 8.

A systematic treatment of the foundations of homological algebra is contained in [Hilton and Stammbach 61 (1971)]. The classical work in this area is the book of H. Cartan and Eilenberg [22 (1956)], but this is written more abstractly. The theory of group cohomology is contained in Brown [20 (1982)] in a spirit similar to that of this book. The general notions of the theory of sheaf cohomology are treated in [Tennison 102 (1975)]. Hirzebruch's book [62 (1956)] is a classical textbook, devoted in the main to applications to the Riemann-Roch theorem.

The part of K-theory treated in this book is covered for the most part by two surveys: topological K-theory by Atiyah [4 (1967)], and algebraic K-theory by Milnor [85 (1971)]. For the questions of algebraic K-theory treated at the end of §§ 22 we note the survey of Suslin [100 (1984)], although this is not written for a general audience.

Finally, since on many occasions we have used topological notions and results (especially in the sections on category theory and homological algebra), we give some topological references. Textbooks written from a point of view similar to that of § 20 of this book are Dold [36 (1980)] and Switzer [101 (1975)]. But in places where a more important role is played by geometrical intuition, for example in connection with the topology of surfaces, the old book of Seifert and Threlfall [93 (1934)] remains irreplaceable. The theory of differentiable manifolds and integrating differential forms on them can be found in the books of Chevalley [25 (1946)] and de Rham [92 (1955)].

We now proceed to the literature referring to the detailed workings of individual examples. Perhaps the theme surfacing throughout the book most richly illustrated with examples is the 'duality' between the functional and the algebraic point of view, the intuition of the elements of a ring as 'functions' on the set of its (maximal or prime) ideals, the analogy between numbers and functions. This is a very old complex of ideas. Properly speaking, the idea of analytically continuing functions from the real line to the complex plane already raises the question of some 'natural' set on which a function should be considered. A big step forward in this direction was the creation of the idea of a Riemann surface. In the article of Dedekind and Weber [31 (1982)], the Riemann surface of an algebraic function field in 1 variable (in our notation, the field $K(C)$, where C is an algebraic curve) is defined in a purely algebraic way as a set of 'homomorphisms' of $K(C)$ into K (with a symbol ∞ adjoined to K). The article of Kronecker [76 (1982)], published in the same issue of the journal, develops a programme for constructing a theory which unifies algebraic numbers and algebraic functions in any number of variables. A discussion of the idea of the parallelism between numbers and functions can be found in Klein's 'Lectures' [73 (1926)]. A treatment of the theory of algebraic functions of 1 variable along ideas of the article of Dedekind and Weber is given in Chevalley's book [26 (1951)].

In connection with questions of point-set topology and logic, it was proved that a Boolean algebra is representable as the ring of continuous functions with values in $\mathbb{F}_2$ on a certain type of topological space (for this see [Birkhoff 10 (1940)]). One can learn about the same ideas applied to rings of continuous real- or complex-valued functions in [Gel'fand, Raikov and Shilov 47 (1960)]; for rings of C^∞ functions see [Bröcker 19 (1975)], and for analytic functions [Hoffman 65 (1962)]. Finally, the concept of scheme, embracing both number theory and algebraic geometry and allowing geometric intuition to be applied to number-theoretical questions was developed by Grothendieck. For this see his survey article [Grothendieck 52 (1960)], the lectures of Manin [83 (1970)] and Chapter 5 of the book [Shafarevich 96 (1972)]. Carrying over infinitesimal methods into the area of number theory, in particular the construction of the p-adic numbers comes under this heading. For an elementary introduction (also to the theory of rings of algebraic integers) see [Borevich and Shafarevich 15 (1964)], for the deeper theory, the book [Weil 106 (1967)].

Another theme running through the book is 'coordinatisation', in the narrow sense of introducing coordinates in the plane and in projective spaces. For this (in particular for the role of Desargues' and Pappus' axioms) see Hilbert's book [59 (1930)], and in a more algebraic form, the books [E. Artin 3 (1967)] and [Baer 7 (1952)]. Continuous geometries are the subject of von Neumann's book [87 (1960)].

The finite fields, which we have frequently encountered, were discovered by Galois; the complete theory of these is already contained in his works [45 (1951)]. Their applications (especially the applications of algebraic geometry over finite fields) to coding theory is treated in the surveys [Goppa 48 (1984)] and [Manin and Vlehduts 82 (1984)].

Algebraic methods in the theory of commuting differential operators began with the results treated in § 5.5 of the book [Ince 69 (1927)]. These results were forgotten and rediscovered several decades later. For a modern survey see [Mumford 86 (1977)].

For ultraproducts see [Barwise and others 8 (1970)].

Tensor products and exterior and symmetric powers of modules are defined in [Kostrikin and Manin 75 (1980)] and [Bourbaki 16 (1942–1948)]. Properties of completion are treated in [Atiyah and Macdonald 5 (1969)].

Clifford algebras appear in a large number of examples. These were introduced by Clifford in the 19th century (see his collected works [29 (1982)]) and rediscovered (in a particular case) by Dirac in the 20th century [35 (1930)], in connection with the attempt to represent a second order linear differential operator as the square of a first order operator with matrix coefficients. A detailed modern treatment can be found in [Bourbaki 17 (1959)].

We proceed to the examples concerning the notion of group. A discussion of the notion of symmetry is the subject of H. Weyl's book [110 (1952)]. For the relation between symmetries and conservation laws in mechanics (E. Noether's theorem), see [Courant and Hilbert 30 (1931)] or [Arnol'd 2 (1974)]. The symmetries of physical laws are discussed in the interesting lectures of Feynman [41 (1965)].

As for examples of groups not realised as transformation groups, the $\mathrm{Ext}(A, B)$ are discussed in any of the courses in homological algebra quoted, the Brauer group in [Deuring 33 (1935)], and the ideal class group in [Atiyah and Macdonald 5 (1969)].

Platonic solids and their connection with finite groups of motion are considered in detail in Hadamard's book [54 (1908)]. Finite subgroups of the group of fraction-linear transformations of the complex plane are treated in another book of Hadamard [55 (1951)]. Symmetries of lattices are analysed in [Klemm 74 (1982)].

A detailed analysis of finite groups generated by reflections is contained in Bourbaki [18 (1968)]. Amazingly enough, the same diagrams given in § 13, in terms of which these groups are classified, also turn up in a whole series of other classification problems (the most important of these being the classification of simple compact or complex Lie groups). A survey of these connections is given in [Hazewinkel and others 58 (1977)].

Geometric crystallography is the subject of the book of Delone, Padurov and Aleksandrov [32 (1934)]. A more modern treatment is contained in [Klemm 74 (1982)], where the groups of ornaments and n-dimensional crystallography are also considered. A chapter of Hilbert and Cohn-Vossen's book [60 (1932)] is also devoted to this. A complete list of ornaments which characterise all of the 17 plane groups can be found in the survey of Mal'tsev [80 (1956)].

All the crystallographic groups were classified by E.S. Fëdorov in 1889 and by Schoenflies in 1890 (independently of one another). In the following year, Fëdorov classified all the groups of plane ornaments [Fëdorov 40 (1891)], displaying what is (for a crystallographer) a highly nontrivial understanding of the geometrical character of the problem. It is amazing that a mathematician as widely read as H. Weyl could write ([110 (1952)], p. 103–4) '... the mathematical notion of a group of transformations was not provided before the nineteenth century; and only on this basis is one able to prove that the 17 symmetries already implicitly known to the Egyptian craftsmen exhaust all possibilities. Strangely enough, the proof was carried out only as late as 1924 by G. Pólya, now teaching at Stanford'. Even more strange is the fact that the above quotation from Weyl has recently been the subject of a series of articles in several issues of the Mathematical Intelligencer [Pedersen and others 89 (1983–1984)]. However, the subject under discussion was the assertion that Egyptian craftsmen knew all 17 symmetries, and none of the participants paid any attention to the untrue assertion as to who solved the mathematical problem as to their classification.

For discrete groups of motion of the Lobachevsky plane and their connections with the theory of Riemann surfaces see the book of Hadamard [55 (1951)]. For the fundamental group, the univeral covering and the group of a knot we refer to the old-fashioned, but geometrical, book [Seifert and Threlfall 93 (1934)].

For the relations between algorithmic problems of group theory and topology, see [Fomenko 42 (1983)].

Braid groups (but without the name) were first considered by Hurwitz, both in the geometrical form (as they are now usually defined), and as fundamental groups, and were subsequently rediscovered, much later, separately in each of their realisations. For this see [Chandler and Magnus 24 (1982)].

For the role of toruses in Liouville's theorem see Arnol'd's book [2 (1974)]. The classical compact groups are carefully worked out in [Chevalley 25 (1946)]. See [Kostrikin and Manin 75 (1980)] for examples of other Lie groups and important relations between them in special dimensions.

For algebraic groups and their relations with discrete groups see the survey of A. Borel [13 (1963)].

Helmholtz-Lie theory, given as an example in § 17 in connection with representation theory, is the subject of an attractive, although slightly difficult, book of H. Weyl [108 (1923)].

For the examples related to the representation of O(4) and the curvature tensor of a 4-dimensional Riemaniann manifold, see [Besse 9 (1981)].

The representations of SU(2) and their relation with quantum mechanics are treated in H. Weyl's book [107 (1928)].

For the examples given as applications of group theory: a treatment of Galois theory is given in [van der Waerden 104 (1930, 1931)]. A brief introduction to differential Galois theory is the book of Kaplansky [71 (1957)]. An example of the groups arising in the Galois theory of extensions of p-adic number fields (the so-called Dëmushkin groups) and having a mysterious parallel with the fundamental groups of surfaces are treated in the book [Cassels and Fröhlich 23 (1967)]. For the applications to invariant theory, see [Dieudonné and Carrell 34 (1971)].

As far as applications of group representations to the classification of elementary particles are concerned, the author is only able to list the references from which he has got to know the subject. The main one is the lectures of Bogolyubov [11 (1967)]. An interesting introduction is the survey of Dyson [37 (1964)]. The Appendix III to Zhelobenko's book [111 (1970)] is also useful.

For the interpretation of the equations of motion of a rigid body in terms of Lie groups and Lie algebras and generalisations of these relations, see [Arnol'd 2 (1974)] and [Fomenko 42 (1983)]. A more complete survey of the theory of formal groups is provided by the book [Hazewinkel 57 (1978)].

A more detailed consideration of the topological constructions given as examples in connection with category theory can be found in [Dold 36 (1980)] and [Switzer 101 (1975)].

For the homology and cohomology groups of a complex see [Hilton and Stammbach 61 (1971)]. De Rham cohomology and a proof of de Rham's theorem is contained in [de Rham 92 (1955)], although de Rham's theorem can now most simply be proved by means of sheaf theory.

The basic example in sheaf cohomology is the Riemann-Roch theorem. This is the subject of the book [Hirzebruch 62 (1956)].

The basic example for topological K-theory is the index theorem. A beautiful introduction to this is provided by Hirzebruch's survey [63 (1965)]. A complete exposition of the proof can be found in [Palais 88 (1965)].

In algebraic K-theory, the theorem on the relation between K_2 and the Brauer group of a field is due to Merkur'ev and Suslin [Suslin 100 (1984)]. Results on the computation of the orders of K_n for finite fields are due to Quillen [91 (1972)]. For conjectures and results on the order of K_n for rings of integers see Soulé [97 (1979)].

To get an impression of the history of the development of algebra and its interaction with the whole of mathematics, an invaluable source is Klein's 'Lectures' [73 (1926)]. Many interesting observations are to be found among the historical remarks in the Bourbaki books. An interesting study, although devoted to the history of a specialised problem, is the book of Chandler and Magnus [24 (1982)].

References*

1. *The life of plants*, vol. 5, part 2, Prosveshchenie, Moscow, 1981 (in Russian)
2. Arnol'd, V.I.: Mathematical methods of classical mechanics, Nauka, Moscow 1974; English translation: Grad. Texts Math. 60, Springer-Verlag, Berlin Heidelberg New York 1978. Zbl. 386.70001
3. Artin, E.: Geometric algebra, Interscience, New York 1957. Zbl. 77, 21; Russian translation: Moscow, Nauka 1969. Zbl. 174, 294
4. Atiyah, M.F.: *K*-theory, Benjamin, New York Amsterdam 1967. Zbl. 159, 533
5. Atiyah, M.F., Macdonald, M.F.: Introduction to commutative algebra, Addison-Wesley, Reading, Mass. 1969. Zbl. 175, 36
6. Atiyah, M.F. and others: Representation theory of Lie groups, Lond. Math. Soc. Lect. Notes *34*, Cambridge Univ. Press, Cambridge New York 1979. Zbl. 426.22014
7. Baer, R.: Linear algebra and projective geometry, Pure Appl. Math. Vol. II, Academic Press, New York 1952. Zbl. 49, 381
8. Barwise, J. and others: Handbook of mathematical logic, Stud. Logic Found. Math. *90*, North-Holland, Amsterdam 1978. Zbl. 443.03001
9. Besse, A. ed.: Géométrie riemannienne en dimension 4, Séminaire Arthur Besse 1978/1979, CEDIC, Paris 1981. Zbl. 472.00010
10. Birkhoff, G.: Lattice theory, Amer. Math. Soc., New York 1940. Zbl. 33, 101
11. Bogolyubov, N.N.: The theory of symmetry of elementary particles, in High-energy physics and the theory of elementary particles, Naukova Dumka, Kiev 1967, 5–112 (in Russian)
12. Borel, A.: Linear algebraic groups, Benjamin, New York Amsterdam 1969. Zbl. 186, 332
13. Borel, A.: Arithmetic properties of linear algebraic groups, in Proc. Int. Congr. Math., Stockholm, 1962, Inst. Mittag-Leffler, Djursholm 1963, 10–22. Zbl. 134, 165
14. Borel, A., Serre, J-P.: Le théorème de Riemann-Roch, Bull. Soc. Math. Fr. *86*, 97–136 (1959). Zbl. 91, 330

---

*For the convenience of the reader, references to reviews in Zentralblatt für Mathematik (Zbl.), compiled using the MATH database, and Jahrbuch über die Fortschritte der Mathematik (Jrb.) have, as far as possible, been included in this bibliography.

Translator's note: After consulting the author and the editors at Springer-Verlag, I have translated the references without modification (except for renumbering them into Latin alphabetical order). As Shafarevich writes, a book in translation must keep its original spirit, or what's the point of translating it at all?

I have also decided against adding a detailed list of references in the style of modern algebra textbooks—the book makes very well the point that algebra is the birthright of all mathematicians and scientists, and that its exposition is too important to be entrusted entirely to professional algebraists. In addition, when I suggested that references to classics of the subject might seem old-fashioned to modern students, Shafarevich's reply was that just because we know of other people's bad habits, it doesn't follow that we should encourage them, does it?.

However, I have taken the liberty of adding a small number of references to texts that have recently appeared; a reader needing further technical references on topics in algebra may also consult the references given in these, and in the Bourbaki series on algebra [16 (1942–1948), 17 (1959), 18 (1968)].

15. Borevich, Z.I., Shafarevich, I.R.: Number theory, Nauka, Moscow 1964. Zbl. 121, 42; English translation: Academic Press, New York London 1966. Zbl. 145, 49
16. Bourbaki, N.: Éléments de mathématiques, Algèbre, Chap. 1–3, Hermann, Paris 1942–1948. I: 1942, Zbl. 60, 68. III: 1948, Zbl. 33, 259
17. Bourbaki, N.: Éléments de mathématiques, Algèbre, Chap. 9, Hermann, Paris 1959. Zbl. 102, 255
18. Bourbaki, N.: Éléments de mathématiques, Groupes et algèbres de Lie, Chap. 4–6, Hermann, Paris 1968. Zbl. 186, 330
19. Bröcker, Th.: Differentiable germs and catastrophes, Cambridge Univ. Press, Cambridge New York 1975. Zbl. 302.58006
20. Brown, K.S.: Cohomology of groups, Grad. Texts Math. 87, Springer-Verlag, Berlin Heidelberg New York 1982. Zbl. 584.20036
21. Burnside, W.: Theory of groups of finite order, Cambridge Univ. Press, Cambridge 1897; Reprint: Dover, New York 1955. Zbl. 64, 251
22. Cartan, H., Eilenberg, S.: Homological algebra, Princeton Univ. Press, Princeton, New Jersey 1956. Zbl. 75, 243
23. Cassels, J.W.S., Fröhlich, A. eds.: Algebraic number theory, Academic Press, London 1967. Zbl. 153, 74
24. Chandler, B., Magnus, W.: The history of combinatorial group theory, Stud. Hist. Math. Phys. Sci. *9*, Springer-Verlag, Berlin Heidelberg New York 1982. Zbl. 498.20001
25. Chevalley, C.: Theory of Lie groups, Princeton Mathematical Series 8, Princeton Univ. Press, Princeton, New Jersey 1946
26. Chevalley, C.: Introduction to the theory of algebraic functions of one variable, Am. Math. Soc., New York 1951. Zbl. 45, 323
27. Chevalley, C.: Classification des groupes de Lie algébriques, Séminaire C. Chevalley 1956–58, Secrétariat mathématique, Paris 1958
28. Chevalley, C.: La théorie des groupes algébriques, Proc. Int. Congr. Math., Edinburgh 1958, Cambridge Univ. Press, Cambridge New York 1960, 53–68. Zbl. 121, 378
29. Clifford, W.K.: Mathematical papers, Macmillan, London 1882
30. Courant, R., Hilbert, D.: Methoden der mathematischen Physik, Bd. I, Springer-Verlag, Berlin, 1931. Zbl. 1, 5; English translation: Methods of mathematical physics, Vol. I, Interscience, New York 1953. Zbl. 53, 28
31. Dedekind, R., Weber, H.: Theorie der algebraischen Funktionen einer Veränderlichen, Crelle J. Reine Angew. Math. *92*, 181–291 (1882). Jrb. 14, 352
32. Delone, B., Padurov, N., Aleksandrov, A.: The mathematical foundations of crystal structure analysis and the definition of the fundamental parallelipiped of repetition by X-ray diffraction, ONTI-GTTI, Moscow-Leningrad 1934 (in Russian)
33. Deuring, M.: Algebren, Springer-Verlag, Berlin 1935. Zbl. 11, 198
34. Dieudonné, J.A., Carrell, J.B.: Invariant theory, old and new, Academic Press, New York London 1971. Zbl. 258.14011
35. Dirac, P.A.M.: The principles of quantum mechanics, Oxford Univ. Press, Oxford 1930. Jrb. 56, 745
36. Dold, A.: Lectures on algebraic topology, Grundlehren Math. Wiss. 200, Springer-Verlag, Berlin Heidelberg New York 1980. Zbl. 434.55001
37. Dyson, F.J.: Mathematics in the physical sciences, Scientific American *211*, 129–146 (1964)
38. Eilenberg, S., MacLane, S.: Natural isomorphisms in group theory, Proc. Natl. Acad. Sci. USA *28*, 537–543 (1942). Zbl. 61, 92
39. Eilenberg, S., MacLane, S.: General theory of natural equivalence, Trans. Am. Math. Soc. *58*, 231–294 (1945). Zbl. 61, 92
40. Fëdorov, E.S.: Symmetries in the plane, *Zapiski imperatorskogo Sankt-Peterburgskogo mineralogicheskogo obshchestva* (Notes of the imperial Saint Petersburg mineralogical society), *28* (2), 345–390 (1891) (in Russian). Jrb. 23, 539
41. Feynman, R.P.: The character of physical laws, Messenger Lectures, Cornell 1964, Cox and Wyman (BBC publications), London 1965

42. Fomenko, A.T.: Differential geometry and topology, Complementary chapters, Moscow Univ. publications, Moscow 1983 (in Russian). Zbl. 517.53001
43. Freudenthal, H.: Octaven, Ausnahmegruppen und Octavengeometrien, Math. Inst. Rijksuniversitet Utrecht 1951. Zbl. 56, 259
44. Frobenius, G.: Gesammelte Abhandlungen, Bd. 1–3, Springer-Verlag, Berlin Heidelberg New York 1968. Zbl. 169, 289
45. Galois, E.: Œuvres mathématiques d'Évariste Galois, Gauthier-Villars, Paris 1951. Zbl. 42, 4
46. Gauss, C.F.: Disquisitiones arithmeticae, Werke, Bd. 1, Springer-Verlag, Berlin 1870; English translation: Yale, New Haven, Conn. London 1966. Zbl. 136, 323
47. Gel'fand, I.M., Raikov, D.A., Shilov, G.E.: Commutative normed rings, Fiz. Mat. G. Iz., Moscow 1960 Zbl. 134, 321; English translation: Chelsea, New York 1964
48. Goppa, V.D.: Codes and information, Usp. Mat. Nauk *39* No. 1, 77–120 (1984); English translation: Russ. Math. Surv. *39* No. 1, 87–141 (1984). Zbl. 578.94011
49. Gorenstein, D.: Finite simple groups, An introduction to their classification, Univ. Series. in Math., Plenum Press New York, London 1982. Zbl. 483.20008
50. Goto, M., Grosshans, F.D.: Semisimple Lie algebras, Lect. Notes. Pure Appl. Math. 38, Marcel Dekker, New York Basel 1978. Zbl. 391.17004
51. Grothendieck, A.: Sur quelques points d'algèbre cohomologique, Tôhoku Math. J. II. Ser. *9*, 119–221 (1957). Zbl. 118, 261
52. Grothendieck, A.: The cohomology theory of abstract algebraic varieties, Proc. Int. Congr. Math., Edinburgh 1958, Cambridge Univ. Press, Cambridge New York 1960, 103–118. Zbl. 119, 369
53. Grothendieck, A., Dieudonné, J.: Éléments de géométrie algébrique, III, Étude cohomologique des faisceaux cohérents, Publ. Math. Inst. Hautes Études Sci. *11* (1961), 1–167 (1962). Zbl. 118, 362
54. Hadamard, J.: Leçons de géométrie élémentaire, II, Géométrie dans l'espace, Armand Colin, Paris 1908
55. Hadamard, J.: Non-Euclidean geometry and the theory of automorphic functions, Gos. Izdat. Tekhn.-Teor. Lit., Moscow Leningrad 1951 (in Russian). Zbl. 45, 361
56. Hamermesh, M.: Group theory and its application to physical problems, Addison-Wesley, Reading, Mass. London 1964. Zbl. 151, 341
57. Hazewinkel, M.: Formal groups and applications, Pure Appl. Math. 78, Academic Press, New York London 1978. Zbl. 454.14020
58. Hazewinkel, M., Hesselink, W., Siersma, D., Veldkamp, F.D.: The ubiquity of Coxeter-Dynkin diagrams (an introduction to the A-D-E problem), Nieuw Arch. Wisk. III. Ser. *25*, 257–307 (1977). Zbl. 377.20037
59. Hilbert, D.: Grundlagen der Geometrie, Teubner, Leipzig Berlin 1930. Jrb. 56, 481
60. Hilbert, D., Cohn-Vossen, S.: Anschauliche Geometrie, Springer-Verlag, Berlin 1932. Zbl. 5, 112; English translation: Geometry and the imagination, Chelsea, New York 1952. Zbl. 47, 388
61. Hilton, P.J., Stammbach, U.: A course in homological algebra, Grad. Texts Math. 4, Springer-Verlag, Berlin Heidelberg New York 1971. Zbl. 238.18006
62. Hirzebruch, F.: Neue topologische Methoden in der algebraischen Geometrie, Springer-Verlag, Berlin Heidelberg New York 1956. Zbl. 70, 163; English translation: Topological methods in algebraic geometry, Springer-Verlag, Berlin Heidelberg New York 1966. Zbl. 138, 420
63. Hirzebruch, F.: Elliptische Differentialoperatoren auf Mannigfaltigkeiten, Arbeitsgemeinschaft für Forschung des Landes Nordrhein-Westfalen *33*, 563–608 (1965). Zbl. 181, 103
64. Hochschild, G.: The structure of Lie groups, Holden-Day, San Francisco London Amsterdam 1965. Zbl. 131, 27
65. Hoffman, K.: Banach spaces of analytic functions, Prentice-Hall, Englewood Cliffs, New Jersey 1962. Zbl. 117, 340
66. Humphreys, J.E.: Linear algebraic groups, Grad. Texts Math. 21, Springer-Verlag, Berlin Heidelberg New York 1975. Zbl. 325.20039

67. Huppert, B.: Endliche Gruppen, Bd.I, Grundlehren Math. Wiss. 134, Springer-Verlag, Berlin Heidelberg New York 1967. Zbl. 217, 72
68. Huppert, B., Blackburn, N.: Finite groups, Vol. II–III, Springer-Verlag, Berlin Heidelberg New York 1982. II: Grundlehren Math. Wiss. 242. Zbl. 477.20001. III: Grundlehren Math. Wiss. 243. Zbl. 514.20002
69. Ince, E.L.: Ordinary differential equations, Longmans-Green, London 1927. Jrb. 53, 399
70. Jordan, C.: Traité des substitutions des équations algébriques, Gauthier-Villars, Paris 1870
71. Kaplansky, I.: An introduction to differential algebra, Hermann, Paris 1957. Zbl. 83, 33
72. Kirillov, A.A.: Elements of the theory of representations, Nauka, Moscow 1972. Zbl. 264.22011; English translation: Grundlehren Math. Wiss. 220, Springer-Verlag, Berlin Heidelberg New York 1976. Zbl. 342.22001
73. Klein, F.: Vorlesungen über die Entwicklung der Mathematik im 19. Jahrhundert, Grundlehren Math. Wiss. 24, Springer-Verlag, Berlin 1926. Jrb. 52, 22
74. Klemm, M.: Symmetrien von Ornamenten und Kristallen, Hochschultext, Springer-Verlag, Berlin Heidelberg New York 1982. Zbl. 482.20034
75. Kostrikin, A.I., Manin, Yu.I.: Linear algebra and geometry, Moscow Univ. publications, Moscow 1980. Zbl. 532.00002; English translation: Gordon and Breach, New York London 1989
76. Kronecker, L.: Grundzüge einer arithmetischen Theorie der algebraischen Grössen, J. Reine Angew. Math. *92*, 1–123 (1882). Jrb. 14, 38
77. Kurosh, A.G.: The theory of groups, Gos. Izdat. Teor.-Tekn. Lit., 1944; English translation: Vol. I, II, Chelsea, New York 1955, 1956. Zbl. 64, 251
78. Lawson, H.B. Jr.: Surfaces minimales et la construction de Calabi-Penrose, Séminaire Bourbaki, Exp. *624*, Astérisque *121/122*, 197–211 (1985)
79. Lie, S., Engel, F.: Theorie der Transformationsgruppen, Bd. I–III, Teubner, Leipzig, I 1883, II 1888, III 1893
80. Mal'tsev, A.I.: Groups and other algebraic systems, in Mathematics, its contents, methods and meaning, vol. 3, 248–331 Academy of Sciences of the USSR, Moscow 1956 (in Russian)
81. Manin, Yu.I.: A course in mathematical logic, Grad. Texts Math. 53, Springer-Verlag, Berlin Heidelberg New York 1977. Zbl. 383.03002
82. Manin, Yu.I., Vlehduts, S.G.: Linear codes and modular curves, in Itogi Nauki Tekh., Ser. Sovrem. Probl. Mat. *25*, 209–257 (1984). Zbl. 629.94013; English translation: J. Sov. Math. *30*, 2611–2643 (1985)
83. Manin, Yu.I.: Lectures on algebraic geometry, Part I, Affine schemes, Moscow Univ. publications, Moscow 1970 (in Russian)
84. Michel, L.: Symmetry defects and broken symmetry configurations, hidden symmetry, Rev. Mod. Phys. *52*, 617–652 (1980)
85. Milnor, J.: Introduction to algebraic K-theory, Ann. Math. Stud., Princeton Univ. Press, Princeton, New Jersey 1971. Zbl. 237.18005
86. Mumford, D.: An algebro-geometric construction of commuting operators and of solutions to Toda lattice equations, Korteweg-de Vries equations and related non-linear equations, in Proc. int. symp. on algebraic geometry Kyoto 1977 Kinokuniya, Tokyo, 115–153, 1977, Zbl. 423.14007
87. von Neumann, J.: Continuous geometry, Princeton Univ. Press, Princeton, New Jersey 1960. Zbl. 171, 280
88. Palais, R.S.: Seminar on the Atiyah-Singer index theorem, Ann. Math. Stud. *57*, Princeton Univ. Press, Princeton, New Jersey 1965. Zbl. 137, 170
89. Pedersen, J.: Geometry: The unity of theory and practice, Math. Intell. *5* No. 4, 37–49, (1983), Zbl. 536.54008. B. Grünbaum, The Emperor's new clothes: full regalia, G string, or nothing? Math. Intell. *6* No. 4, 47–53 (1984) Zbl. 561.52014, and P. Hilton and J. Pedersen, Comments on Grünbaum's article, Math. Intelli. *6* No. 4, 54–56 (1984). Zbl. 561.52015
90. Pontryagin, L.S.: Topological groups, Redak. tekh.-teor. lit., Moscow Leningrad 1938; English translation: Oxford Univ. Press, London Milford Haven Conn. Zbl. 22, 171

91. Quillen, D.: On the cohomology and K-theory of the general linear groups over a finite field, Ann. Math. (2) *96*, 552–586 (1972). Zbl. 249.18022
92. de Rham, G.: Variétés différentiables, Formes, courants, formes harmoniques, Hermann, Paris 1955. Zbl. 65, 324
93. Seifert, H., Threlfall, W.: Lehrbuch der Topologie, Teubner, Leipzig Berlin 1934. Zbl. 9, 86
94. *Séminaire Sophus Lie*, Théorie des algèbres de Lie, Topologie des groupes de Lie, Paris 1955
95. Serre, J-P.: Représentations linéaires des groupes finis, Hermann, Paris 1967; English translation: Linear representations of finite groups, Grad. Texts Math. 42, Springer-Verlag, Berlin Heidelberg New York 1977. Zbl. 355.20006
96. Shafarevich, I.R.: Basic algebraic geometry, Nauka, Moscow 1972; English translation: Grundlehren Math. Wiss. 213, Springer-Verlag, Berlin Heidelberg New York 1974. Zbl. 284.14001
97. Soulé, K.: K-théorie des anneaux d'entiers de corps de nombres et cohomologie étale, Invent. Math. *55*, 251–295 (1979). Zbl. 437.12008
98. Speiser, A.: Die Theorie der Gruppen von endlicher Ordnung, Springer-Verlag, Berlin 1937. Zbl. 17, 153
99. Springer, T.A.: Linear algebraic groups, Prog. Math. 9, Birkhäuser, Boston 1981. Zbl. 453.14022
100. Suslin, A.A.: Algebraic K-theory and the norm residue homomorphism, in Itogi Nauki Tekh., Ser. Sovrem. Probl. Mat. *25*, 115–207 (1984). Zbl. 558.12013; English translation in J. Sov. Math. *30*, 2556–2611 (1985)
101. Switzer, R.M.: Algebraic topology—homotopy and homology, Grundlehren Math. Wiss. 212, Springer-Verlag, Berlin Heidelberg New York 1975. Zbl. 305.55001
102. Tennison, B.R.: Sheaf theory, Lond. Math. Soc. Lect. Notes 20, Cambridge Univ. Press, Cambridge New York 1975. Zbl. 313.18010
103. Tits, J.: Groupes simples et géométries associées, in Proc. Int. Congr. Math., Stockholm, 1962, Inst. Mittag-Leffler, Djursholm 1963, 197–221. Zbl. 131, 265
104. van der Waerden, B.L.: Moderne Algebra, Bd. 1, 2, Springer-Verlag, Berlin 1930, 1931; I: Jrb. 56, 138. II: Zbl. 2, 8. English translation: Algebra, Vol. I, II, Ungar, New York 1970.
105. Weber, H.: Lehrbuch der Algebra, Bd. 1, 2, Vieweg, Braunschweig 1898, 1899
106. Weil, A.: Basic number theory, Grundlehren Math. Wiss. 144, Springer-Verlag, Berlin Heidelberg New York 1967. Zbl. 176, 336
107. Weyl, H.: Gruppentheorie und Quantenmechanik, Hirzel, Leipzig 1928; English translation: The theory of groups and quantum mechanics, Princeton 1930. Jrb. 54, 954
108. Weyl, H.: Mathematische Analyse des Raumproblems, Springer-Verlag, Berlin 1923. Jrb. 49, 81. Reprinted Wissenschaftliche Buchgesellschaft, Darmstadt 1977
109. Weyl, H.: The classical groups, Princeton Univ. Press, Princeton, New Jersey 1939. Zbl. 20, 206
110. Weyl, H.: Symmetry, Princeton Univ. Press, Princeton, New Jersey, 1952. Zbl. 46, 4
111. Zhelobenko, D.P.: Compact Lie groups and their representations, Nauka, Moscow 1970; English translation: Am. Math. Soc., Providence 1973. Zbl. 228.22013

Added in translation:

112. Cohn, P.M.: Algebra, vol. 1, 2, J. Wiley, London New York Sidney 1974, 1977. 1: Zbl. 272.00003. 2: Zbl. 341.00002, 2nd ed.: J. Wiley, 1982, Zbl. 481.00001
113. Curtis, C.W., Reiner, I.: Methods of representation theory, vol. I, II, Pure and Applied Mathematics, J. Wiley, London New York Sidney 1981, 1987. I: Zbl. 469.20001. II: Zbl. 616.20001
114. Jacobson, N.: Lectures in abstract algebra, vol. I, II, III, Van Nostrand, Princeton, New Jersey 1951, 1953, 1964, I: Zbl, 44, 260. II: Zbl. 53, 212. III: Zbl. 124, 270
115. Matsumura, H.: Commutative ring theory, C.U.P, Cambridge New York Melbourne, 1986. Translation from the Japanese (Kyoritsu Tokyo 1980) Zbl. 603.13001

Index of Names

Subject Index

《国外数学名著系列》(影印版)

(按初版出版时间排序)

1. 拓扑学 I：总论　S. P. Novikov (Ed.)　2006.1
2. 代数学基础　Igor R. Shafarevich　2006.1
3. 现代数论导引 (第二版)　Yu. I. Manin　A. A. Panchishkin　2006.1
4. 现代概率论基础 (第二版)　Olav Kallenberg　2006.1
5. 数值数学　Alfio Quarteroni　Riccardo Sacco　Fausto Saleri　2006.1
6. 数值最优化　Jorge Nocedal　Stephen J. Wright　2006.1
7. 动力系统　Jürgen Jost　2006.1
8. 复杂性理论　Ingo Wegener　2006.1
9. 计算流体力学原理　Pieter Wesseling　2006.1
10. 计算统计学基础　James E. Gentle　2006.1
11. 非线性时间序列　Jianqing Fan　Qiwei Yao　2006.1
12. 函数型数据分析 (第二版)　J. O. Ramsay　B. W. Silverman　2006.1
13. 矩阵迭代分析 (第二版)　Richard S. Varga　2006.1
14. 偏微分方程的并行算法　Petter Bjørstad　Mitchell Luskin(Eds.)　2006.1
15. 非线性问题的牛顿法　Peter Deuflhard　2006.1
16. 区域分解算法：算法与理论　A. Toselli　O. Widlund　2006.1
17. 常微分方程的解法 I：非刚性问题 (第二版)　E. Hairer　S. P. Nørsett　G. Wanner　2006.1
18. 常微分方程的解法 II：刚性与微分代数问题 (第二版)　E. Hairer　G. Wanner　2006.1
19. 偏微分方程与数值方法　Stig Larsson　Vidar Thomée　2006.1
20. 椭圆型微分方程的理论与数值处理　W. Hackbusch　2006.1
21. 几何拓扑：局部性、周期性和伽罗瓦对称性　Dennis P. Sullivan　2006.1
22. 图论编程：分类树算法　Victor N. Kasyanov　Vladimir A. Evstigneev　2006.1
23. 经济、生态与环境科学中的数学模型　Natali Hritonenko　Yuri Yatsenko　2006.1
24. 代数数论　Jürgen Neukirch　2007.1
25. 代数复杂性理论　Peter Bürgisser　Michael Clausen　M. Amin Shokrollahi　2007.1
26. 一致双曲性之外的动力学：一种整体的几何学的与概率论的观点　Christian Bonatti　Lorenzo J. Díaz　Marcelo Viana　2007.1
27. 算子代数理论 I　Masamichi Takesaki　2007.1
28. 离散几何中的研究问题　Peter Brass　William Moser　János Pach　2007.1
29. 数论中未解决的问题 (第三版)　Richard K. Guy　2007.1
30. 黎曼几何 (第二版)　Peter Petersen　2007.1
31. 递归可枚举集和图灵度：可计算函数与可计算生成集研究　Robert I. Soare　2007.1
32. 模型论引论　David Marker　2007.1

33. 线性微分方程的伽罗瓦理论　Marius van der Put　Michael F. Singer　2007.1
34. 代数几何 II：代数簇的上同调，代数曲面　I. R. Shafarevich (Ed.)　2007.1
35. 伯克利数学问题集 (第三版)　Paulo Ney de Souza　Jorge-Nuno Silva　2007.1
36. 陶伯理论：百年进展　Jacob Korevaar　2007.1

37. 同调代数方法 (第二版)　Sergei I. Gelfand　Yuri I. Manin　2009.1
38. 图像处理与分析：变分，PDE，小波及随机方法　Tony F. Chan　Jianhong Shen　2009.1
39. 稀疏线性系统的迭代方法　Yousef Saad　2009.1
40. 模型参数估计的反问题理论与方法　Albert Tarantola　2009.1
41. 常微分方程和微分代数方程的计算机方法　Uri M. Ascher　Linda R. Petzold　2009.1
42. 无约束最优化与非线性方程的数值方法　J. E. Dennis Jr.　Robert B. Schnabel　2009.1
43. 代数几何 I：代数曲线，代数流形与概型　I. R. Shafarevich (Ed.)　2009.1
44. 代数几何 III：复代数簇，代数曲线及雅可比行列式　A. N. Parshin　I. R. Shafarevich (Eds.)　2009.1
45. 代数几何 IV：线性代数群，不变量理论　A. N. Parshin　I. R. Shafarevich (Eds.)　2009.1
46. 代数几何 V：Fano 簇　A. N. Parshin　I. R. Shafarevich (Eds.)　2009.1
47. 交换调和分析 I：总论，古典问题　V. P. Khavin　N. K. Nikol'skij (Eds.)　2009.1
48. 复分析 I：整函数与亚纯函数，多解析函数及其广义性　A. A. Gonchar　V. P. Havin　N. K. Nikolski (Eds.)　2009.1
49. 计算不变量理论　Harm Derksen　Gregor Kemper　2009.1
50. 动力系统 V：分歧理论和突变理论　V. I. Arnol'd (Ed.)　2009.1
51. 动力系统 VII：可积系统，不完整动力系统　V. I. Arnol'd, S. P. Novikov (Eds.)　2009.1
52. 动力系统 VIII：奇异系统 II：应用　V. I. Arnol'd (Ed.)　2009.1
53. 动力系统 IX：带有双曲性的动力系统　D. V. Anosov (Ed.)　2009.1
54. 动力系统 X：旋涡的一般理论　V. V. Kozlov　2009.1
55. 几何 I：微分几何基本思想与概念　R. V. Gamkrelidze (Ed.)　2009.1
56. 几何 II：常曲率空间　E. B. Vinberg (Ed.)　2009.1
57. 几何 III：曲面理论　Yu. D. Burago　V. A. Zalgaller (Eds.)　2009.1
58. 几何 IV：非正规黎曼几何　Yu. G. Reshetnyak (Ed.)　2009.1
59. 几何 V：最小曲面　R. Osserman (Ed.)　2009.1
60. 几何 VI：黎曼几何　M. M. Postnikov　2009.1
61. 李群与李代数 I：李理论基础，李交换群　A. L. Onishchik (Ed.)　2009.1
62. 李群与李代数 II:李群的离散子群,李群与李代数的上同调　A. L. Onishchik　E. B. Vinberg (Eds.)　2009.1
63. 李群与李代数 III：李群与李代数的结构　A. L. Onishchik　E. B. Vinberg (Eds.)　2009.1
64. 经典力学与天体力学中的数学问题　Vladimir I. Arnold　Valery V. Kozlov　Anatoly I. Neishtadt　2009.1
65. 数论 IV：超越数　A. N. Parshin　I. R. Shafarevich (Eds.)　2009.1
66. 偏微分方程 IV：微局部分析和双曲型方程　Yu. V. Egorov　M. A. Shubin (Eds.)　2009.1
67. 拓扑学 II：同伦与同调，经典流形　S. P. Novikov　V. A. Rokhlin (Eds.)　2009.1